Frontiers of Physics and Plasma Science (FPPS-2016)

Journal of Physics: Conference Series Volume 836

Ujjain, India
7 – 8 November 2016

ISBN: 978-1-5108-4182-6
ISSN: 1742-6588

Printed from e-media with permission by:

Curran Associates, Inc.
57 Morehouse Lane
Red Hook, NY 12571

Printed by Curran Associates, Inc. (2017)

For permission requests, please contact the Institute of Physics
at the address below.

Institute of Physics
Dirac House, Temple Back
Bristol BS1 6BE UK

Phone: 44 1 17 929 7481
Fax: 44 1 17 920 0979

techtracking@iop.org

Additional copies of this publication are available from:

Curran Associates, Inc.
57 Morehouse Lane
Red Hook, NY 12571 USA
Phone: 845-758-0400
Fax: 845-758-2633
Email: curran@proceedings.com
Web: www.proceedings.com

Table of contents

Volume 836

Frontiers of Physics and Plasma Science7–8 November 2016, Ujjain Engineering College, Ujjain, India

Accepted papers received: 11 April 2017Published online: 31 May 2017

Preface

Papers

Plasma Physics

General Physics

Preface to the conference proceedings

We are very pleased to introduce the proceeding of FPPS-2016; the international conference "**Frontiers of Physics and Plasma Science**" that took place on 7 and 8 November, 2016 in the campus of Ujjain Engineering College, Ujjain (India). The goal of the meeting was to provide a broad prospective to the plasma science emphasizing physics with a new plasma technologies. The scientific program of the conference focused on the advancement of the all branches of physics in achieving all applications of the plasma science. The conference spans a wide range of topics, reporting experiments, techniques and ideas that advance the plasma science worldwide.

There were 20 invited lectures and 04 oral presentations covering the different area of the conference. The keynote lecture was delivered by Dr. Rajdeep Singh Rawat (NTU, Singapore) on "Density plasma focus: novel high energy density plasma device". Prof. Y.C. Saxena (IPR, Gandhinagar, Ahmedabad), Prof. R. P. Sharma (IIT, New Delhi), Prof. Fernando Haas (Brazil), Prof. Davoud Dorranian (Tehran, Iran), Dr. Raju Khanal (Tribhuwan University, Nepal), Prof. Avinash Khare (IIT, New Delhi), Dr. Navin Dwivedi (Israel), Prof. V.K. Tripathi (IIT New Delhi), Dr. J. Ghosh (IPR, Gandhinagar, Gujarat), Dr. Devendra Sharma (IPR, Gandhinagar, Gujarat), Prof. R.K. Thareja (IIT Kanpur), Dr. Vipul Arora (RRCAT, Indore), Prof. M. P. Bora (Gauhati University, Guwahati) and many more have delivered their lectures in the field of plasma science and its applications. The program was chaired in a professional and efficient way by the session chairmen who were selected for their international standing in the subject.

The 165 abstracts that were presented in two days (during parallel poster session) formed a heart of the conference and provided ample opportunity for the discussions. The 170 participants, 110 of whom were students had many fruitful discussions and exchange the ideas that contributed to the success of the conference. A panel of experts (Prof. R.P. Sharma, Prof. T. S. Gill, Prof. N.S. Saini, Prof. K. P. Maheshwari, Dr. A.P. Misra, and Dr. V. Ganeshan) was there to judge the presented posters and 6 best posters were awarded prize. Participants from six countries (India, Nepal, Iran, Singapore, Brazil and Israel) made the conference truly international in scope. Of the total number of presented abstracts, 54 of these are included in this proceedings volume. These papers maintained the high promise suggested by the written abstracts.

Finally, we would like to extend our special thanks to Dr. Umesh Pendharkar, Principal, Ujjain Engineering College, Ujjain for his constant motivation provided during the preparation of the conference. We appreciate the project fellow Shweta Jain (MPCST, Bhopal), Archana Patidar (UGC, New Delhi), Tripti Rimza (SERB-DST, New Delhi) and Karishma Jena (MPCST, Bhopal) for their hard work and enthusiasm for making this conference a grand success. We also gratefully acknowledge the contribution of Prof. B. D. Shrivastava, Vikram University, Ujjain in assembling the conference proceeding. We would like to thank all of the participants attending the conference and also to the committees for their valuable contribution. It is our pleasure to acknowledge the Directorate Technical Education (Bhopal), SERB-DST (New Delhi) and MPCST (Bhopal) for their financial support.

Dr. Prerana Sharma
Department of Physics, Ujjain engineering college, Ujjain, – 456010, India
Editor

IOP Conf. Series: Journal of Physics: Conf. Series **836** (2017) 011002 doi:10.1088/1742-6596/836/1/011002

Patron Council

Shri Deepak Joshi,
Honorable Minister, Tech. Edu. & Skill Development Deptt. MP Govt., Bhopal.
Mrs. Kalpana Shrivastava,
Principal Secretary, Tech. Education, MP Govt., Bhopal.
Dr. Ashish Dongre,
Director, Tech. Edu. & Skill Development Deptt. MP Govt., Bhopal.
Chief Mentor
Dr. Umesh Pendharkar,
Principal, Ujjain Engineering College, Ujjain (M.P.)

Coordinator

Dr. Prerana Sharma
Asso. Prof. & Head, Physics Deptt.

Co- Coordinators

Dr. H.K. Patel,
Head, Mathematics Deptt.
Prof. Y.K. Joshi,
Asst. Prof., Comp. Sc. Deptt.
Dr.R.K. Rai,
Asso. Prof., Physics Deptt.
Dr. V.K. Hinge,
Asst. Prof., Physics Deptt.

Fellow Associates

Shweta Jain, Project Fellow MPCST, Bhopal (M.P.) India

Archana Patidar, Project Fellow UGC New Delhi, India

Tripti Rimza, Project Fellow SERB-DST New Delhi, India

Karishma Jena, Project Fellow MPCST, Bhopal (M.P.) India

IOP Conf. Series: Journal of Physics: Conf. Series **836** (2017) 011002 doi:10.1088/1742-6596/836/1/011002

International Scientific Advisory Committee

Prof Swadesh Mahajan, Univ of Texas, Austin, USA
Prof. B. Eliasson, Strathclyde University, UK
Dr. Ioannis Kourakis, Queen's Univ. Belfast, UK
Prof Dieter HH Hoffmann, Darmstadt, Germany
Prof. R. Bingham, Strathclyde University, UK
Prof. G. Brodin, Umea University, Sweden
Prof. F. Haas, Instituto de Fisica, Brazil
Prof. S.S. Pandey, VC, Vikram Univ. Ujjain, India
Prof. A.K. Gwal, VC, AISECT, Bhopal, India
Prof. S. Chaturvedi, Director IPR, Ahmedabad, India
Prof. Dhiraj Bora, IPR, Ahmedabad, India
Prof. P.A. Naik, Director RRCAT, Indore, India
Dr. J.A. Chakera, RRCAT, Indore, India
Prof. Avinash Khare, DU, Delhi, India
Prof. B.P. Pandey, Macquarie University, Australia
Prof. Y.C. Saxena, IPR Ahmedabad, India
Prof. Abhijit Sen, IPR, Ahmedabad, India
Prof. R.P. Sharma, IIT, Delhi, India
Prof. V.K. Tripathi, IIT, Delhi, India
Prof. Basudev Ghosh, Jadavpur Univ., Kolkata, India
Prof. Vinod Krishan, IIA, Bangalore, India
Prof. Raju Khannal, TU, Kathmandu, Nepal
Prof. J. Ghosh, IPR, Ahmedabad, India
Prof. Devendra Sharma, IPR Ahmedabad, India

Conference Advisory Committee

Prof. S.K. Ghosh, VU, Ujjain, India
Prof. K.P. Joshi, DAVV, Indore, India
Prof. M.S. Tiwari, DSHGSU, Sagar, India
Prof. Ashutosh Mishra, DAVV, Indore, India
Prof. R.K. Chhajlani, VU, Ujjain, India
Prof. B.D. Shrivastava, VU, Ujjain, India
Prof. G.K. Upadhyay, VU, Ujjain, India
Prof. S.B. Shirivastava, VU, Ujjain, India
Prof. P.K. Sen, SGSITS, Indore, India
Dr. Vipul Arora, RRCAT, Indore, India
Prof. G. Mishra, DAVV, Indore, India
Dr. Poornima Khare, RGPV, Bhopal, India
Dr. Swati Dubey, VU, Ujjain, India
Dr. Nishchhal Yadav, VU, Ujjain, India
Dr. G. Ahirwar, VU, Ujjain, India
Dr. S. Shastri, GPC, Ujjain, India
Dr. M.K. Rathore, MPCST, Bhopal, India
Dr. J. Parashar, SATI, Vidisha, India
Dr. Y. Choyal, DAVV, Indore, India
Dr. R.P. Prajapati, GGCU, Bilaspur, India
Dr. Praveen Sharma, BU, Bhopal, India

IOP Conf. Series: Journal of Physics: Conf. Series **836** (2017) 011002 doi:10.1088/1742-6596/836/1/011002

Local Organizing Committee

Dr. S.K. Jain, Registrar, RGPV, Bhopal
Dr. J.K. Shrivastava, Head, Chem. Engg. Deptt., UECU
Prof. B.L. Gupta, Head, Electrical Engg. Deptt., UECU
Dr. A.K. Sthapak, Head, Civil Engg. Deptt., UECU
Prof. R. Nagaich, Head, Mech. Engg/. Deptt., UECU
Prof. Y. S. Thakur, Head, Elx. & Comm. Deptt., UECU
Prof. K.S. Chandel, Head, Chemistry Deptt., UECU
Prof. A.K. Mishra, Chemistry Deptt., UECU
Dr. B.K. Singh, Controller Examination, UECU
Dr. Sunil Punjabi, Dean Academic, UECU
Dr. J. Palsania, Dean, Student Welfare, UECU
Dr. Savita Maru, Dean Planning, UECU
Dr. A.K Dwivedi, DDO, UECU
Dr. V.K. Sukhwani, Mechanical Engg. Deptt., UECU
Prof. A. Bhardwaj, Head Phys. Deptt. P.G.C, Biaora

IOP Conf. Series: Journal of Physics: Conf. Series **836** (2017) 011003 doi:10.1088/1742-6596/836/1/011003

01. SERB-DST, New Delhi, India

Scientific & Engineering Research Board- Department of Science & Technology, New Delhi

SERB INDIA

Science & Engineering Research Board
Department of Science & Technology
Government of India

02. DTE, Bhopal, (M.P.) India

Directorate of Technical Education, Bhopal

IOP Conf. Series: Journal of Physics: Conf. Series **836** (2017) 011004 doi:10.1088/1742-6596/836/1/011004

Peer review statement

All papers published in this volume of *Journal of Physics: Conference Series* have been peer reviewed through processes administered by the proceedings Editors. Reviews were conducted by expert referees to the professional and scientific standards expected of a proceedings journal published by IOP Publishing.

Terahertz pulses in a periodic plasma channel via modulated electron beam

Pratibha Malik[a], Suresh C Sharma, Rinku Sharma

Department of Applied Physics, Delhi Technological University, Delhi-110042, India

Email: [a)]pratibhamalik27@gmail.com

Abstract. The proposed scheme offers a promising way to study the effect of modulated relativistic electron beam in a periodic plasma channel (wiggler). The fluid equations model is used to find out the numerical behaviour of the radiation wave. In this scheme, a modulated relativistic electron beam makes possible to generate and amplify the high power terahertz (THz) radiation pulses. Using this model, we have analyzed and calculated the nonlinear dispersion relation of the radiation wave. The radiation frequency of THz wave can be tuned by varying the wiggler wavelength and beam energy. The excitation of the radiation wave is being developed with the resonance between the modulated frequency of electron beam and radiation wave in the periodic plasma channel (wiggler field) via Cerenkov interaction along the axial direction of channel, which provides the maximum enhancement in the growth rate of the radiation wave. Additionally, the growth rate of the unstable mode of the radiation wave depends on beam energy and modulation index, and becomes larger as modulation index approaches unity.

1. Introduction

Terahertz far-infrared radiation (THz) is considered to be the significant cutting edge of the electromagnetic spectrum for the last few decades, with various unique utility in high resolution imaging with medical application, security inspection, and remote sensing [1-4]. The THz radiation can be generated using various schemes such as ultra-short laser pulses [5] and quantum cascade lasers [6].

The conventional electromagnetic wiggler based FEL has been verified an effective source of generating short wavelength electromagnetic radiation but due to difficulties occurring in the designing of the practical implementations of such wigglers are limited. To overcome this limitation new plasma wigglers[7- 8] based FELs are introduced.The injection of an electron beam in a periodically system is a crucial example for the development of narrative radiation sources such as free electron lasers (FEL). In this paper, we present a schemeto generate and amplify Terahertz radiation using under-denserippled plasma wiggler.

2. Model

Consider a density modulated relativistic electron beam of velocity $v_b^0\hat{z}$ and density $n_{b1}(n_b^0+\Delta n_b^0\exp[-i\omega t+i\omega z/v_b^o])$ is being introduced in ripple plasma wiggler of density $n_w(n_p^0+n_w^p\exp[i\vec{w}.\vec{r}])$ with wave vector $\vec{w}$ .Here, the terms n_p^0 and n_w^p are the density of plasma electrons at equilibrium and perturbation, respectively. This non-linear interaction excites and amplifies the THz electromagnetic wave $(\omega_1 = 2\gamma_0^2 w v_b^0, \vec{k}_1)$ co-propagating with beam wave($(\omega_1, \vec{k}_b)$. The response offered by beam electrons and plasma electrons on account of THz radiation and beam wave is governed by the equation of motion

$$\frac{\partial}{\partial t}(\gamma_1 \vec{v}_{b1}) + (\vec{v}_{b1}.\nabla)\gamma_1 \vec{v}_{b1} = -\frac{e\vec{E}_T^1}{m} - \frac{e\vec{E}_b^1}{m} - \frac{e}{mc}(\vec{v}_{b1} \times \vec{B}_T), \quad (1)$$

$$\frac{\partial n_{b1}}{\partial t} + \nabla.(n_{b1}\vec{v}_{b1}) = 0. \quad (2)$$

After solving the equations (1) and (2) further, we get the perturbed velocity and density of the beam and plasma electrons, respectively. Consequently, this produces the nonlinear current density for THz radiation wave and beam wave. This nonlinear current density stimulate the excitation of the unstable radiation wave

$$\vec{J}_n^l = (\vec{J}_n^l)_T + (\vec{J}_n^l)_{beam} \quad (3)$$

Furthermore, using the Maxwell's third and fourth equations, we solve the wave equations and get the nonlinear dispersion relation and growth rate in unstable mode of radiation wave as follows

$$\lambda_w = \frac{4\pi c\omega_1(\gamma_o^2 - 1)\cos(\psi)}{\omega_1^2 + \gamma_o^2\omega_p^2}, \quad (4)$$

$$\beta = \left[\frac{3}{16}\right]^{\frac{1}{2}} \left[\frac{\omega_b^2\omega_w^4 w_x^2}{\gamma_o^3 c^2(\omega_p^2 + k_1^2c^2)^{1/2}(k_b^2 - k_1^2)^2}\left((1+\Delta+(1+\Delta)\gamma_o^2\frac{w_x^2}{k_{bz}^2}) + \Delta\frac{k_o}{k_{bz}}\right)\right]^{\frac{1}{3}}. \quad (5)$$

where the terms ω_b , ω_w and ω_p are represented as beam plasma frequency, plasma wiggler frequency and electron plasma frequency, respectively in equation (5). This relation shows that the growth rate in the unstable mode of THz radiation wave depends on density of beam electrons or beam current, plasma electrons and inversely proportional to radiation frequency of the wave. Additionally, the above results can be extended to find out the efficiency of the radiation system.

3. Results and Discussions

The results show that the amplification in growth rate of the unstable mode of wave is the sensitive function of the plasma density, beam current and modulation index. Using equation (4),the variation of radiation frequency versus plasma wiggler period is shown in figure 1.This figure illustrates that the wiggler period initially increases and then sharply starts decreasing as frequency increases further. In addition, as the plasma density increases more, the wiggler period becomes shorter and it could be more resonant to generate and amplify the coherent THz radiation. Using equation (5), figure 2 depicts the variation of growth rate of the radiation wave with beam density. This shows that growth rate increases as beam density increases. This result is in good agreement with the results obtained by[9-10]. From figure 3, it can be seen that the growth rate of the wave increases with modulation index but decreases as beam energy increases.

Using equation (4), in figure 4, we observe the variation of wiggler period with ripple angle and plasma density. This figure illustrates that the wiggler period decreases as ripple angle and plasma density increases and becomes sharper to amply the THz radiation emission.

4. Conclusions

In summary, the interaction of pre-bunched relativistic electron beam with plasma kind of wiggler has been considered as one of the important alternative scheme to amplify the THz radiation emission. The present scheme reveals that the amplification of THz radiation wave depends on the beam energy, beam density, plasma density and modulation index and the desired radiation frequency can be tuned using plasma wiggler period and beam energy.

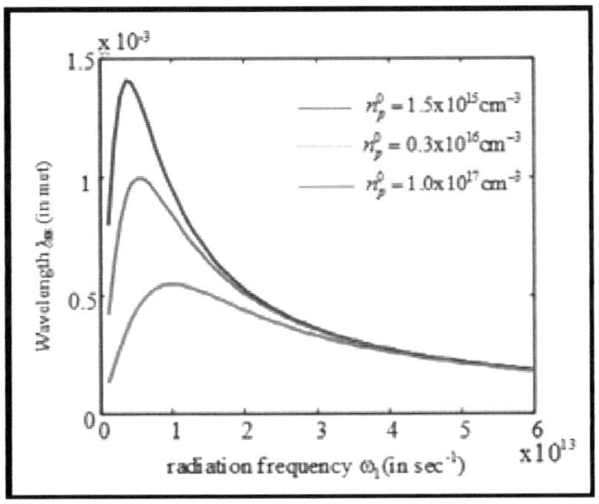

Figure 1. Figure showing that the periodic plasma wavelength decreases as plasma density increases and becomes sharper for THz radiation

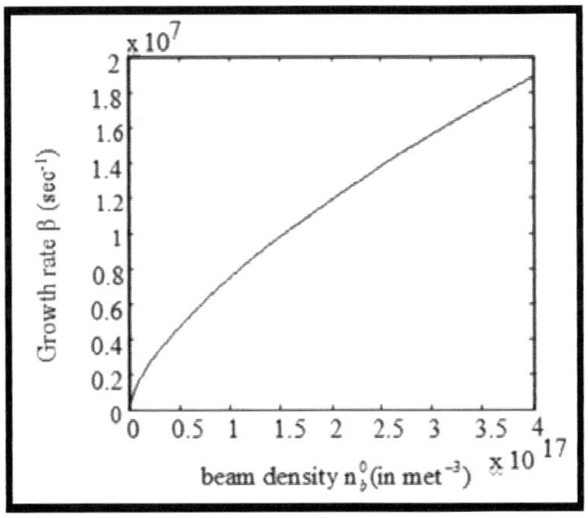

Figure 2. Variation of growth rate β of radiation wave with beam density n_b^o.

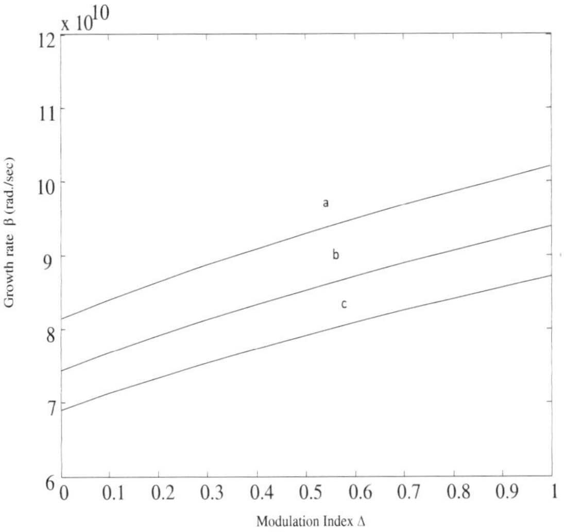

Figure 3. Variation of growth rate (β) with modulation index (Δ) at beam energy (a) E_b=1.0 MeV (b) E_b=2.0 MeV and (c) E_b=3.0 MeV

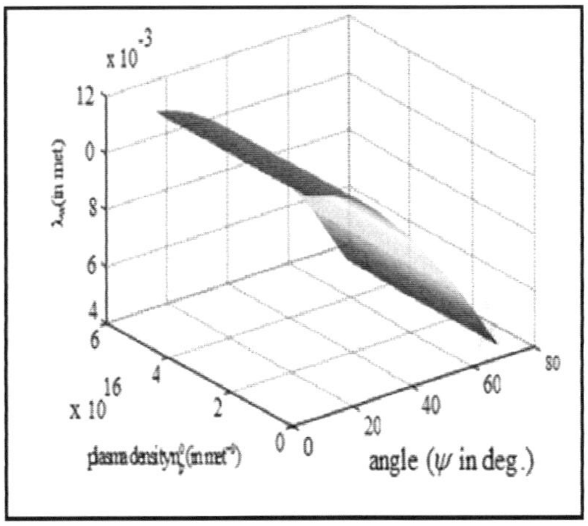

Figure 4. Variation of periodic plasma plasma wavelength λ_w with angle ψ and density n_p^o.

References

[1] Ferguson B and Zhang XC 2002 *Nat. Mater.* **1** 26
[2] Tonouchi M 2007 *Nat. Photon.* **1** 97
[3] Leemans W P, Tilborg J Van, Faure J, Geddes C G R, Toth C, Schroeder C B , Esarey E, Fubiani G and Dugan D 2004 *Phys. Plasmas* **11** 899
[4] HashimshonyD, Zigler A and Papadopoulos K 2001 *Phys. Rev. Lett.* **86** 2806

[5] Zhang X C, Bhu, Darrow J and Auston D 1990 *Appl. Phys. Lett.* **56** 1011
[6] Faist J, Capasso F, Sivco D L, Sirtori C, Hutchinson AL and Cho AY 1994 *Science* **264** 553
[7] Joshi C, Katsouleas T, Dawson J M, Yan YT and Slater J M1987 *IEEE Trans. J. Quant. Elect.* **23** 1571
[8] Miano R G and Vaccaro V G 1990 *Phys. Scr.* **30** 192
[9] Namiot V A and Shchurova LY 2012 *Phys. Lett.* A **375** 2759
[10] Zhang S C 2009 *Phys. Plasmas* **16** 093107

IOP Conf. Series: Journal of Physics: Conf. Series **836** (2017) 012002 doi:10.1088/1742-6596/836/1/012002

Nonlinear eigen-mode structures in complex astroclouds

P K Karmakar[a] and A Haloi

Department of Physics, Tezpur University, Napaam-784028, Tezpur, Assam, India

E-mail: [a)]pkk@tezu.ernet.in

Abstract. The evolutionary dynamics of strongly nonlinear waves (of arbitrary amplitude)in an inhomogeneous complex astrophysical viscous cloud is investigated without recourse to any kind of swindle. It consists of warm lighter electrons and ions (Boltzmanian); and cold massive bi-polar dust grains (inertial fluids) alongside vigorous neutral dynamics in quasi-neural hydrodynamic equilibrium. Application of the Sagdeev pseudo-potential method transforms the analytic model into a conjugated pair of intermixed non-integrable energy integral laws. A natural excitation of electrostatic quasi-monotonic compressive dispersive shock-like eigen-modes is numerically demonstrated. In contrast, the self-gravitational waves grow purely as non-monotonic compressive oscillatory shock-like structures. The unique features of both the distinct classes are depicted. Their non-trivial significance in the astro-context is emphasized.

1. Introduction

The nonlinear astrophysical eigen-mode dynamics amid gravito-electrostatic interplay have been an interesting area of research due to its diversified roles played in the interstellar space and cosmic environments. Their significance in the transport processes of fluid material in star and bounded equilibrium structure formation mechanisms, via the re-distribution of involved dynamical properties leading to astro cloud fragmentation into clumpy substructures, is well known [1]. Such eigen-patterns cause kinetic energization of astro particles responsible for different phenomenological tremendous effects on the Jet-induced (triggered) mode of star formation evolutions yet to be well explored [1-3]. At this backdrop, we herein develop an analytic model to see the strongly nonlinear gravito-electrostatic eigen-modes in a complex inhomogeneous bi-polar astro cloud in the Sagdeev pseudo-potential framework [4] on the relevant astro fluid scales of space and time. The dynamics and relevancy of the explored shock-like structures in the astrophysical contexts are presented.

2. Model formulation

We consider an inhomogeneous viscous astrophysical cloud model consisting of warm lighter electrons and ions (Boltzmannians); and cold massive bi-polar dust grains with partial ionization (fluids)in quasi-neural hydrodynamic equilibrium in a flat geometry (1-D).The multi-grains have equal polytropic indices of $\gamma_{d+} = \gamma_{d-} = \gamma_{dn} = \gamma = 3$ [2, 5].We, for analytic simplicity, ignore the complications, such as turbulence, non-thermal distributions, cosmic ray interactions, etc. The electronic and ionic dynamics are governed by the Boltzmann distribution laws in normalized form with all the generic notations [3] respectively as

$$N_e = N_{e0} \exp(\Phi) \quad (1)$$

$$N_i = N_{i0} \exp(-\Phi). \quad (2)$$

The cloud dust dynamics are dictated by the modified normalized equations of continuity, momentum balance, adiabatic pressure and coupling electro-gravitational Poisson potential respectively cast as

$$\left(\partial N_{dj}/\partial T\right)+\left[\partial\left(N_{dj}M_{dj}\right)/\partial X\right]=0, \tag{3}$$

$$\frac{\partial M_{dj}}{\partial T}+M_{dj}\frac{\partial M_{dj}}{\partial X}=-\delta_{d-,dj}\left(\frac{q_{dj}}{e}\right)\frac{\partial\Phi}{\partial X}-3\delta_{d-,dj}\left(\frac{T_{dj}}{T_p}\right)N_{dj}\frac{\partial N_{dj}}{\partial X}-\frac{\partial\Psi}{\partial X}+\kappa_{dj}\frac{\partial^2 M_{dj}}{\partial X^2}, \tag{4}$$

$$\frac{\partial P_{dj}}{\partial T}+M_{dj}\frac{\partial P_{dj}}{\partial X}+\gamma P_{dj}\frac{\partial M_{dj}}{\partial X}=0, \tag{5}$$

$$\frac{\partial^2\Phi}{\partial X^2}=\beta\left[n_{e0}N_e-n_{i0}N_i+Z_{d-}n_{d-0}N_{d-}-Z_{d+}n_{d+0}N_{d+}\right], \tag{6}$$

$$\frac{\partial^2\Psi}{\partial X^2}=\frac{1}{\rho_{d0}}\left[m_{d-}n_{d-0}N_{d-}+m_{d+}n_{d+0}N_{d+}+m_{dn}n_{dn0}N_{dn}\right]. \tag{7}$$

The independent spatiotemporal coordinates, *X* and *T*, are normalized by the Jeans length λ_J and Jeans time $\omega_J^{-1}=\left(c_{ss}/\lambda_J\right)^{-1}$; respectively. N_e, N_i and N_{dj} are the normalized concentrations of electrons, ions and dust species. The normalization is by the respective equilibrium values n_{e0}, n_{i0} and n_{dj0}, with *j* = '+'for positively charged grains), '-' for negative grains and '*n*' for neutral grains. Next, $q_{dj}=Z_{dj}|e|$ is the grain charge in terms of the charge number Z_{dj} and electronic charge *e*. $\delta_{d-,dj}=m_{d-}/m_{dj}$ denotes the mass ratio of the negative to the jth dust species and $\beta=e^2/(\rho_{d0}m_{d-}G)$.The universal gravitational constant, through which gravitating matter interacts, is denoted by *G* (= 6.67×10^{-11} m^3 kg^{-1} s^{-2}). κ_{dj} is the normalized(by Jeans value) kinematic viscosity of the jth dust fluid. The fluid velocity M_{dj} is normalized by the dust acoustic phase speed, $c_{ss}=\left(T_p/m_{d-}\right)^{1/2}$, with $T_e \sim T_i = T_p$ ($>>T_{dj}$, the jth species temperature) as the plasma temperature (in *e*V). $P_{dj}=p_{dj}/p_{dj0}=N_{dj}^{\ \gamma}$ is the dust adiabatic pressure normalized by the equilibrium dust isothermal pressure, $p_{dj0}=n_{dj}T_{dj}$. Lastly, the electrostatic and gravitational potentials, Φ and Ψ, are normalized by T_p/e and $c_{ss}^{\ 2}$; respectively.

3. Eigen-mode structure equations

To apply the well-known Sagdeev pseudo-potential method [4],all the physical dependent variables in equations (1)-(7) are transformed into a time-independent form by the Galilean coordinate transformation, $\xi=X-\mu T$, with μ denoting the normalized (by c_{ss}) reference frame velocity. We introduce two integral functions as, $f_{dj}(\Phi)=\int_0^{\Phi}N_{dj}\,d\Phi$, approximating pure electrostatic case, and $g_{dj}(\Psi)=\int_0^{\Psi}N_{dj}\,d\Psi$, assuming pure self-gravitational one. We analytically solve for N_{dj} under the appropriate boundary conditions as, $N_e\to1$, $N_i\to1$, $N_{dj}\to1$, $M_j\to0$, $\Phi\to0$, $\Psi\to0$, $\partial\Phi/\partial\xi\to0$ and $\partial\Psi/\partial\xi\to0$ at $\xi\to\pm\infty$ for local disturbances(with $\kappa_{dj}\neq0$). It leads to the inter-coupled energy integral laws as

$$(1/2)(\partial\Phi/\partial\xi)^2+V_E(\Phi,\Psi)=0, \tag{8}$$

$$(1/2)(\partial\Psi/\partial\rho)^2+V_G(\Phi,\Psi)=0, \tag{9}$$

where the electrostatic and gravitational Sagdeev potentials are respectively given as

$$V_E(\Phi,\Psi)=\beta\left[\left\{n_{e0}+n_{i0}+Z_{d-}n_{d-0}f_{d-}(\Phi)\Big|_{\Phi=0,\frac{\partial\Phi}{\partial\xi}=0}-Z_{d+}n_{d+0}f_{d+}(\Phi)\Big|_{\Phi=0,\frac{\partial\Phi}{\partial\xi}=0}\right\}\right.$$

$$-\left\{n_{e0}e^{\Phi}+n_{i0}e^{-\Phi}+Z_{d-}n_{d-0}f_{d-}(\Phi)-Z_{d+}n_{d+0}f_{d+}(\Phi)\right\}\Big], \tag{10}$$

$$V_G(\Phi,\Psi)=\frac{1}{\rho_0}\left[\left\{m_{d+}n_{d+0}g_{d+}(\Psi)\big|_{\Psi=0,\frac{\partial\Psi}{\partial\xi}=0}+m_{d-}n_{d-0}g_{d-}(\Psi)\big|_{\Psi=0,\frac{\partial\Psi}{\partial\xi}=0}+m_{dn}n_{dn0}g_{dn}(\Psi)\big|_{\Psi=0,\frac{\partial\Psi}{\partial\xi}=0}\right\}\right.$$
$$\left.-\left\{m_{d+}n_{d+0}g_{d+}(\Psi)+m_{d-}n_{d-0}g_{d-}(\Psi)+m_{dn}n_{dn0}g_{dn}(\Psi)\right\}\right] \tag{11}$$

We perform analytical tests to check the existential conditions for nonlinear coherent structures to exist. It is seen that equation (10)-(11) satisfy the following conditions for compressive shock-like structures as

$$V_{E(G)}(\Phi,\Psi)=0,\ \left[\partial V_{E(G)}(\Phi,\Psi)/\partial\Phi(\Psi)\right]\neq 0,\text{at } \Phi=0,\ \Psi=0 \tag{12a}$$

$$\left[\partial^2 V_{E(G)}(\Phi,\Psi)/\partial\Phi^2(\Psi^2)\right]<0,\ \text{at } \Phi=0,\ \Psi=0 \tag{12b}$$

$$V_{E(G)}(\Phi,\Psi)\neq 0,\ \text{at } \Phi=\Phi_{max}\ (\Psi=\Psi_{max}) \tag{12c}$$

$$V_{E(G)}(\Phi,\Psi)<0,\ \text{at } 0<|\Phi|<|\Phi_{max}|\,(0<|\Psi|<|\Psi_{max}|) \tag{12d}$$

In the above equations except in the arguments, maps the gravitational counterparts. So, $V_E(\Phi,\Psi)$and$V_G(\Phi,\Psi)$fulfil all the conditions equation (12), for the phylogenes is of compressive shock-like patterns.

4. Results and discussions

We numerically solve equations (8)-(9) for exact eigen-structure characterization with the fourth-order Runge-Kutta method [6] in judicious plasma parameter windows[2, 7]to get the results (figures 1-2).

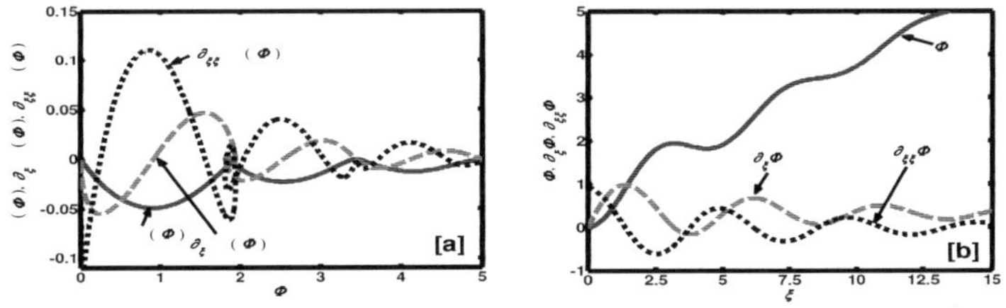

Figure 1. Profile of the normalized electrostatic [a] Sagdeev potential [$V_E(\Phi,\Psi)$], gradient [$\partial_\xi V_E(\Phi,\Psi)$] and differential curvature [$\partial_{\xi\xi}V_E(\Phi,\Psi)$]; and [b] physical potential [Φ], gradient [$\partial_\xi\Phi$] and differential curvature [$\partial_{\xi\xi}\Phi$] for μ = 2.98. Fine details are discussed in the text.

Figure 1 shows profile of the normalized electrostatic [a] Sagdeev potential [$V_E(\Phi,\Psi)$,deflated by dividing with 10^{-17}, blue solid line], its gradient [$\partial_\xi V_E(\Phi,\Psi)$,by dividing with 10^{-17}, red dashed line] and its differential curvature [$\partial_{\xi\xi}V_E(\Phi,\Psi)$,by dividing with 10^{-17},black dotted line]; and [b] physical potential [Φ , rescaled by dividing with 10^{-9},blue solid line], its gradient [$\partial_\xi\Phi$,by dividing with 10^{-9}, red dashed line]and its differential curvature [$\partial_{\xi\xi}\Phi$,by dividing with 10^{-9},black dotted line]for μ = 2.98.Different inputs are $(\xi)_i = 1.00\times10^{-2}$ with $\Delta\xi = 1.00\times10^{-2}$, $(\Phi)_i = 2.00\times10^{-9}$, $(\Phi_\xi)_i = 1.00\times10^{-11}$, $(\Psi)_i = 1.00\times10^{-4}$, and $(\Psi_\xi)_i = 1.00\times10^{-3}$. The other parameters kept fixed are $n_{e0} = 5.00\times10^3$ m^{-3}, $n_{i0} = 5.00\times10^3$ m^{-3}, $n_{d-0} = 7.00\times10^{-1}$ m^{-3}, $n_{d+0} = 1.00\times10^{-1}$ m^{-3}, $n_{dn0} = 9.00\times10^{-1}$ m^{-3}, $Z_{d-} = 1.50\times10^2$, $Z_{d+} = 1.00\times10^2$, $m_{d-} = 2.80\times10^{-8}$ kg, $m_{d+} = 1.00\times10^{-8}$ kg, $m_{dn} = 1.00\times10^{-11}$ kg, $\alpha_1 = 1.10\times10^{-2}$, $\alpha_2 = 1.20\times10^{-2}$, $\alpha_3 = 1.00\times10^{-2}$, $\kappa_{d-} = 2.00\times10^{-2}$, $\kappa_{d+} = 2.00\times10^{-2}$, and $\kappa_{dn} = 1.00\times10^{-2}$. The electrostatic Sagdeev potential, $V_E(\Phi,\Psi)$, evolves as a rhythmic chain of rarefactive disturbances with gradually decreasing amplitude (Fig. 1[a]). It is seen numerically as well that $V_E(\Phi,\Psi)$ satisfies all the approximate analytic conditions (see(12)) for the evolution of compressive shock-like structures. The

Sagdeev field, $\partial_\xi V_E(\Phi,\Psi)$, shows damped oscillatory propagation as hybrid inter-mixture of gradually attenuated rarefactive and compressive disturbances (figure 1(a)). The Sagdeev potential curvature, $\partial_{\xi\xi} V_E(\Phi,\Psi)$, likewise depicts the phase trajectory as an admixture of compressive and rarefactive patterns with decreasing wave amplitude. The physical potential Φ, evolves as quasi-monotonic compressive dispersive shock-like pattern for μ =2.98 (figure 1(b)). The field, $\partial_\xi \Phi$ propagates as damped oscillatory compressive disturbances; whereas, the curvature $\partial_{\xi\xi}\Phi$ shows analogous features as figure1(a).It shows that the deviation from exact global quasi-neutrality is more pronounced near the vicinity of the cloud centre than elsewhere.

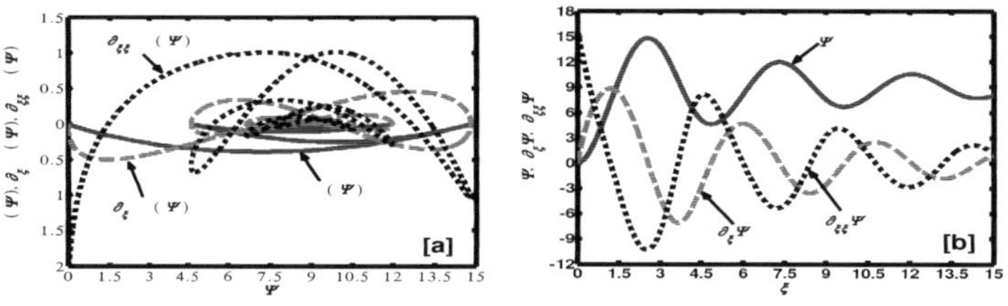

Figure 2. Same as figure 1, but for the gravitational wave dynamics.

Figure 2 depicts the normalized self-gravitational [a] Sagdeev potential [$V_G(\Phi,\Psi)$,rescaled by multiplying with 10^{-2},blue solid line], its gradient [$\partial_\xi V_G(\Phi,\Psi)$, by multiplying with 10^{-2}, red dashed line]and its differential curvature [$\partial_{\xi\xi} V_G(\Phi,\Psi)$, by multiplying with 10^{-2}, black dotted line]; and [b] physical potential [Ψ], its gradient [$\partial_\xi \Psi$] and its differential curvature [$\partial_{\xi\xi}\Psi$]as figure 1.Clearly, $V_G(\Phi,\Psi)$ evolves as oscillatory rarefactive disturbance with similar attenuation (figure2[a]). It fulfils the analytic conditions(see equation(12)) for compressive shock-like shapes. The Sagdeev field, $\partial_\xi V_G(\Phi,\Psi)$, shows oscillatory propagation of gradually attenuated rarefactive and compressive disturbances (figure 2(a)). The Sagdeev potential curvature, $\partial_{\xi\xi} V_G(\Phi,\Psi)$, shows samefeatures as Fig. 1[a].The self-gravitational potential, Ψ, evolves as non-monotonic compressive oscillatory shock-like structures (figure 2(b)). Its field $\partial_\xi\Psi$ and curvature $\partial_{\xi\xi}\Psi$ show almost the same as figure 1(b).

5. Conclusions

The strongly nonlinear behaviour of gravito-electrostatic waves in inhomogeneous self-gravitating viscous multi-fluidic complex plasma are studied. It is methodologically executed in the Sagdeev pseudo-potential framework. The analysis depicts quasi-monotonic compressive dispersive shock-like fluctuations (electrostatic) and non-monotonic compressive oscillatory shock-like fluctuations (self-gravitational) to exist in the astro cloud. The results may be useful in understanding diverse nonlinear wave activities in cosmic, interstellar space and astrophysical environments contributing to the formation mechanisms of bounded equilibrium astro-structures via dynamic cloud collapse processes.

Acknowledgement
The financial support of DST, GoI, Grant No. SR/FTP/PS-021/2011), is thankfully recognized.

References

[1] Bodenheimer P H 2011 *Principles of Star formation* (New York: Spinger)
[2] Verheest F 2000 *Waves in Dusty Space Plasmas* (Dordrecht: Kluwer)
[3] Haloi A and Karmakar P K 2015 *Astrophys. Space Sci.* **358** 41
[4] Sagdeev R Z 1966 *Rev. Plasma Phys.* **4** 23
[5] Ahmad Z, Mushtaq A and Mamun A A 2013 *Phys. Plasmas* **20** 032302
[6] Kiusalaas J 2005 *Numerical Methods in Engineering with MATLAB* (New York: Cambridge)
[7] Spitzer L 1978 *Physical Processes in the Interstellar Medium* (New York: John Wiley)

IOP Conf. Series: Journal of Physics: Conf. Series **836** (2017) 012003 doi:10.1088/1742-6596/836/1/012003

Effect of lighter and heavier pair ions on solitary waves

S Sijo, V Anu, M Manesh, G Sreekala, T W Neethu, D E Savithri and C Venugopal[a]

School of Pure & Applied Physics, Mahatma Gandhi University, Priyadarshini Hills, Kottayam 686 560, Kerala, India

E-mail: [a)]cvgmgphys@yahoo.co.in

Abstract. We investigate solitary waves in a multicomponent plasma of positively and negatively charged lighter and heavier ionsand electrons ofsolar and cometary origin; the electron components are modeled by kappa distributions. The solitary wave solution of the Zakharov-Kuznetsov (ZK) equation is plotted for parameters of comet Halley; physical variables like densities and temperatures of negatively charged lighter ions significantly affect the width and amplitude of the solitary waves.

1. Introduction

A cometary plasma is composed of many species of ions including O^+, H_2^+, He^+, etc[1]. Apart from these ions, presence of solar wind protons and electrons and photo-electronsmakes it a truemuti-ion plasma [2-5].

In addition tothe above positive ions, negative ions [6] and co-existence of both positively and negatively charged ions are also possible [7].

As regards electrons, the spacecraft ICE observed photo-electrons in the tail of comet Giacobini-Zinner[8]. The deviations from typical Maxwellian distribution make the use of kappa distribution due to a highly energetic tail [9]. And finally, Voelzke and Izaguirreanalyzeda number of images of nonlinear structures of comet Halley and some solitary structures were clearly identified [10].

We are thus interested in solitary waves in a sixcomponent plasma. The lighter and heavier pair ions are treated as cold, while both the electrons are described by a Boltzmann-like distribution given by

$$n_s = n_{s0}\left[1+e_s\psi\left(k_B T_s(\kappa_s-3/2)\right)^{-1}\right]^{-\kappa_s+1/2} \tag{1}$$

where s $=se$(for solar electrons) or ce(cometary photo-electrons). $n$denotes the density (the subscript 0 denotes the equilibrium value), e_s the charge, T_s the temperature and κ_s the spectral index for the species 's'. k_B is the Boltzmann's constant and ψ , the potential.

The normalized form of the equations of continuity and motion for the lighter and heavier pair ions along with the Poisson's equation are given by

$$\partial_t n_j + \nabla.(\mathrm{n}_j \mathrm{u}_j) = 0 \tag{2}$$

$$\partial_t u_j + (u_j.\nabla)u_j = q\alpha_k\beta_k\nabla\psi + q(u_{jz}\hat{y} - u_{jy}\hat{z})\Omega_j - 5/3\,\beta_{+,-,}\sigma_{+,-}n_{+,-}^{-1/3}\nabla n_{+,-} \tag{3}$$

$$q=\pm;\; k=+,-,2;\; j=H^+,H^-,1,2$$

$$\nabla^2\psi = n_1 - (1 - z_{H+}\mu_{H+} + z_{H-}\mu_{H-} + \mu_{se} + \mu_{ce})n_2 - (1 - z_{H+}\mu_{H+} - z_2\mu_2 + \mu_{se} + \mu_{ce})n_{H-} - (1 - z_2\mu_2 + z_{H-}\mu_{H-} + \mu_{se} + \mu_{ce})n_{H+} + \mu_{se}(1 - \psi\left(\sigma_{se}(\kappa_{se} - 3/2)\right)^{-1})^{(-\kappa_{se}+1/2)} + \mu_{ce}(1 - \psi\left(\sigma_{ce}(\kappa_{ce} - 3/2)\right)^{-1})^{(-\kappa_{ce}+1/2)} \quad (4)$$

In (2) to (4) n_j and u_j are the number densities and fluid speeds for the species j (j= H^+, H^-, 1(negative heavier ions) and 2 (positive heavier ions)), normalized by their equilibrium values n_{j0} and $\left(z_1 k_B T_1 (m_1)^{-1}\right)^{1/2}$ respectively. ψ, which is now the normalized electric potential, is normalized by $k_B T_1 (e)^{-1}$. Space and time (t) coordinates are respectively normalized by the Debye length $\lambda_{D1} = \left(z_1 k_B T_1 \left(4\pi z_1^2 e^2 n_{10}\right)^{-1}\right)^{1/2}$ and the inverse of the plasma frequency $\omega_{p1}^{-1} = \left(m_1 \left(4\pi z_1^2 e^2 n_{10}\right)^{-1}\right)^{1/2}$. Also $\alpha_i = z_i z_1^{-1}$, $\alpha_2 = z_2 z_1^{-1}$, $\beta_i = m_1 m_i^{-1}$, $\beta_2 = m_1 m_2^{-1}$, $\mu_j = n_{j0}\left(z_1 n_{10}\right)^{-1}$ and $\sigma_j = T_j T_1^{-1}$. T_j and T_1 are the temperatures of species j and negative heavier ions respectively. m_i, m_1 and m_2 are, respectively, the masses of hydrogen ions, negatively and positively charged heavier ions while z_i, z_1 and z_2 are the corresponding charge numbers.

To derive ZK equation, we introduce new variables as:

$$X = \varepsilon^{1/2}(x - \lambda t), \quad Y = \varepsilon^{1/2} y, \quad Z = \varepsilon^{1/2} z, \quad \tau = \varepsilon^{3/2} t$$

All physical quantities are expressed asymptotically as a power series as:

$$n_j = 1 + \varepsilon n_j^{(1)} + \varepsilon^2 n_j^{(2)} + ...; \qquad u_{xj} = \varepsilon u_{xj}^{(1)} + \varepsilon^2 u_{xj}^{(2)} + ...$$

$$u_{y,zj} = \varepsilon^{3/2} u_{y,zj}^{(1)} + \varepsilon^2 u_{y,zj}^{(2)} + ...; \qquad \psi = \varepsilon\psi^{(1)} + \varepsilon^2\psi^{(2)} + ...$$

By above substitutions and equating different powers of ε, we arrive at the ZK equation as,

$$\partial_\tau \psi^{(1)} + A\psi^{(1)}\partial_X \psi^{(1)} + \partial_X\left(B\partial^2_X \psi^{(1)} + C(\partial^2_Y + \partial^2_Z)\psi^{(1)}\right) = 0 \quad (5)$$

where $A = A1(A0)^{-1}$, $B = \lambda^4 (A0)^{-1}$ and $C = A2(A0)^{-1}$

$$A0 = 2\lambda(1 + (1 - z_{H+}\mu_{H+} + z_{H-}\mu_{H-} + \mu_{se} + \mu_{ce})\alpha_2\beta_2 - 9\alpha_{H-}\beta_{H-}\left(3 - \frac{5\beta_{H-}\sigma_{H-}}{\lambda^2}\right)^{-2} \times (1 - z_{H+}\mu_{H+} - z_2\mu_2 + \mu_{se} + \mu_{ce}) + 9\alpha_{H+}\beta_{H+}\left(3 - \frac{5\beta_{H+}\sigma_{H+}}{\lambda^2}\right)^{-2}(1 - z_2\mu_2 + z_{H-}\mu_{H-} + \mu_{se} + \mu_{ce}))$$

$$A1 = -3 + (1 - z_{H+}\mu_{H+} + z_{H-}\mu_{H-} + \mu_{se} + \mu_{ce})3\alpha_2^2\beta_2^2 + (1 - z_{H+}\mu_{H+} - z_2\mu_2 + \mu_{se} + \mu_{ce})3\alpha_{H-}^2\beta_{H-}^2 \frac{27 - \frac{5\beta_{H-}\sigma_{H-}}{\lambda^2}}{\left(3 - \frac{5\beta_{H-}\sigma_{H-}}{\lambda^2}\right)^3} + (1 - z_2\mu_2 + z_{H-}\mu_{H-} + \mu_{se} + \mu_{ce})3\alpha_{H+}^2\beta_{H+}^2 \frac{27 - \frac{5\beta_{H+}\sigma_{H+}}{\lambda^2}}{\left(3 - \frac{5\beta_{H+}\sigma_{H+}}{\lambda^2}\right)^3} - \lambda^4\left(\frac{\mu_{ce}}{\sigma_{ce}^2}\frac{(\kappa_{ce} - 1/2)(\kappa_{ce} + 1/2)}{(\kappa_{ce} - 3/2)^2} + \frac{\mu_{se}}{\sigma_{se}^2}\frac{(\kappa_{se} - 1/2)(\kappa_{se} + 1/2)}{(\kappa_{se} - 3/2)^2}\right)$$

$$A2 = \lambda^4 (1 + \frac{1}{\Omega_1^2} + (1 - z_{H+}\mu_{H+} + z_{H-}\mu_{H-} + \mu_{se} + \mu_{ce})\frac{\alpha_2\beta_2}{\Omega_2^2} - (1 - z_{H+}\mu_{H+} - z_2\mu_2 + \mu_{se} + \mu_{ce})\frac{9\alpha_{H-}\beta_{H-}}{\Omega_{H-}^2\left(3 - \frac{5\beta_{H-}\sigma_{H-}}{\lambda^2}\right)^2}$$

$$+ (1 - z_2\mu_2 + z_{H-}\mu_{H-} + \mu_{se} + \mu_{ce}) 9\alpha_{H+}\beta_{H+}\left(\Omega_{H+}^2\left(3 - \frac{5\beta_{H+}\sigma_{H+}}{\lambda^2}\right)^2\right)^{-1})$$

The solution of the equation (5) can be written as $\psi = \psi_m Sech^2\left(\xi (W)^{-1}\right)$ where $\xi = l_x X + l_y Y + l_z Z - u_0 \tau$ is the transformed coordinate, $\psi_m = 3u_0 \left(l_x A\right)^{-1}$ is the amplitude and $W = 2\sqrt{\left(Bl_x^3 + Cl_x(l_y^2 + l_z^2)\right)u_0^{-1}}$ is the width of the soliton. l_x, l_y and l_z are direction cosines.

2. Results and discussions

We use parameters observed at comet Halley for our computations: the density of lighter hydrogen ions $n_{i0} = 4.95\,cm^{-3}$;(temperature $T_i = 8\times10^4 K$) and the solar electron temperature $T_{se} = 2\times10^5 K$ [11]. The temperature of cometary photo-electrons $T_{ce} = 2\times10^4 K$. The number densities of negatively and positively charged heavier oxygen ions were set at $n_{10} = 0.05\,cm^{-3}$ and $n_{20} = 0.5\,cm^{-3}$ respectively. The heavier oxygen pair ions were detected at an energy of the order of $1\,eV$ [6]. The other parameters for the figures are $\kappa_{ce} = 7/2$, $\kappa_{se} = 11/2$, $z_2 = 1$ and $z_1 = 2$.

Figure 1 is a plot of the width W of the solitary wave vs. number density of negatively charged lighter ions for different phase velocities. From the plots, it is seen that as number densities of negatively charged lighter ions increase, the width of the solitary wave increases (left panel) or decreases (middle and right panels); i.e., the values of phase velocities play a significant role in the shape of the solitary waves.A plot, similar to figure 1, of the amplitude ψ_m versus the density of the negatively charged lighter ions n_{H-} , shows that the amplitude ψ_m is also affected by n_{H-} .

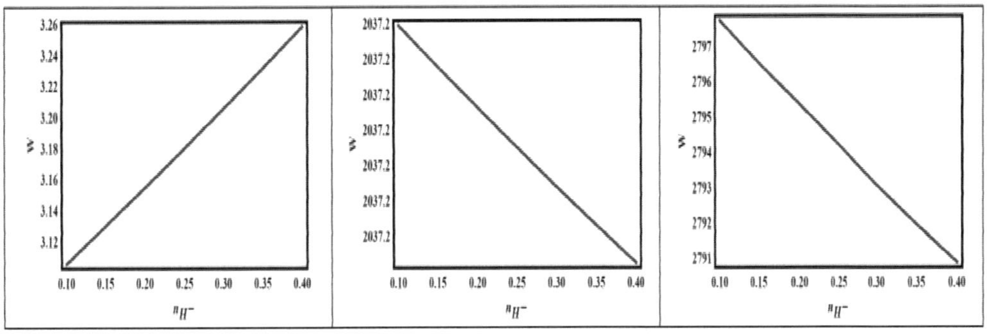

Figure 1. Width W of the solitary waves vs. densities n_{H-} for different phase velocities.

The plot of the width W of the solitary wave vs. temperatures of negatively charged lighter ions for different phase velocities is depicted in figure 2. From the plots, it is seen that as temperatures of negatively charged lighter ions increases, the width of the solitary wave decreases (left panel) or increases (middle and right panels); i.e. the values of phase velocities and temperatures of H^- ions play a significant role in the shape of the solitary waves.

Finally, variations of amplitude of the solitary waves vs. temperatures of negatively chargedlighter ions for different phase velocities was also studied in figure 3. Unlike figure 2, amplitude of the solitary waves decreased (corresponding to left and middle panels of figure 2) or increased (right panel) as temperatures of negatively charged lighter ions increased. Also, the middle panel corresponds to a rarefactive solitary wave.

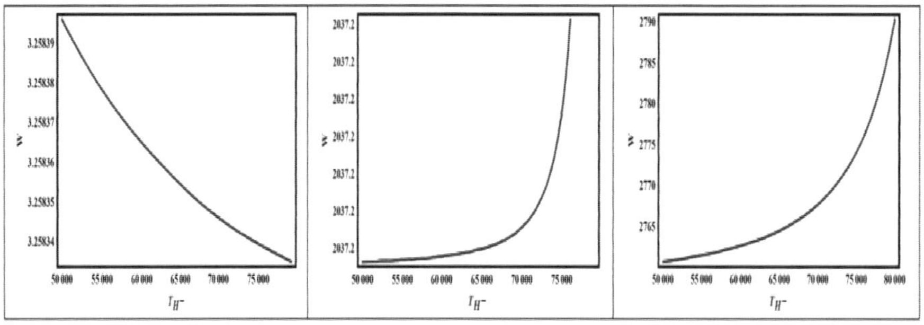

Figure 2. Width W of the solitary waves vs. temperature T_{H-} for different phase velocities.

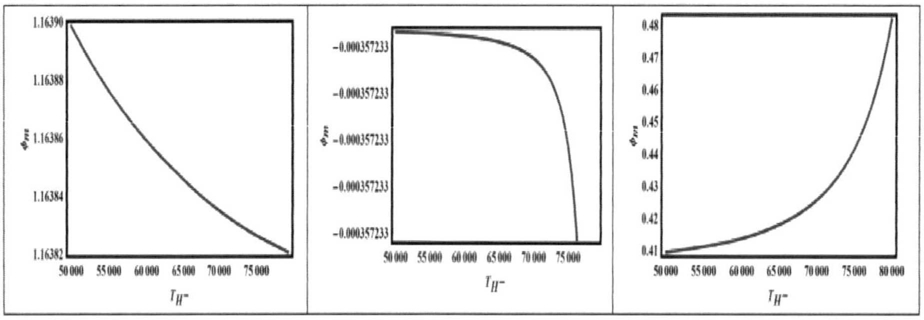

Figure 3. Amplitude ψ_m of the solitary waves vs. temperature T_{H-} for different phase velocities.

3. Conclusions

We have investigated the influence of lighter and heavier pair ions on solitary waves in a multi component plasma. The components in the plasma are lighter and heavier positively and negatively charged ions, hot electrons of solar origin and photo-electrons of cometary origin. We find that for different values of densities and temperatures of negatively charged lighter ions affects the width and amplitude of the solitary waves. Also polarities of the solitary waves are dependent on the phase velocities.

References

[1] Balsiger H *et al.* 1986 *Nature* **321** 330
[2] Rubin M, Hansen K C, Gombosi T I, Combi M R, Altwegg K and Balsiger H 2009 *Icarus* **199** 505
[3] Ipavich F M *et al.* 1986 *Science* **232** 366
[4] Neugebauer M *et al.* 2007 *Astrophys. J.* **667** 1262
[5] Kharchenko V *et al.* 2003 *Astrophys. J. Lett.* **585** L73
[6] Chaizy P *et al.* 1991 *Nature* **349** 393
[7] Ellis T A and Neff J S 1991 *Icarus* **91** 280
[8] Zwickl R D, Baker D N, Bame S J, Feldman W C, Fuselier S A, Huebner W F, McComas D J and Young D T 1986 *Geophys. Res. Lett.* **13** 401
[9] Vasyliunas V M 1968 *J. Geophys. Res.* **73** 2839
[10] Voelzke M R and Izaguirre L S 2012 *Planet. Space. Sci.* **65** 104
[11] Brinca A L and Tsurutani B T 1987 *Astron. Astrophys.* **187** 311

IOP Conf. Series: Journal of Physics: Conf. Series **836** (2017) 012004 doi:10.1088/1742-6596/836/1/012004

Dynamics of streaming instability with quantum correction

H P Goutam and P K Karmakar[a]

Department of Physics, Tezpur University, Napaam-784028, Tezpur, Assam, India

E-mail: [a)]pkk@tezu.ernet.in

Abstract. A modified quantum hydrodynamic model (m-QHD) is herein proposed on the basis of the Thomas-Fermi (TF) theory of many fermionic quantum systems to investigate the dynamics of electrostatic streaming instability modes in a complex (dusty) quantum plasma system. The newly formulated m-QHD, as an amelioration over the existing usual QHD, employs a dimensionality-dependent Bohmian quantum correction prefactor, $\gamma = [(D\text{-}2)/3D]$, in the electron quantum dynamics, where D symbolizing the problem dimensionality under consideration. The normal mode analysis of the coupled structure equations reveals the excitation of two distinct streaming modes associated with the flowing ions (against electrons and dust) and the flowing dust particulates (against the electrons and ions). It is mainly shown that the γ-factor introduces a new source of stability and dispersive effects to the ion-streaming instability solely; but not to the dust counterparts. A non-trivial application of our investigation in electrostatic beam-plasma (flow-driven) coupled dynamics leading to the development of self-sustained intense electric current, and hence, of strong magnetic field in compact astrophysical objects (in dwarf-family stars) is summarily indicated.

1. Introduction

The existence of streaming instability in various quantum plasma fluids has recently gathered immense attraction of interdisciplinary researchers because of a number of fundamental reasons realizable in both laboratory and compact astrophysical environments [1]. This kind of instability, driven purely by plasma flow currents, acts as a natural source mechanism to the excitation process of various coherent structures in the form of solitons, shocks, vortices, and so forth [1-2]. It can play a significant role also in the generation mechanism of highly energized particles, self-sustained intense electric field and strong magnetic field in compact astro-objects (dwarf-family stars) [3].

A complex (dusty) quantum plasma has long been known to support two distinct types of such instabilities, ion-streaming instability and dust-streaming instability, depending purely on the dynamical scales of space and time [4]. The involved technique has applied a normal quantum hydrodynamic model (QHD). However, application of normal quantum hydrodynamic model (QHD) in the description of evolutionary dynamics of flow-driven instabilities in quantum plasmas in diversified situations is inadequate and improper [5]. The electrostatic instability mechanisms operative in a complex viscoelastic quantum plasma in the light of a modified quantum hydrodynamic model (m-QHD) has recently been investigated. The modification here has implemented a quantum mechanical correction prefactor (γ) in the front of the Bohm potential associated with the electronic quantum dynamics. The m-QHD has been derived on the basis of the Thomas-Fermi (TF) theory of many fermionic quantum systems with the spin-dynamics correctly considered. It has non-trivially been shown that the correction prefactor acts as a stabilizer for the ion-stream case only.

The main motivation of this paper is to illuminate an analytic forum to understand the instability growth dynamics and associated features in the parametric domains previously remaining unexplored. We apply the m-QHD formalism in our new complex quantum plasma system with the viscoelastic effects absolutely ignored. The model is devised with a main focus to see the instability dynamics in a pure natural form. The effective role of the correction prefactor on the instability properties and further applicability in compact astrophysical objects are specifically highlighted.

2. Model and formulation

We consider a complex quantum plasma system composed of three components. The constituents include the degenerate quantum electrons; non-degenerate classical ions and dust particulates. The global quasi-neutrality is assumed to pre-exist in hydrodynamic equilibrium. It is anticipated that the quantum plasma dynamics is governed by the TF-based m-QHD model [5]. It includes the basic equations as the continuity, momentum, and closing electrostatic Poisson equations as

$$\partial_t n_j + \partial_x (n_j v_j) = 0, \tag{1}$$

$$m_j n_j D_j v_j = -q_j n_j \partial_x \varphi - \partial_x p_j + \gamma (\hbar^2/2m_j) \partial_x \left(\partial_x^2 \sqrt{n_j} / \sqrt{n_j} \right), \tag{2}$$

$$\partial_x^2 \varphi = -4\pi \sum_j q_j n_j . \tag{3}$$

The notations n_j, v_j, m_j, q_j, and p_j represent the population density, flow velocity, mass, charge and Fermi pressure of jth species ($j = e$ for electrons, i for ions, d for dust grains). The Fermi pressure here is given by, $p_j = (m_j v^2_{Fj}/n^2_{j0}) n^3_j$, where, $v_{Fj} = \sqrt{(2k_B T_{Fj}/m_j)}$ is the Fermi velocity with k_B as the Boltzmann constant, T_{Fj} the Fermi temperature and n_{j0} the equilibrium plasma population density; respectively. The symbol ϕ represents the self-consistent plasma electrostatic potential. Also, γ and $\hbar$ respectively represent the dimensional quantum-correction prefactor and the reduced (angular) Planck constant.

A Fourier-based normal mode analysis over equations (1) – (3) against a slight perturbation about the equilibrium yields the following linear generalized dispersion relation as

$$1 - \frac{\omega_{pe}^2}{(\omega - kv_{e0})^2 - \Omega_e^2} - \frac{\omega_{pi}^2}{(\omega - kv_{i0})^2 - \Omega_i^2} - \frac{\omega_{pd}^2}{(\omega - kv_{d0})^2 - \Omega_d^2} = 0 . \tag{4}$$

Here, $\omega_{pj} = \sqrt{(4\pi q^2_j n_{j0}/m_j)}$ denotes the plasma oscillation frequency corresponding to the jth specie and $\Omega_j = \sqrt{(k^2 v^2_{Fj} + \gamma k^4 \hbar^2/4m^2_j)}$ signifies the effective frequency of quantum fluctuations. The following special cases arising from equation (4) are in order.

2.1. Ion-streaming instability.

We assume that the ions are streaming against the electrons and dust grains. Applying the conditions $|\omega - kv_{e0}| << \Omega_e$, $|\omega - kv_{i0}| >> \Omega_i$ and $\omega >> kv_{d0}$ in equation (4), one gets

$$1 + \frac{\omega_{pe}^2}{\Omega_e^2} - \frac{\omega_{pi}^2}{(\omega - kv_{i0})^2} - \frac{\omega_{pd}^2}{\omega^2} = 0 . \tag{5}$$

Assuming $\omega = kv_{i0} + \upsilon$, $\upsilon << kv_{i0}$ and $kv_{i0} \approx \omega_{pd}\Omega_e/\sqrt{(\Omega^2_e + \omega^2_{pe})}$ in equation (5), we get the desired dispersion relation in normalized (standard) form as

$$\Omega = \left(\frac{-1 + i\sqrt{3}}{2^{4/3}} \right) \left[KV_{i0} \left(K^2 + \gamma K^4 H^2 \right) \left(K^2 + \Omega_{pe}^2 + \gamma K^4 H^2 \right)^{-1} \right]^{1/3} . \tag{6}$$

here, the various variables are normalized with all usual notations as follows: $\Omega = \omega/\omega_{pi}$, $K = kv_{\mathrm{Fe}}/\omega_{pi}$, $V_{i0} = v_{i0}/v_{\mathrm{Fe}}$, $H = \hbar\omega_{pi}/2m_e v^2_{\mathrm{Fe}}$ and $\Omega_{pe} = \omega_{pe}/\omega_{pi}$.

2.2. Dust-streaming instability.
In the case of dust streaming against the electrons and ions, we apply the assumption $|\omega\text{-}kv_{e0}| << \Omega_e$, $\omega >> kv_{i0}$ and $|\omega\text{-}kv_{d0}| >> \Omega_d$ in equation (4). Proceeding in the same way as in the ion-streaming case, we get the normalized form of the dust-streaming frequency as

$$\Omega = \left(\frac{-1+i\sqrt{3}}{2^{4/3}}\right)\left[KV_{d0}\left(K^2+\gamma K^4H^2\right)\left(K^2+\Omega_{pe}^2+\gamma K^4H^2\right)^{-1}\right]^{1/3}. \tag{7}$$

The normalizations are carried out with respect to the dust-streaming parameters as: $\Omega = \omega/\omega_{pd}$, $K = kv_{\mathrm{Fe}}/\omega_{pd}$, $V_{d0} = vd_0/v_{\mathrm{Fe}}$, $H = \hbar\omega_{pd}/2m_e v^2_{\mathrm{Fe}}$ and $\Omega_{pe} = \omega_{pe}/\omega_{pd}$.

3. Results and discussions

The above dispersion relations [equations (6) - (7)] are solved numerically with incremental search method [6] with judicious parametric values as : $n_{e0} \approx n_{i0} \sim 10^{25}$ - 10^{26} cm^{-3} ; dust grains (silicate, glass with magnesium-iron-silicate stoichiometry, forsterite-olivine admixture, etc) density, $n_d \approx 3\times10^{14}$ – 3×10^{16} cm^{-3} [7-8]; dust mass, $m_d \sim 10^{-14}$ – 10^{-16} g [7-8]; plasma temperature, $T \sim 10^3$ - 10^4 K; Fermi temperature, $T_{Fe} = \hbar^2(3\pi^2 n_{e0})^{2/3}/2k_B m_e = 9\times10^6$ K and Fermi velocity, $v_{Fe} = (2k_B T_{Fe}/m_e)^{1/2} \sim 10^9$. Such high-density and low-temperature degenerate quantum plasmas exist in white dwarfs, such as GD56, GD362, G29-38, WD2326+049, and so forth [7-8].

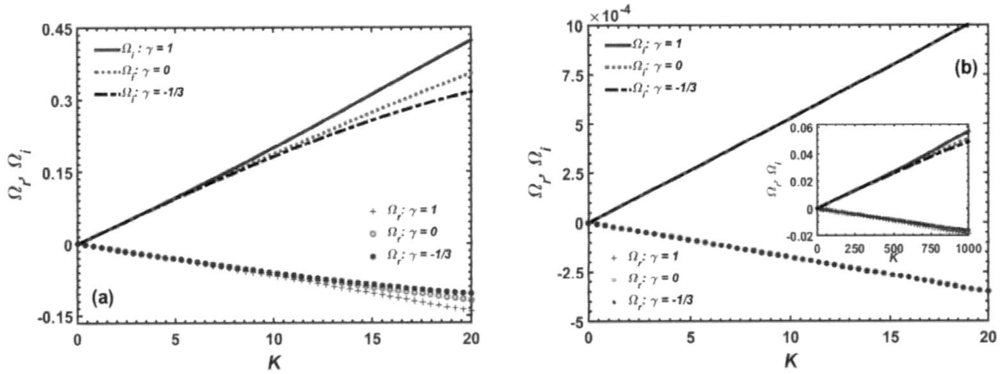

Figure 1. Profile of the normalized growth rate [Ω_i, upper set of lines] and real frequency [Ω_r, lower set of lines] for the (a) ion-streaming and (b) dust-streaming instability with the normalized wave number [K]. Various input values are in the text. The inset panel (right) shows the instability behavior in the short-wavelength domain.

Figure 1(a) shows the normalized real and imaginary parts of the ion-streaming instability frequency against the normalized wave number. As in the previous work [5], the growth rate in the Wigner-Poisson (WP) formalism (γ = 1, blue solid line) increases in comparison with the pure classical case (γ = 0, red dotted line). In contrast, the growth rate decreases in the TF formalism (γ = -1/3, black dashed-dotted line) against the classical case. We speculate that the real frequency comes out to be negative due to the higher value of the plasma flow velocity in comparison with the ion-streaming phase velocity. The ion-streaming wave is the modified form of the dust ion acoustic wave (DIAW) in the presence of flow. It is further noted that, due to the presence of the quantum pressure, there is noticeable quantitative changes in the instability growth rate. If the streaming is in 1-D, as herein, the quantum pressure reduces the growth (as per the TF formalism [5]). Also, the quantum effects are more prominent in the short-wavelength domain. It is, moreover, seen that, there is minute deviation from the linear behaviour in the dispersion characteristics (Ω_r-profiles). Thus, it implicates that the quantum pressure provides weak dispersive effects to the ion-streaming wave dynamics.

Figure 1(b) depicts the normalized real and imaginary parts of the dust-streaming instability frequency with the normalized wave number. The effect of quantum pressure is not noticeable in the long wavelength domain. The quantum pressure becomes effective beyond $K = 700$, which thus, has the corresponding wavelength much smaller than that of the dust acoustic wave (DAW).

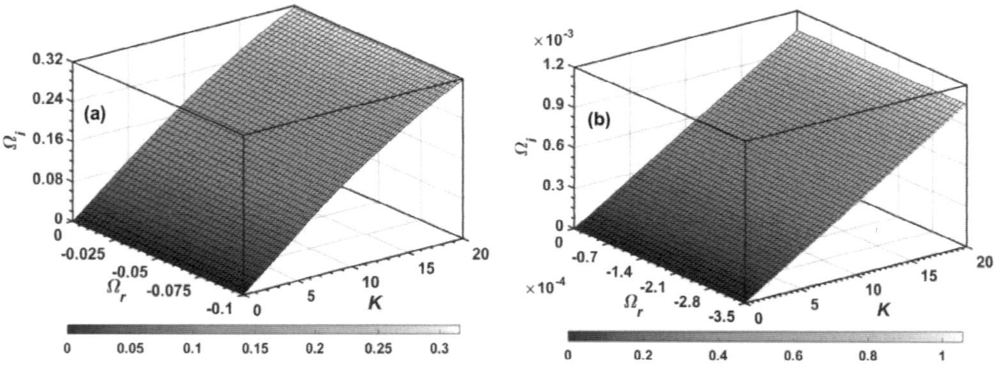

Figure 2. Profile of the normalized growth rate [Ω_i] with the normalized real frequency [Ω_r] and normalized wave number [K] for $\gamma = -1/3$. Various inputs are the same as in figure 1. The colorbars indicate the colormaps of spectral variations in the Ω_i-values in the ion- and dust-streaming cases; respectively.

Figure 2 portrays the 3-D profiles of the instability parameters ($\gamma = -1/3$) for a better display of the instability behaviors in both of the above cases in a phase plan spanned by the normalized real frequency and normalized wave number. It is seen that the ion-streaming instability profile admits a quasi-linear variation along the K-axis, while the dust-streaming profile is perfectly linear. It further confirms the fact that the quantum pressure is effective only in the ion-streaming dynamics.

4. Conclusions

In summary, we have carried out an instability analysis of the streaming quantum dusty plasma with the degenerate electrons. Two distinct classes of instability, corresponding to ion- and dust-flows, are studied. It is shown that the quantum-correction effects are prominent in the ion-streaming dynamics only; but, not so for the dust case due to the large difference in the length-scales. It is concluded that the adopted quantum correction γ-factor introduces a new source of stability and dispersive effects to the ion-streaming instability solely; but not to the dust counterparts.

Finally, we hope that the analysis can be useful to understand the electrostatic beam-plasma coupled dynamics and flow-driven instabilities responsible for generating the intense electric currents and strong magnetic fields in the compact astrophysical objects of dwarf-family.

References

[1] Cap F 1994 *Waves and Instabilities in Plasmas* (New York: Springer)
[2] Haas F 2011 *Quantum Plasmas: A Hydrodynamic Approach* (New York: Springer)
[3] Fortov V E 2016 *Extreme States of Matter-High Energy Density Physics* (Switzerland: Springer)
[4] Ali S and Shukla P K 2007 *Euro Phys. J. D* **41** 319
[5] Karmakar P K and Goutam H P 2016 *Phys. Plasmas* **23** 112121
[6] Kiusalaas J 2005 *Numerical Methods in Engineering with MATLAB* (New York: Cambridge)
[7] Jura M, Farihi J and Zuckerman B 2007 *Astrophys. J.* **663** 1285
[8] Jura M, Farihi J Zuckerman B and Becklin E E 2007 *Astron. J.* **133** 1927

IOP Conf. Series: Journal of Physics: Conf. Series **836** (2017) 012005 doi:10.1088/1742-6596/836/1/012005

Terahertz radiation generation by using modulated electron beam with plasma wave wiggler

J Panwar[a], S C Sharma and R Sharma

Department of Applied Physics, Delhi Technological University, Delhi-110 042, India

E-mail: [a)]panwar.jyots2010@gmail.com

Abstract. The generation of high power terahertz radiation has been studied by the interaction of density modulated relativistic electron beam (REB) with the plasma pump wave. The REB is pre bunched by the two laser beams which later translate to density modulation on passing through the drift space. The modulated beam enters the plasma medium and excites the electron plasma wave which leads to the oscillatory velocity of the electron beam. This transverse velocity couples with the modulated beam density and gives rise to nonlinear current density that leads to the emission of radiation of the order of terahertz. In results, the power of the radiation is evaluated and analysed. The power varies as square of the electron beam radius and field of the pump wave. The radiation thus obtained has numerous scopes in various fields.

1. Introduction

The generation of terahertz radiation using relativistic electron beams in free electron lasers has become an active area of research due to its wide applications in diverse fields including medical, security, spectroscopy, defence etc [1-2]. The emission of tunable terahertz radiation using pre bunched relativistic electron beam by two lasers has been studied by Kumar and Tripathi [3]. They noticed that the output power and amplitude of the radiation increases with the frequency. The frequency of the radiation was also found to be tuned by lasers wavelength. Plasma wave helps in obtaining higher output efficiency due to its short wavelength. Joshi et al.[4] have employed plasma density wave as a purely electric wiggler and studied the effect of plasma on the free electron radiation. They concluded that strong wiggler strength and short wiggler wavelength of plasma wave leads to emission of radiation at modest electron beam energy. Son et al.[5] have studied the emission of terahertz radiation using plasma wave as an undulator by forward Raman scattering and observed that the beam energy required reduces to a considerable amount and the energy spread reduces due to space charge effect. In this paper, we have studied the effect of relativistic electron beam and the pump wave on the emission of THz radiation, emitted by the interaction of density modulated relativistic electron beam with the space charge wave excited in the plasma medium.

2. Model

A relativistic density modulated electron beam is obtained following Kumar and Tripathi [3]. The two laser beams leads to the energy modulation of the REB by exerting a pondermotive force on it in the modulator.

$v_m = -\lambda v_b \sin(\omega t_0 + \theta_b)\hat{z}$, where $\lambda = \frac{kP}{2\gamma^4}\frac{c^2}{v_b^2}\frac{e^2 A_a A_b^*}{m^2 c^2 \omega_a \omega_b}\frac{\sin(\theta_b)}{\theta_b}$, $\theta_b = \frac{(\omega - kv_b)P}{2v_b}$ is the phase angle of modulator, (ω, k) is the frequency and wave vector of modulated beam, P is the length of interaction region, v_b is the initial beam velocity, A_a, A_b and ω_a, ω_b are the electric field amplitudes and frequencies of the laser beams, respectively, γ is the relativistic gamma factor. The energy modulated beam on travelling through the drift space gets density modulated. $n_m = -2n_b J_1\left(\frac{\omega L\lambda}{v_b}\right)\cos(\omega t_3 - \psi)$, where n_b is the initial beam density.

The beam on interaction with the electrostatic pump wave acquires an oscillatory velocity $\vec{v_0} = \frac{e\vec{E_0}}{im\omega_0\gamma^3}$, where $\vec{E_0}$ is the pump electric field, ω_0 is the frequency of space charge pump wave. This velocity couples with the modualted beam density and leads to non-linear current density at phase (ω', k), where $\omega' = \omega + \omega_0$.

$$\vec{J}_{\omega',k} = \frac{2n_b e^2 A_0}{im\omega_0\gamma^3} J_1\left(\frac{\omega L\lambda}{v_b}\right) e^{-i(\omega' t - kz)} \tag{1}$$

This acts as an antenna and helps in evaluating the retarded vector potential. The power of the radiation is obtained using the poynting vector, calculated using the vector potential. The average power of the emitted THz radiation is evaluated as

$$4\pi r^2 \overline{S_{av}} = \frac{8\pi n_b^2 \omega'^2 e^4 A_0^2 r_e^4 l_e^2 J_1^2\left(\frac{\omega L\lambda}{v_b}\right)}{c^3 \varepsilon_0 m^2 \omega_0^2 \gamma^6}, \tag{2}$$

where ω' is the frequency of output radiation, l_e is the electron bunch length, r_e is the radius of beam. The normalized power is given as

$$\text{Normalized power} = \frac{2n_b^2 \omega'^2 e^4 r_e^4 l_e^2 J_1^2\left(\frac{\omega L\lambda}{v_b}\right)}{l_s^2 c^4 \varepsilon_0^2 m^2 \omega_0^2 \gamma^6}, \tag{3}$$

l_s is the length of space charge wave wiggler.

3. Results and discussions

The parameters used for calculating the power of the output radiation are: $v_b = 0.93c$, $n_b = 3\times10^{15} cm^{-3}$, $\gamma = 2.7$, $r_e = 50\mu m$, $l_e = 0.35cm$, $l_s = 0.4cm$. Using Eq. (3), we have plotted figure 1. The figure indicates that the normalized power of the output radiation first increases with the frequency, attains a maximum value and then decreases. This behaviour is observed due to the Bessel's function character of the density modulation. At a fixed value of frequency, the power is found to decrease with the wiggler length.

Figure 2 shows the same behaviour and for a particular value of frequency, the output power increases with the increase in beam density.

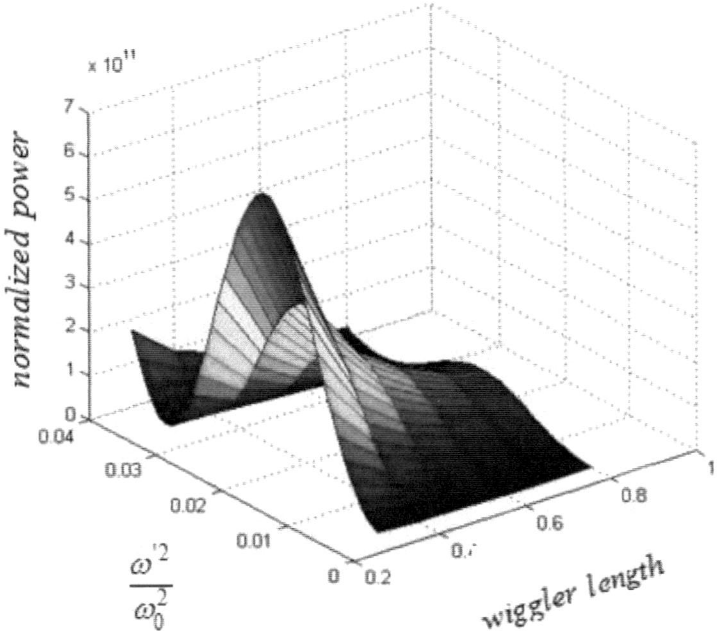

Figure 1. Figure showing the dependence of normalized power on the normalized frequency and wiggler length.

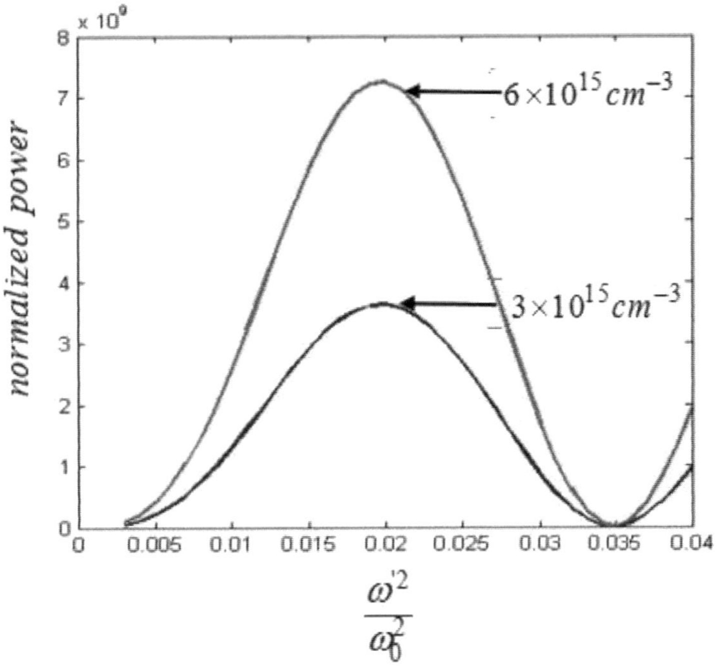

Figure 2. Figure showing the variation of normalized power with the normalized frequency for beam densities $n_b = 3\times10^{15}cm^{-3}$, $6\times10^{15}cm^{-3}$.

4. Conclusions

A pre bunched relativistic electron beam proves to be a great source for the generation of high power THz radiation. The power scales as square of the beam radius and beam density. It varies inversely as square of the wiggler length. The radiation can thus be tuned by varying the parameters of the REB and pump wave. The output radiation obtained can be used for practical applications where high power is required.

References

[1] Taylor Z D, Singh R S, Culjat M O, Suen J Y, Grundfest W S, Lee H and Brown E R 2008 *Opt. Lett.* **33** 1258
[2] Walther M, Fischer B M, Ortner A, Bitzer A, Thoman A and Helm H 2010 *Anal. Bioanal.* Chem. **397** 1009
[3] Kumar M and Tripathi V K 2012 *Phys. Plasmas* **19** 073109
[4] Joshi C, Katsouleas T, Dawson J M, Yan Y T and Slater J M 1987 *IEEE J Quant. Elec.* **23** 1571
[5] Son S, Moon S J and Park J 2012 *Opt. Lett.* **37** 5172

Normal mode behaviours in solar prominence plasmas

M Gohain and P K Karmakar[a]

Department of Physics, Tezpur University, Napaam-784028, Tezpur, Assam, India

E-mail: [a)]pkk@tezu.ernet.in

Abstract. The evolutionary dynamics of normal modes, non-adiabatic magneto-hydro-dynamic (MHD) waves, in the prominence plasmas in the presence of kinematic viscosity and inertia-dependent species stratification is herein explored. It judiciously implements a new energy equation devised to describe optically thin radiation losses, thermal conduction, viscosity, heating mechanisms, and so forth. The dispersion analysis demonstrates that the thermal conduction plays as a primary sink for both the wave classes of fast and slow spectral groups. Finally, asynoptic indication to further refinements and scopes is concisely stressed.

1. Introduction

The solar prominence plasmas, which are composed of partially ionized gases supported against the solar gravity by the coronal magnetic field, have recently acquired rapidly growing interest among astro-plasma communities. The evolutionary prominence dynamics, via the excitation of collective waves and exotic oscillations, have extensively been investigated [1-3]. The prominence oscillations can be of both small and large amplitudes. These small amplitude oscillations have been detected mainly in the velocity field describable in the framework of magneto-hydro-dynamic (MHD) waves [4, 5].Besides, the coronal gravity and electromagnetic effects in dictating the viscous prominence kinetics with proper thermodynamic functions are yet to be well explored [6-8].The real picture of non-adiabatic prominence waves (eigen-modes) can be obtained by considering different unavoidable mechanisms, such as radiative losses, heating, electro-thermal conduction, and so forth.

The inclusion of the dissipative agencies (e.g., viscosity, conductions, radiative cooling, etc.) could explain the spatiotemporal damping [4, 5]. A refined exploration would require a proper energy-transport law integrating all the phenomenological effects. The wave damping mechanisms in viscous stratified prominence plasma have hitherto been remaining an open long-sought goal as far as seen. This paper is motivated by the need for realizing the realistic prominence mode behaviors. We herein propose a theoretical analysis to see mainly the damping factors and behaviors. The results are important in the intriguing coronal seismology and in in-situ mensuration of the coronal atmosphere.

2. Physical and mathematical models

We consider an unbounded prominence plasma configuration in the presence of magnetic field acting along the x-direction in homogeneous stratified MHD equilibrium. It includes all the possible realistic effects, such as optically thin radiative losses, viscosity, gravity, etc. A schematic is sketched in figure 1.

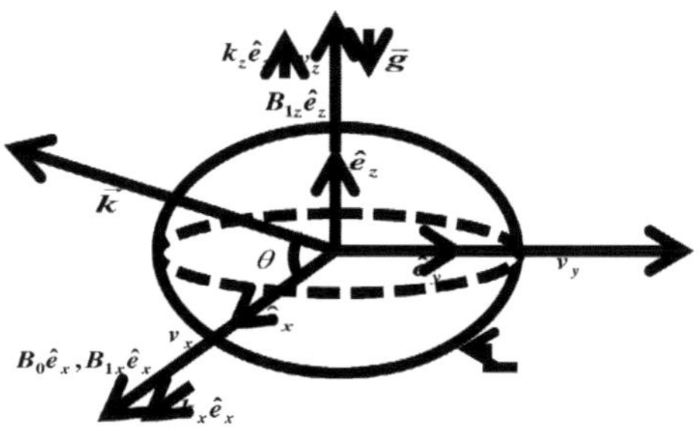

Figure 1. Schematic diagram of the solar prominence plasma configuration envisaged in the proposed model. The sensible dynamical variables and their resolved counterparts are directionally highlighted with different arrows.

The model is setup by a closed set of non-adiabatic MHD equations with all the generic notations [5, 7].If no source or sink, it includes the following respective conservation laws of mass, force, etc.

$$D\rho/Dt + \rho\vec{\nabla}.\vec{v} = 0 \tag{1}$$

$$\rho(D\vec{v}/Dt) = -\vec{\nabla}p + (1/\mu)(\vec{\nabla}\times\vec{B})\times\vec{B} + \rho\vec{g} + \rho\upsilon\{(4/3)\vec{\nabla}(\vec{\nabla}.\vec{v}) - \vec{\nabla}\times\vec{\nabla}\times\vec{v}\} \tag{2}$$

$$Dp/Dt - (\gamma p/\rho)(D\rho/Dt) + (\gamma-1)\left[\rho L(\rho,T) - \vec{\nabla}.(K\vec{\nabla}T)\right] = 0 \tag{3}$$

$$\partial\vec{B}/\partial t = \vec{\nabla}\times(\vec{v}\times\vec{B}) \tag{4}$$

$$\vec{\nabla}.\vec{B} = 0 \tag{5}$$

$$p = \rho RT/\tilde{\mu} \tag{6}$$

here, $D/Dt = \partial/\partial t + v.\vec{\nabla}$ represents the material (Lagrangian or convective) derivative associated with the mean MHD flow dynamics and ρ is the mass density. The symbol R denotes the ideal gas (molar) constant and $\tilde{\mu}$ is the plasma degree of ionization, defined as $\tilde{\mu} = 1/1+\xi_i$, with $\xi_i = n_i/(n_i + n_n)$, as the relative population density of the ionized species with respect to the total plasma population density. The non-adiabatic effects act via a conduction term, $\vec{\nabla}.(K\vec{\nabla}T)$, where K is the thermal conductivity and T is the temperature (in K). γ is the adiabatic index and L the heat-loss function[4]. In a homogeneous MHD equilibrium [4, 5], where the temperature and density are uniform, one finds

$$L(\rho_0, T_0) = 0. \tag{7}$$

The function L measures the difference between an arbitrary heat input and radiative heat loss. Such losses are described by optically thin radiative loss function [4, 5] expressed as

$$L(\rho,T) = \chi^*\rho T^\alpha - h\rho^a T^b \tag{8}$$

Here, χ^* and α are the temperature-dependent piecewise functions accounting for the different radiative loss regimes. Different heating scenarios are specified by the distinct exponents a and b. Applying the Fourier-based normal mode analysis over basic structure equations (1)-(6),with all the relevant perturbations evolving as plane waves of the form $\sim \exp i(\omega t - k.r)$ under no boundary effects considered, we finally get the linear quintic dispersion relation after decomposition exercises as

$$a_5\omega^5 + a_4\omega^4 + a_3\omega^3 + a_2\omega^2 + a_1\omega^1 + a_0 = 0 \tag{9}$$

where the various involved coefficients in the presence of viscous and gravitational effects are as

$$a_0 = \left(\frac{i}{\mu\rho_0^3}\right)\left[B_{0x}^2 p_0 k_x^2(-AT_0 + H\rho_0)\left\{k_x^2 + i\left(-ik_z + \frac{1}{2H_B}\right)^2\right\}\right] \tag{10}$$

$$a_1 = \left(\frac{1}{\mu\rho_0^2}\right)\left[B_{0x}^2 c_s^2 p_0 k_x^2\left\{k_x^2 + i\left(-ik_z + \frac{1}{2H_B}\right)^2\right\} + \frac{4}{3}i\mu p_0\upsilon k_x^2\left(-ik_z + \frac{1}{2H_B}\right)^2(AT_0 - H\rho_0) - i\mu p_0\upsilon\left(-ik_z + \frac{1}{2H_B}\right)^4\right.$$
$$(AT_0 - H\rho_0) + \frac{1}{3}\mu gAT_0\rho_0\upsilon k_x^2\left(-ik_z + \frac{1}{2H_B}\right) + \frac{4}{3}AT_0B_{0x}^2\upsilon k_x^2 - \frac{7}{3}AT_0B_{0x}^2\upsilon k_x^2\left(-ik_z + \frac{1}{2H_B}\right)^2 + \frac{4}{3}i\mu gAT_0\rho_0\upsilon k_x^2\left(-ik_z + \frac{1}{2H_B}\right)$$
$$+ AT_0B_{0x}^2\upsilon\left(-ik_z + \frac{1}{2H_B}\right)^2 + i\mu gAT_0\rho_0\upsilon\left(-ik_z + \frac{1}{2H_B}\right)^3 - \frac{4}{3}\mu p_0\upsilon k_x^2\left(-ik_z + \frac{1}{2H_B}\right)^2(AT_0 - H\rho_0) + \mu p_0\upsilon k_x^4(AT_0 - H\rho_0), \tag{11}$$

$$a_2=\left(\frac{1}{\mu\rho_0^2}\right)\left[\begin{array}{l} i\left(AB_{0x}^2T_0+\mu p_0\right)\left(AT_0-H\rho_0\right)\left\{k_x^2+i\left(-ik_z+\frac{1}{2H_B}\right)^2\right\}-\mu gAT_0\rho_0\left(-ik_z+\frac{1}{2H_B}\right)\\ +i\mu c_s^2p_0\rho_0\upsilon\left\{k_x^4+i\left(-ik_z+\frac{1}{2H_B}\right)^4\right\}-\mu gp_0\rho_0\upsilon k_x^2\left(-ik_z+\frac{1}{2H_B}\right)-\frac{24}{9}i\mu AT_0\rho_0\upsilon^2k_x^2\left(-ik_z+\frac{1}{2H_B}\right)^2\\ +\frac{1}{3}iB_{0x}^2p_0\upsilon k_x^2\left\{k_x^2+i\left(-ik_z+\frac{1}{2H_B}\right)^2\right\}+\mu gp_0\rho_0\upsilon\left(-ik_z+\frac{1}{2H_B}\right)^3+\frac{4}{3}\mu AT_0\rho_0\upsilon^2\left\{k_x^4+i\left(-ik_z+\frac{1}{2H_B}\right)^4\right\}\end{array}\right], \quad (12)$$

$$a_3=\left(\frac{1}{\mu\rho_0^2}\right)\left[\begin{array}{l} -p_0\left(B_{0x}^2+\mu c_s^2\rho_0\right)\left\{k_x^2+i\left(-ik_z+\frac{1}{2H_B}\right)^2\right\}-i\mu gp_0\rho_0\left(-ik_z+\frac{1}{2H_B}\right)\\ +\frac{24}{9}\mu p_0\rho_0\upsilon^2k_x^2\left(-ik_z+\frac{1}{2H_B}\right)^2-\frac{4}{3}\mu\rho_0\rho_0\upsilon^2\left\{k_x^4+\left(-ik_z+\frac{1}{2H_B}\right)^4\right\}-\frac{7}{3}\mu AT_0\rho_0\upsilon\left\{k_x^2-\left(-ik_z+\frac{1}{2H_B}\right)^2\right\}\end{array}\right], \quad (13)$$

$$a_4=-\frac{i}{\rho_0}\left[AT_0+\frac{7}{3}p_0\upsilon\left\{k_x^2-\left(-ik_z+\frac{1}{2H_B}\right)^2\right\}\right], \quad (14)$$

$$a_5=\frac{p_0}{\rho_0} \quad (15)$$

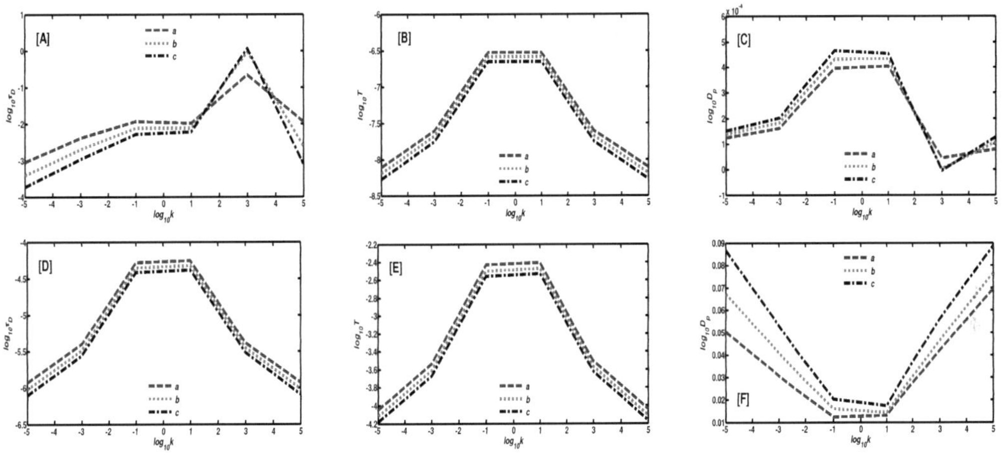

Figure 2. Profile of the logarithmic damping time ($\log_{10}\tau_D$), logarithmic period ($\log_{10}T$), and logarithmic damping per period ($\log_{10}D_P$) with the logarithmic wave number ($\log_{10}k$) for the *fast* ([A]-[C]), and *slow* ([D]-[F]) modes in different conditions. The constant heating mechanism is defined by $a=b=0$. The dashed (blue), dotted (red), and dashed-dotted (black) lines link to the three distinct prominence conditions: (a) T = 6000 K, (b) T = 7000 K and (c) T = 8000 K (black); respectively. The fixed inputs are $\chi^*=2.02\times10^{-15}$, $\alpha=8.06$, and $\rho=5.00\times10^{-11}$ kg m^{-3}, R = 8.3 m^2 kg s^{-2} K^{-1} mol^{-1}, $\tilde{\mu}$ = 0.8, B = 10^{-3} T, $\mu=4\pi\times10^{-7}$ H m^{-1}, $\gamma=5/3$, and $\theta=45^0$; respectively.

It is interesting to note here that the coefficients (a_0 - a_5)in equation(9)procreate all the previous results[4], but as special cases with the roles of gravity and viscosity entirely turned off. Here, the difference lies in the gravitational effects via a_2 and a_3; and dissipative effects via a_1, a_2, a_3 and a_4. Clearly, equation(9) is a fifth-degree polynomial in ω with a strong dispersive dependency via k. The principal goal lies in the time damping of MHD waves, and hence, k is considered as a real, but $\omega = \omega_R + i\omega_I$ as a complex. Out of all the five roots of equation (9) obtained numerically, only one root is purely imaginary, indicating non-propagatory thermal or condensation mode [9]. The rest of the roots are two conjugate pairs, one for the slow mode and the other the fast mode. Thus, the damping time is $\tau_D = 1/\omega_I$, the period of the waves is $T = 2\pi/\omega_R$ (hereafter, T as period) and the damping per period is $D_P = \omega_I/\omega_R$.

3. Results

A numerical analysis based on of the dispersion relation (9), yields the results as follows. As in figure 2, with increase in temperature, τ_D of the fast wave at first decreases followed by a rapid rise to a peak (at $k \sim 10^3$ m^{-1}), and again decreases. We speculate that τ_D of the slow wave decreases in magnitude with increase in temperature. This means, higher the temperature, the more is the time required to damp the waves; and vice versa. Likewise, T decreases in a similar fashion with respect to τ_D of the slow wave, but D_P, in contrast, increases with rise in temperature. Thus, the thermal conduction plays an important role in the damping mechanism. The changes in the damping behaviour is due to the coupled interplay of the intrinsic MHD field ($B = 10^{-3}$ T) and gravito-electro-magnetically induced field ($B_{GEM} = -(1/c_A)E_{GIP} \sim 10^{-8}$ T) as per the well-known electromagnetic Faraday-Lenz action[7].

4. Conclusions

We have studied the damping behaviour of the normal eigen-modes (MHD waves) in the solar prominence plasmas considering all the realistic non-adiabatic terms in an appropriate energy law. A new generalized dispersion relation (quintic) is derived by employing linear normal mode analysis. It enables us to characterize two distinct classes of modes: the fast wave and slow ramified wave. It is seen that the damping time, τ_D, of the fast and slow waves decrease with increase in temperature. We further see that the thermal conduction plays an important role in the damping mechanisms. Thus, the joint effects of non-adiabaticity sourced by thermal conduction, radiative losses and heating; alongside viscosity and gravitational stratification, are able to damp both the fast and slow modes. The model formalism presented here may be useful for basic study of the prominence seismology for characterizing the solar structure, morphology and coronal behavior. In a broader sense, it can also be extended to explore the MHD phenomena prevalent in different stellar and astrophysical configurations. For more concretized implications and applications, we finally admit that further refinements are necessary with the proper inclusion of all other prevailing relevant effects [7-9], such as ion-neutral damping, phase mixing, geometric curvature effects, tunnel-induced wave leakage, etc.

Acknowledgement
The financial support from DST, GoI, Grant No. SR/FTP/PS-021/2011), is thankfully recognized.

References

[1] Balthasar H, Weihr E, Schleicher H and Wohl H 1993 *Astron. Astrophys.* **277** 635
[2] Oliver R and Ballester J L 2002 *Sol. Phys.* **206** 45
[3] Terradas J, Molowny-Horas R, Wiehr E, *et al.* 2002 *Astron. Astrophys.* **393** 637
[4] Carbonell M, Oliver R and Ballester J L 2004 *Astron. Astrophys.* **415** 739
[5] Terradas J, Carbonell M, Oliver R and Ballester J L 2005 *Astron. Astrophys.***434** 741
[6] Goedbloed H P and Poedts S 2004 *Principles of Magnetohydrodynamics* (Cambridge)
[7] Priest E 2014 *Magnetohydrodynamics of the Sun* (New York: Cambridge Univ. Press)
[8] Vial J C and Engvold O 2015 *Solar Prominences* (New York: Springer)

[9] Field G B 1965 *Astrophys. J.* **142** 531

IOP Conf. Series: Journal of Physics: Conf. Series **836** (2017) 012007 doi:10.1088/1742-6596/836/1/012007

Oblique ion-acoustic shock waves in a magnetized, multi-species plasma

Manesh Michael, Anu Varghese, Neethu Theresa Wilington, Sijo Sebastian, Savithri Devi E, Sreekala G and Chandu Venugopal[a]

School of Pure & Applied Physics, Mahatma Gandhi University, Kottayam 686560, Kerala, India.

E-mail: [a)]cvgmgphys@yahoo.co.in

Abstract. Ion-Acoustic (IA) shock waves are studied in a magnetized, five component cometary plasma consisting of positively and negatively charged oxygen ions, kappa described hydrogen ions, hot solar electrons and slightly colder cometary electrons. The Korteweg–de Vries–Burgers (KdVB) equation is derived and the variation of the shock profile is studied. It is found that the obliqueness of shock wave propagation significantly varies the basic properties like amplitude and width of the shock.

1. Introduction

In nature, nonlinear structures such as solitons, shocks and double layers arise due to nonlinearity, dispersion and dissipation. Space environments like the auroral zones, magnetospheres, rings of planets and the solar wind are perfect laboratories for investigating these structures.

The nonlinearity of the ion acoustic wave has been studied extensively the first theoretical formulation was given by Sagdeev [1] and Ikezi et al. [2] provided its first experimental observation. The space crafts Viking and Freja found electrostatic solitary waves in the magnetosphere as density depressions.

Solitons are a result of nonlinearity being balanced by dispersion effects. However, a medium, which is both dispersive and dissipative, supports a shock wave instead of a soliton. Shock waves are described by the Korteweg de-Vries Burgers (KdVB) equation in 1D and the Kadomstev-Petviashivili-Burgers (KPB) in 2D. The dissipative Burger term in both these equations is due to dissipative mechanisms such as wave-particle interactions, multi-ion streaming, dust charge fluctuations, anomalous viscosity, etc. [3-5]. When wave breaking is balanced by dispersion and dissipation, a monotonic or oscillatory dispersive shock wave is generated [6]. IA shock waves have been extensively studied recently [7-11].

This paper is an extension of our previous investigation of IA shock waves in a five component unmagnetized plasma of kappa described electrons and ions [12]. Here, we investigate IA shock waves in similar, but magnetized plasma.

2. Theoretical Modelling and basic equations

We consider a five component magnetized cometary plasma consisting of positively and negatively charged oxygen ions, kappa described hydrogen ions, hot solar electrons and slightly colder cometary electrons. The external magnetic field is along the z-axis $(B_0\hat{z})$. The charge neutrality of the system is described using the equation, $n_{ce0} + n_{se0} + Z_1 n_{10} = n_{H0} + Z_2 n_{20}$

Here n_{ce0}, n_{he0} represent the equilibrium densities of colder cometary and hotter solar electrons respectively while n_{10}, n_{20}, n_{H0} are the equilibrium densities of negatively charged oxygen (O^-) ions (charge number Z_1), positively charged oxygen (O^+) ions (charge number Z_2)and hydrogen ions respectively. The general expression for a kappa distribution of species 's' is given by,

$$n_s = n_{s0}\left[1 + e_s\varphi\left(k_B T_s(\kappa_s - 3/2)\right)^{-1}\right]^{-\kappa_s + 1/2} \tag{1}$$

In (1), s = H for hydrogen, $= se$ for solar electrons and $= ce$ for cometary photo-electrons. n_s denotes the density (with the subscript '0' denoting the equilibrium value), e_s the charge, T_s the temperature and κ_s, the spectral index for species 's'. k_B is the Boltzmann's constant and φ, the potential. The dynamics of the heavier ions (j=1, 2) are described using following normalised equations

$$\frac{\partial N_j}{\partial \tau} + \nabla.(N_j V_j) = 0 \tag{2}$$

$$\frac{\partial V_j}{\partial \tau} + (V_j.\nabla)V_j = \alpha_j \nabla\phi + \eta_j \nabla^2 V_j + \varepsilon_j \Omega_j (V_{jy}\hat{i} - V_{jx}\hat{j}) \tag{3}$$

where $\alpha_1 = 1$, $\alpha_2 = -(Z_2/Z_1)(m_1/m_2)$, $\varepsilon_1 = -1$ and $\varepsilon_2 = 1$.

We normalize the equations using the parameters of O^- ions according to, $\phi = e\varphi\left(k_B T_1\right)^{-1}, V_j = v_j c_s^{-1}$, $\tau = t\omega_{p1}$ where $c_s = \left(Z_1 k_B T_1 m_1^{-1}\right)^{1/2}$ and $\omega_{p1} = \left(4\pi Z_1^2 e^2 n_{10} m_1^{-1}\right)^{1/2}$. The variable x is normalized using $\lambda_{D1} = \left((Z_1 k_B T_1)(4\pi Z_1^2 e^2 n_{10})^{-1}\right)^{1/2}$ while $N_j = n_j (n_{j0})^{-1}$. The gyro-frequencies are normalised using ω_{p1}. Thus N_1, N_2 represent the normalised densities of O^- and O^+ ions respectively; V_1 and $V_2$are their corresponding velocities. Ω_1 and Ω_2 represent their gyro-frequencies; η_1 and η_2 are their normalized kinematic viscosities. The normalised Poisson equation is given by

$$\nabla^2\phi = N_1 - N_2\left(1 + \mu_{ce} + \mu_{se} - \mu_H\right) + \sum_s \varepsilon_s \mu_s \left(1 - \frac{\varepsilon_s \phi}{\sigma_s\left(\kappa_s - 3/2\right)}\right)^{-(\kappa_{ce} - 1/2)} \tag{4}$$

where s = 'se', 'ce' or H, $\varepsilon_{ce} = \varepsilon_{se} = +1$ and $\varepsilon_H = -1$

Also $\mu_{ce} = \dfrac{n_{ce0}}{Z_1 n_{10}}$, $\mu_{se} = \dfrac{n_{se0}}{Z_1 n_{10}}$, $\mu_H = \dfrac{n_{H0}}{Z_1 n_{10}}$, $\sigma_{ce} = \dfrac{T_{ce}}{T_1}$, $\sigma_{se} = \dfrac{T_{se}}{T_1}$ and $\sigma_H = \dfrac{T_H}{T_1}$

Equations (2) to (4) are solved by the standard reductive perturbation method [12]. Finally we obtained the KdVB equation which describes nonlinear wave propagation as

$$\partial_\tau \phi^1 + A\phi^1 \partial_\xi \phi^1 + B\partial_{\xi\xi\xi}\phi^1 - C\partial_{\xi\xi}\phi^1 = 0 \tag{5}$$

In (5) the coefficients A, B, and C are given by

$$A = \left[3 l_z^4 \lambda^{-4}\left(m^2\left(Z_2 Z_1^{-1}\right)^2\left(1 + \mu_{ce} + \mu_{se} - \mu_H\right) - 1\right) - L\right] D^{-1}$$

$$B = \left[1 + (1 - l_z^2)\left(\Omega_1^{-2} + \left(1 + \mu_{ce} + \mu_{se} - \mu_H\right)\Omega_2^{-2} m\ (Z_2 Z_1^{-1})\right)\right] D^{-1}$$

$$C = \left[l_z^4 \lambda^{-3}\left(\eta_{10} + \eta_{20}\left(1 + \mu_{ce} + \mu_{se} - \mu_H\right) m\ (Z_2 Z_1^{-1})\right)\right] D^{-1}$$

where

$$L=\frac{\mu_{ce}\left(\kappa^2{}_{ce}-1/4\right)}{\left(\kappa_{ce}-3/2\right)^2\sigma_{ce}^2}+\frac{\mu_{se}\left(\kappa^2{}_{se}-1/4\right)}{\left(\kappa_{se}-3/2\right)^2\sigma_{se}^2}-\frac{\mu_H\left(\kappa^2{}_H-1/4\right)}{\left(\kappa_H-3/2\right)^2\sigma_H^2}\text{ and } D=2l_z^2\lambda^{-3}\left(1-\alpha_2\left(1+\mu_{ce}+\mu_{se}-\mu_H\right)\right)$$

The solution of the KdVB equation is obtained by adopting the tanh method and is [13]

$$\phi^1=\phi_m\left(1-0.25\left[1+\tanh\left(\omega^{-1}(\varsigma-V\tau)\right)\right]^2\right) \tag{6}$$

In (6) $\phi_m=0.48(AB)^{-1}C^2$ is the shock amplitude, $\omega=10BC^{-1}$ is its width and $V=0.24B^{-1}C^2$ the shock speed.

3. Results

In this paper, we plot solution (6) for the plasma environment of the comet Halley [14]. The observed density of the hydrogen ions was $n_H=4.95\,\text{cm}^{-3}$ (temperature $T_H=8\times10^4\,K$). We set the densities of positively charged oxygen ions at $n_{20}=0.5\,\text{cm}^{-3}$ and that of negatively charged oxygen ions at $n_{10}=0.05\,\text{cm}^{-3}$ [15].The temperature of the solar (or hot) electrons was $T_{se}=2\times10^5\,\text{K}$. The temperature of the colder photo-electron was set at $T_{ce}=2\times10^4\,\text{K}$; the background magnetic field is $8g$ [14].

Figure 1 shows the variation of the shock profile of IA solitary waves with the spectral indices κ. The parameters used for the figures are $n_{10}=0.05\,\text{cm}^{-3}$, $n_{20}=0.5\,\text{cm}^{-3}$, $n_H=4.95\,\text{cm}^{-3}$, $T_{ce}=2\times10^4\,\text{K}$, $T_{se}=2\times10^5\,\text{K}$, $T_1=T_2=1.16\times10^4\,K$, $Z_1=Z_2=2$. Curve (a) is for the spectral index κ = 2, curve (b) for κ = 3 and curve (c) for κ = 4; the amplitude of shock wave increases with increasing κ.

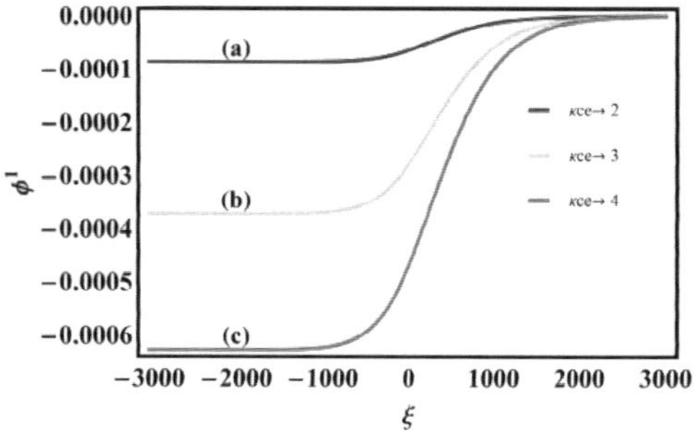

Figure 1. Shock profile as a function of spectral index

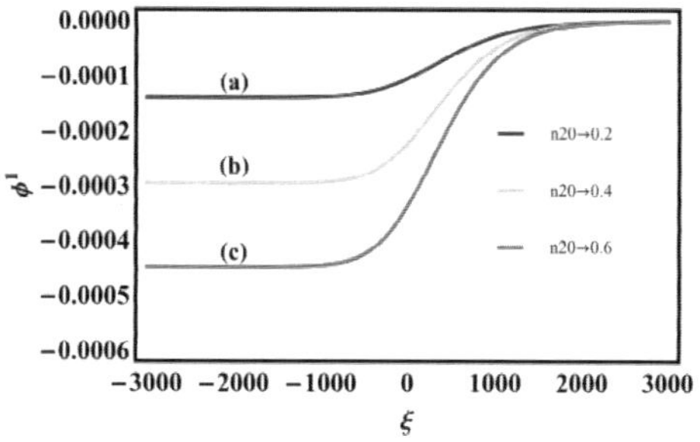

Figure 2. Shock profile as a function of O^+ ion density

Figure 2 is again a plot of the shock profile as a function of the positive oxygen ion densities n_{20}. The spectral indices $\kappa_{ce}=\kappa_{se}=\kappa_H=3$; the other parameters are the same as in figure 1. Curve (a) is for n_{20}=0.2 cm^{-3}, curve (b) is for n_{20}=0.4 cm^{-3} and curve (c) is for n_{20}=0.6 cm^{-3}; the strength of the shock wave increases with an increase of positively charged oxygen ion densities.

In figure 3, we depict the variation of amplitude of shock wave versus kappa as a function of obliqueness of the shock wave. The parameters chosenfor the figures 3 and 4 are the same as in figure 1. In both figures curve (a) is for $l_z=0.2$, curve (b) is for $l_z=0.3$ and curve (c) is for $l_z=0.4$; the magnitude of the amplitude of the shock increases as the obliqueness decreases.The variation of the width (ω) of the shock as a function of obliqueness of the wave is depicted in figure 4. We find that the width of the shock wave increases with an increase of obliqueness.A study similar to that of figure

2, but as a function of the positively charged oxygen ion kinematic viscosity η_{20}, shows that the strength of the shock wave increases with increasing η_{20}.

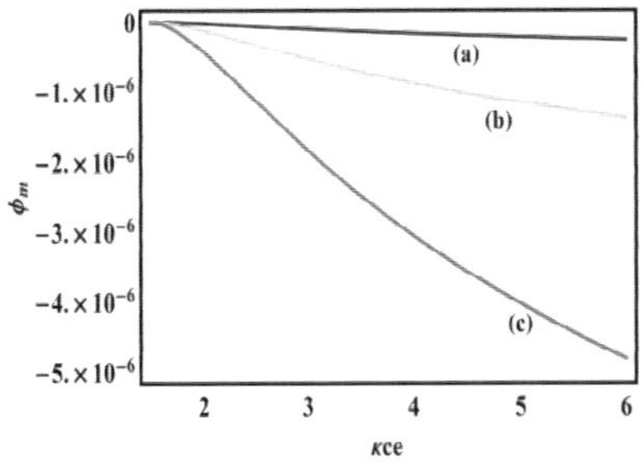

Figure 3. Amplitude of Shock wave Versus kappa as function of l_z kappa as function of l_z.

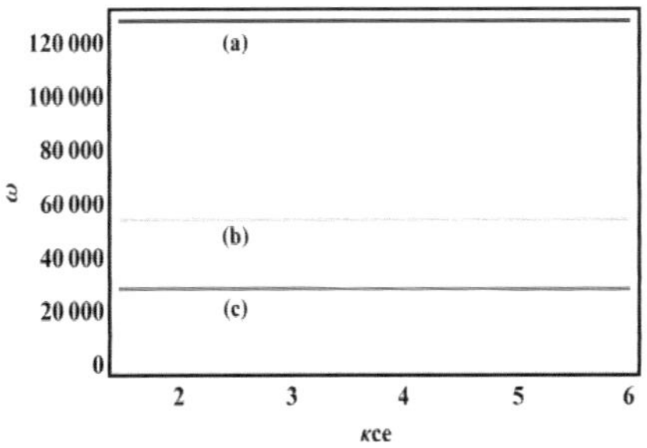

Figure 4. Width of shock wave Versus

4. Conclusions

We have studied IA shock waves in a five component magnetized plasma composed of positively and negatively charged oxygen ions, kappa described lighter hydrogen ions and hot and cold electrons by deriving the KdVB equation. We find that the strength of the shock wave enhances with increasing spectral indices and positively charged oxygen ion densities; it also increases with increasing viscosity of positively charged oxygen ions and temperature of cometary electrons. The obliqueness of the shock wave with the magnetic field significantly varies the characteristics like width and amplitude of the wave.

References

[1] Sagdeev R Z and Leontovich MA 1966 *Rev. Plasma Phys.* **4** 23
[2] Ikezi H, Taylor R and Baker D 1970 *Phys. Rev. Lett.* **25** 11
[3] Mamun A A and Shukla P K 2002 *Phys. Plasmas* **9** 1468
[4] Xue J K 2003 *Phys. Plasmas* **10** 4893
[5] Sahu B and Roychoudhury R 2007 *Phys. Plasmas* **14** 072310
[6] Shukla P K and Mamun A A 2003 *New J. Phys.***5** 17
[7] Shukla P K and Mamun A A 2001 *IEEE Tr. Plasma Sci.* **29** 221
[8] Mamun A A, Eliasson B and Shukla P K 2004 *Phys. Lett. A* **332** 412
[9] Ghosh S and Gupta MR 2005 *Phys. Plasmas* **12** 092306
[10] Pakzad H R 2012 *Ind. J. Phys.* **86** 743
[11] El-Hanbaly A M, El-Shewy E K, Sallah M and Darweesh H F 2005 *J. Theor. Appl. Phys.* **9** 167
[12] Manesh M, Neethu T W, Neethu J, Sijo S, Sreekala G and Venugopal C 2016 *J. Theor. Appl. Phys.* **10** 289
[13] Shahmansouri M and Mamun A A 2013 *Phys. Plasmas* **20** 082122
[14] Brinca A L and Tsurutani B T 1987 *Astron. Astrophys.* **187** 311
[15] Chaizy P *et al.* 1991 *Nature* **349** 393

IOP Conf. Series: Journal of Physics: Conf. Series **836** (2017) 012008 doi:10.1088/1742-6596/836/1/012008

On aspects of electron-LO phonon interaction in magnetized semiconductors in the presence of a hybrid pump field

Swati Dubey, Ayushi Paliwal[a] and S Ghosh

School of Studies in Physics, Vikram University, Ujjain 456010, India

E-mail: [a)]ayushipaliwal.paliwal@gmail.com

Abstract. Most realistic propagation of an intense hybrid pump wave in a magnetized semiconductor plasma has been considered to study some aspects of electron-LO phonon interactions. Hydrodynamic model for one component plasma along with coupled mode theory has been used to study parametric amplification due to polaron mode. Expressions for parametric gain coefficient arising due to parametric instability and threshold field required to incite parametric amplification has been derived. The compound semiconductors of group III-V and II-VI are unique within the universe of simple octet compounds, enable them to dominate higher performance electronics and optoelectronics. Present study aims to compare materials for which favourable magnitudes of parametric gain and threshold value could be obtained with suitable values of external parameters. Numerical estimations were carried out using the data of two different group compound semiconductors namely ZnSe and GaAs. Both the gain coefficients and threshold pump field are found to be strongly dependent on the carrier concentration of the medium. Resonance between plasma frequency and collective excitation frequency affects the process of amplification in both cases. Higher gain is achieved for GaAs which has smaller coupling coefficient as compared to ZnSe. Hybrid pump propagation is found to strengthen the electron-LO phonon coupling.

1. Introduction

In polar materials interaction of electric field takes place with both electrons and optical phonons. Single particle scattering is significant in transport theory. As per the direction of incidence of electromagnetic wave, the collective modes are labelled as longitudinal, transverse, mixed or hybrid, [1]. Present study considers a hybrid pump wave for infinite semiconductor plasma i.e. E_0 must have components that are both parallel and perpendicular to the direction of propagation [2],due to following reasons:-

- So far the response of an electron gas to longitudinal or transverse excitations has been studied by several workers [3-4], but it is not easy to create purely longitudinal or transverse pump.
- When an electromagnetic wave impinges on the surface of crystal at normal incidence, it excites only TO phonons inside the crystal if the provided frequency is matched accurately to that of the phonons. This is so both waves are transverse and consequently they couple together. LO phonons may not be excited in this way, because of the fact that these phonons being longitudinal, do not couple to transverse incident wave. If the incident wave falls obliquely onto the surface, there is a longitudinal electric field at surface which then acts to excite the LO phonons [5].

IOP Conf. Series: Journal of Physics: Conf. Series **836** (2017) 012008 doi:10.1088/1742-6596/836/1/012008

Hence in this study, we discuss the physics of interaction of electrons and LO phonons in semiconductor materials, interacting with and through, hybrid pump electromagnetic field. Previously we reported electromagnetic wave - polar semiconductor interaction in III-V polar semiconductor material under hybrid pump field configuration [6].

In the present study we develop a comparative model and to accomplish purpose of study we chose III-V GaAs and II-VI ZnSe direct band gap polar semiconductors.Motive behind choosing these particular semiconductors is the fact that in contrast to pure semiconductors GaAs and ZnSe are direct band gap ionic semiconductors with lower electron masses and higher radiative recombination coefficients which makes them useful for high speed electronics and optoelectronics[7]. As the materials belonging different groups have different magnitudes of physical parameters which may provide a better insight to understand different aspects of electron-LO phonon interactions. Present study probably for the first time establishes a comparative investigation of amplification characteristics of different group polar semiconductors shined by a hybrid pump field.

2. Theoretical formulation

This section deals with the theoretical formulation of the second order nonlinear optical susceptibility $\chi^{(2)}$ for the polaron mode in magnetized doped polar semiconductor plasmas. We consider the propagation of an hybrid pump wave, periodic in both space and time as, $E_0 = \left(E_{0x}\hat{x} + E_{0y}\hat{y}\right).\exp\left[i(\mathrm{k}_0\,\hat{x} - \omega_0 t)\right]$, in a homogeneous semiconductor plasma subjected to a magnetic field along z axis i.e. $B_0 = B_0\hat{k}$, perpendicular to direction of wave propagation.

In order to develop vital mathematics the basic equations and expressions are adopted from [6] (hereafter referred as paper I) and the procedure adopted from Neogi and Ghosh [8] (here after referred as paper II). We assume that the momentum and energy transfer between the pump, polaron and signal wave satisfy phase matching conditions $k_s = k_0 - k_{pl}$ and $\omega_S = \omega_0 - \omega_{pl}$.

Propagation of pump wave creates density perturbations in the medium. Using the basic equations of paper I and adopting the procedure of paper II, we get equation for the density perturbation by neglecting the Doppler shift under Rotating Wave Approximation as

$$\frac{\partial^2 n_1}{\partial t^2} + 2\Gamma_e \frac{\partial n_1}{\partial t} + \overline{\omega}_p^{2} Z n_1 - \omega_{p,pl}^2 \cdot n_0 A_1 \cdot \frac{\partial R}{\partial x} - 2\Gamma_{ph} \cdot T = -i(k_0 - k_{pl}) n_1 A_2 \overline{E} \tag{1}$$

where, $\overline{E} = (e/m) E_{eff}$ $\quad E_{eff} = -E_{0x} + \upsilon_{0y} B_{0z}$ $\quad \overline{\omega}_p^{2} = \omega_p^2 A_1$

$$\upsilon_{0x} = \overline{E} / 2\Gamma_e - i\omega_0 \quad \upsilon_{0y} = \left(-\frac{e}{m} E_{0y} - \omega_c \upsilon_{0x}\right) / 2\Gamma_e - i\omega_0$$

The above equation describes coupling between polaron mode and the side band waves in the presence of intense pump. Meaning of all symbols is same as given in paper I. Now we may obtain second-order susceptibility via nonlinear polarization $P_{nl} = \varepsilon_0 \chi^{(2)} E_0 E_{pl}^*$ as

$$\chi_{eff} = \frac{-iq(k_0 - k_{pl})}{M_{pl}} \left\{ \frac{\omega_p^2 \omega_{p,pl}^2}{\omega_0 \omega_1} \right\} Q^* G^* \tag{2}$$

Now by setting P_{nl} equal to zero, one may obtain the effective pump amplitude required to incite the parametric process,

$$E_{th} = \left| \left({m_e}\big/{e(k_0 - k_{pl}) A_2} \right) \delta_1 \delta_2 \right| \quad \text{, where } \delta_1 = \left(\overline{\omega}_p^{2} Z - \omega_1^2\right)^{1/2} \text{ and}$$

$$\delta_2 = \left(\overline{\omega}_p^{2} Z - \omega_{0,pl}^2\right)^{1/2}$$

Now the absorption coefficient can be achieved by using the relation,

$$\alpha_{para} = \frac{k_s}{2\varepsilon_1}\mathrm{Im}(\chi_{eff}^{(2)})\left|E_{eff}\right| = \frac{k_s}{2\varepsilon_1}\left\{\frac{-iq(k_0 - k_{pl})}{M_{pl}}\left(\frac{\omega_p^2\omega_{p,pl}^2}{\omega_0\omega_1}\right)\right\}Q^*G^*\left|E_{eff}\right| \tag{3}$$

Where α_{para} is the nonlinear absorption coefficient. Equation (3) is subsequently employed to study the amplification characteristics of the scattered wave in the parametric process. The nonlinear parametric gain of the signal as well as the idler waves can be possible only if α_{para} is negative.

3. Results and discussions

Set of relevant parameters of GaAs is given in paper I and for ZnSe: $\alpha = 0.43$, $\omega_{LO} = 4.814\times10^{13}\,s^{-1}$ $\varepsilon_s = 9.6$, $\varepsilon_{opt} = 6.3$ and $k_{pl} = 4\times10^5\,m^{-1}$. Numerical appreciations of above comparative study are depicted in figures 1-4 in which solid line curve represents GaAs and dotted curves are for ZnSe material. Figure 1 describes the variation of threshold pump field with external magnetic field. Figure depicts that for both the materials, increasing magnitudes of magnetic field reduces the threshold pump field. Pump field required to incite parametric process is slightly lower for GaAs than ZnSe.

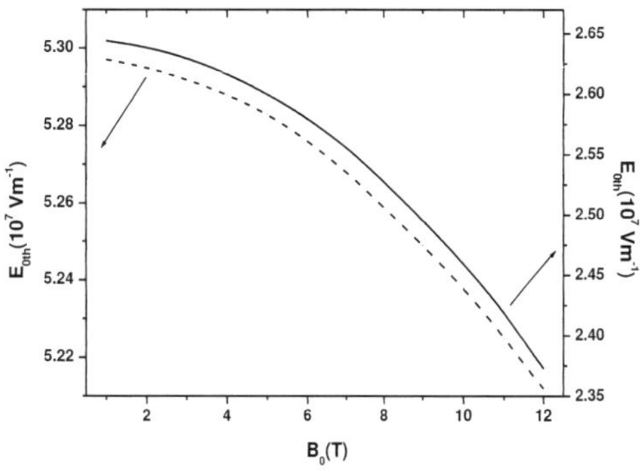

Figure 1. Variation of threshold field E_{th} with magnetic field at $B_0 = 12$ tesla

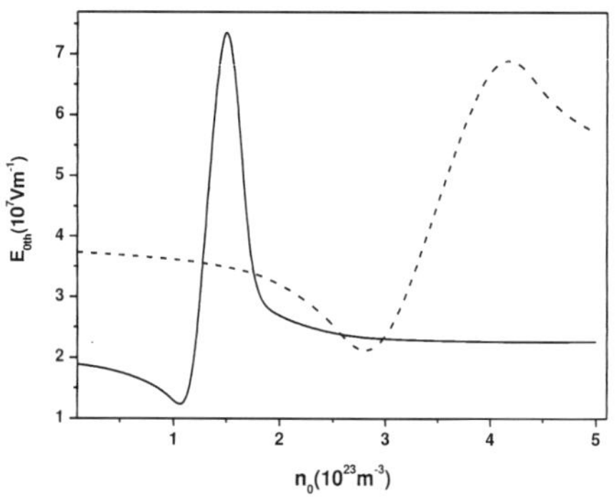

Figure 2. Variation of threshold field E_{th} with carrier concentration at $n_0 = 3\times10^{23}\,m^{-3}$

As depicted in figure 2 carrier concentration significantly affects the threshold field magnitude. Initially increasing carrier concentration leads to a reduction in threshold field till a minimum $\left(E_{th} \approx 1.25\times10^7\,Vm^{-1}\right)$ is touched at $n_0 \approx 1.2\times10^{23}\,m^{-3}$ for GaAs. Minima of threshold for ZnSe $\left(E_{th} \approx 2.1\times10^7\,Vm^{-1}\right)$ shifts towards higher concentration $\left(n_0 \approx 3\times10^{23}\,m^{-3}\right)$. It is clear that GaAs requires relatively lower pump field to incite the amplification process than ZnSe at a lower carrier concentration. Minima of both the curve signify the resonance between plasma frequency and collective cyclotron frequency. Further increment in carrier concentration makes $\omega_p^2 > \omega_{0,pl}^2 - \omega_c^2$, hence increments in threshold field is observed.

Absorption coefficient α_{para} is plotted as a function of magnetic field and carrier concentration in figures 3 and 4 respectively, at pump field $E_0 = 5.5\times10^7\,Vm^{-1}$. Higher magnetic field proves to be favourable resulting into amplification of the signal wave in both the semiconductors with lower threshold pump field. Highest parametric gain obtained is $\approx 8.7\times10^5\,m^{-1}$ for ZnSe and $\approx 1.8\times10^6\,m^{-1}$ for GaAs. Figure 4 shows the variation of absorption coefficient with carrier concentration. Initially absorption of the signal wave is depicted in figure for both the materials. It is observed that in both the cases slight increment in carrier concentration leads to amplification of the signal waves. Gain

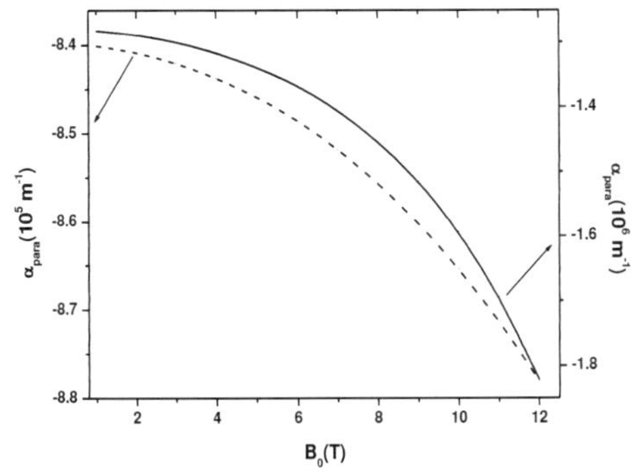

Figure 3. Variation of absorption coefficient with magnetic field at $B_0 = 12$ tesla

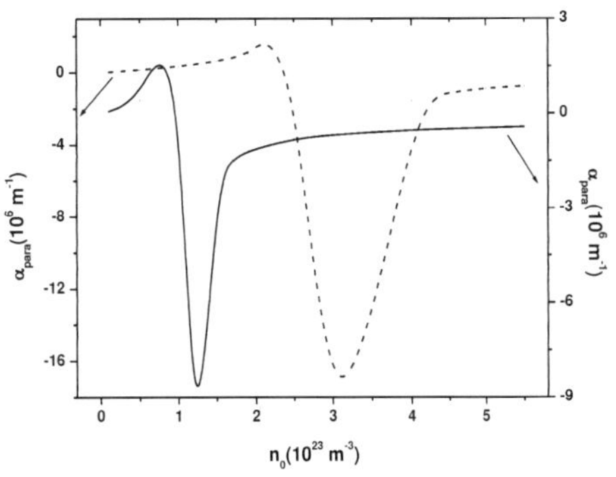

Figure 4. Variation of absorption coefficient with carrier density at $n_0 = 3\times10^{23} m^{-3}$

increases upto $\alpha_{para} \approx 1.7\times10^7 m^{-1}$ (for ZnSe) and $\alpha_{para} \approx 0.9\times10^7 m^{-1}$ (for GaAs).Sign reversal of absorption coefficient is due to the reason that initially plasma frequency $\omega_p^2 < \omega_{0,pl}^2 - \omega_c^2$ causing positive values of α i.e. absorption. A slight increment in carrier concentration gives maximum gain due to resonance between plasma frequency and collective normal mode frequency. Further increments in doping concentration reduce parametric gain because $\omega_p^2 > \omega_{0,pl}^2 - \omega_c^2$. This behaviour could be utilised in the optical switching applications. Figure clearly demonstrated the shifting of the maximum gain position towards the higher magnitude of doping concentration for ZnSe as compared to GaAs. Maximum gain is obtained at the concentrations where minimum threshold reported in figure 1 for both semiconductors.

4. Conclusions

This comparative study proves III-V GaAs more suitable material than II-VI ZnSe for commercial optical devices as far as threshold and parametric amplification characteristics are concerned. Effective mass and Fröhlich coupling coefficient are very significant parameters for polar materials. It is found that material with lower effective mass and Fröhlich coupling coefficient (GaAs) is favourable for the design of optical parametric amplifier. Probably hybrid pump propagation strengthens the electron LO phonon coupling. Thus one may conclude that the threshold and amplification characteristics strongly depend on material parameters. It is hoped that result of this paper may be useful in designing optical devices.

References

[1] Steele M C and Vural B1969 *Wave Interactions in Solid State Plasmas* (New York: Mc Graw Hill) p 2
[2] Dubey S and Ghosh S 1993 *Phys. Stat. Sol. (b)* **180** 275
[3] Agrawal R, Dubey S and Ghosh S 2012 *Int. J. Eng. Sci. Res.* **2** 1886
[4] Comas F, Trallerro C and Cardona M 1997 *Phys. Rev. B* **56** 4115
[5] Omar M A1999 *Elementary Solid State Physics* (Singapore: Pearson education) p 127
[6] Paliwal A, Dubey S and Ghosh S 2015 *AIP Conf. Proc.* **1670** 030020
[7] Tsao J Y 2004 *The World of Compound Semiconductors* p 2
[8] Neogi A and Ghosh S1989 *Phys. Stat. Sol. B* **152** 691

IOP Conf. Series: Journal of Physics: Conf. Series **836** (2017) 012009 doi:10.1088/1742-6596/836/1/012009

Effect of magnetic field on the Rayleigh Taylor instability of rotating and stratified plasma

PK Sharma, Anita Tiwari[a] and Shraddha Argal

University Institute of Technology, Barkatullah University ,Bhopal (M.P.) 462026, India

E-mail: [a)]anitatiwari7987@gmail.com

Abstract. In the present study the effect of magnetic field and rotation have been carried out on the Rayleigh Taylor instability of conducting and rotating plasma, which is assumed to be incompressible and confined between two rigid planes $z = 0$ and $z = h$. The dispersion relation of the problem is obtained by solving the basic MHD equations of the problem with the help normal mode technique and appropriate boundary conditions. The dispersion relation of the medium is analysed and the effect of magnetic field and angular velocity (rotation effect) have been examined on the growth rate of Rayleigh Taylor instability. It is found that the magnetic field and angular velocity (rotation effect) have stabilizing influence on the Rayleigh Taylor instability.

1. Introduction

The Rayleigh Taylor instability has created the attention of authors due to its crucial role in astrophysical and space phenomenon's, e.g. heating of solar corona, supernova implosion, explosion, Crab nebula and crucial role in ICF (inertial confinement fusion) and laser plasma [1, 2]. When heavier fluid is supported by lighter fluid in presence of gravitational force, the instability arises at the interface of fluids is known as Rayleigh Taylor instability [3,4].

In this connection Chandrasekhar [5] has given a good understanding of the classical RT instability of superposed fluids and Goldston and Rutherford [6] have discussed the classical RT instability of stratified fluids. The effect of magnetic field on the RT instability of stratified fluid in presence of suspended particles has been carried out by Chhajlani et al. [7]. Sharma and Chhajlani [8,9] have studied the effect of rotation and FLR correction along with suspended particles on the Rayleigh Taylor instability.

The RT instability of stratified fluids has been discussed by various authors. Cao et al.[10] have studied effect of quantum and magnetic field on the RT instability. Gupta and Singh [11] have investigated stability of stratified rotating viscoelastic fluid in the presence of variable magnetic field. Hoshoudy [12] has discussed Rayleigh Taylor instability in magnetized plasma. The effects of quantum and rotation on the Rayleigh Taylor instability of stratified fluids have been also revealed by Hoshoudy [13].

Keeping in mind the great scientific and applied interest of Rayleigh Taylor instability in convection problems, it is attempt to discuss the influence of magnetic field and angular velocity on the Rayleigh Taylor instability. This problem, to the best of our knowledge, has not been investigated yet.

2. Mathematical model

We consider incompressible plasma of finite thickness confined between two rigid planes z = 0, z = h under action of magnetic field **H** (H_x (z), 0, 0) and gravitational field $\boldsymbol{g}$ (0, 0, -g). Therefore the relevant magneto hydrodynamic equations of problem can be written as

$$\rho \frac{\partial \boldsymbol{U}}{\partial t} = -\nabla p + \rho \boldsymbol{g} + 2\rho(\boldsymbol{U} \times \boldsymbol{\Omega}) + \frac{\mu_e}{4\pi}(\nabla \times \boldsymbol{H}) \times \boldsymbol{H} \tag{1}$$

$$\frac{\partial \rho}{\partial t} + \nabla.(\rho \boldsymbol{U}) = 0 \tag{2}$$

$$\nabla.\boldsymbol{U} = 0 \tag{3}$$

$$\nabla.\boldsymbol{H} = 0 \tag{4}$$

$$\frac{\partial \boldsymbol{H}}{\partial t} = \nabla \times (\boldsymbol{U} \times \boldsymbol{H}) \tag{5}$$

The symbols $\rho, \boldsymbol{U}, p, \mu_e$ represents the fluid density, fluid velocity, pressure, permeability of magnetic field respectively. To examine the stability of system we assume the following perturbation in physical quantities

$$p = p_0 + p_1, \qquad \rho = \rho_0 + \rho_1, \qquad \mathbf{H} = \mathbf{H_0} + \mathbf{h}, \qquad \boldsymbol{U} = \boldsymbol{U_0} + \boldsymbol{u}. \tag{6}$$

In the above equation, symbols p_0 , ρ_0, $\boldsymbol{H_0}$, $\boldsymbol{U_0}$, denote the equilibrium state and p_1, ρ_1, $\boldsymbol{h}$ (h_x, h_y, h_z), $\boldsymbol{u}$ ($u_x u_y u_z$),denote the perturbed state of pressure, fluid density, magnetic field, velocity of fluid and quantum pressure. Now analyzing the disturbance into normal modes we assume the perturbation quantities are of the form

$$f(x, y, z, t) = f(z) exp(ik_x x + ik_y y - i\omega t) \tag{7}$$

where, $i\omega$ is growth rate of perturbation and $k^2 = k_x^2 + k_y^2$ is wave number. We linearized the basic equations of the problem by using normal mode analysis and obtain a general differential equation in the z component of velocity (u_z) as

$$i\omega\left[\frac{\partial}{\partial z}\left(\rho_0 \frac{\partial u_z}{\partial z}\right) - \rho_0 k^2 u_z\right] + \frac{gk^2 u_z}{i\omega}\frac{\partial \rho_0}{\partial z} + \frac{4\Omega^2}{i\omega\left(1 + \frac{\mu_e H_{0x}^2(z) k_x^2}{4\pi\rho_0 i^2 \omega^2}\right)}\frac{\partial}{\partial z}\left(\rho_0 \frac{\partial u_z}{\partial z}\right) + \frac{\mu_e H_{0x}^2(z) k_x^2}{4\pi i\omega}\left(\frac{\partial^2}{\partial z^2} - k^2\right)u_z + \frac{2\mu_e H_{0x}(z) k_x^2}{4\pi i\omega}\frac{\partial H_{0x}(z)}{\partial z}\frac{\partial u_z}{\partial z} = 0 \tag{8}$$

Equation (8) represents the effect of magnetic field and rotation on the Rayleigh Taylor instability of stratified fluids.

3. Dispersion Relation

We have considered that the stratified plasma of finite thickness is bounded by two rigid boundaries z = 0 and z = h. In this case the density and magnetic field have exponential dependence on z, therefore the density and magnetic field distribution at z = 0 are given as

$$\rho_0(z) = \rho_0 exp\left(z(L_D)^{-1}\right), H_{0x}(z) = H_{0x}(z) exp\left(z(2L_D)^{-1}\right) \tag{9}$$

where L_D is a constant. The general solution of equation (8) is obtained as follows

$$u_z = u_0 sin\left(\frac{n\pi z}{h}\right) exp(\lambda z) \tag{10}$$

here $\lambda = -(2L_D)^{-1}$ and n is an integer. Equation (8) is solved by using equations (9-10) and obtained general dispersion relation as

$$\sigma^4+\sigma^2\left[V_A^2k_x^2+\frac{4\Omega^2\left(h^2+4L_D^2n^2\pi^2\right)}{\left(h^2+4L_D^2n^2\pi^2+4k^2h^2L_D^2\right)}-\frac{4gk^2h^2L_D}{\left(h^2+4L_D^2n^2\pi^2+4k^2h^2L_D^2\right)}\right]$$
$$+V_A^2k_x^2\left[V_A^2k_x^2-\frac{4gk^2h^2L_D}{\left(h^2+4L_D^2n^2\pi^2+4k^2h^2L_D^2\right)}\right]=0 \tag{11}$$

where $\sigma = i\omega$ and $V_A^2=\dfrac{\mu_e H_{0x}^2(z)}{4\pi\rho_0}$ is the effect of magnetic field known as Alfven speed.

4. Discussion

4.1 Absence of quantum effect, magnetic field and rotation

We put $\omega_q^2 = 0$, $V_A^2 = 0$, and $\Omega=0$ in the general dispersion relation (11) and obtain

$$\sigma^2-\frac{4gk^2h^2L_D}{h^2+4n^2\pi^2L_D^2+4k^2h^2L_D^2}=0 \tag{12}$$

Equation (12) is a dispersion relation for the classical RT instability which has been discussed by Goldston and Rutherford [5]. The growth rate of classical RT instability is obtained as

$$\sigma_{classical}=\left(\frac{4gk^2h^2L_D}{h^2+4n^2\pi^2L_D^2+4k^2h^2L_D^2}\right)^{1/2} \tag{13}$$

From equation (13) it is clear that system has an exponentially growing perturbation.

4.2 Absence of magnetic field

In the absence of magnetic field the dispersion relation and growth rate becomes as

$$\sigma^2+\frac{4\Omega^2\left(h^2+4L_D^2n^2\pi^2\right)}{\left(h^2+4L_D^2n^2\pi^2+4k^2h^2L_D^2\right)}-\frac{4gk^2h^2L_D}{\left(h^2+4L_D^2n^2\pi^2+4k^2h^2L_D^2\right)}=0 \tag{14}$$

$$\sigma_{Rotation}=\left(\frac{4gk^2h^2L_D}{h^2+4n^2\pi^2L_D^2+4k^2h^2L_D^2}-\frac{4\Omega^2\left(h^2+4L_D^2n^2\pi^2\right)}{\left(h^2+4L_D^2n^2\pi^2+4k^2h^2L_D^2\right)}\right)^{1/2} \tag{15}$$

From equations (12 and 15) we find that $\sigma_{Rotation} < \sigma_{classical}$ this implies that the angular velocity (rotation effect) has stabilizing influence on the Rayleigh Taylor instability of stratified plasma.

To investigate the effect of rotation and magnetic field, we normalize the general dispersion relation by using following parameter, where $\omega_{pe}^2=\rho\ e^2\left(m_e^2\varepsilon_0\right)^{-1}$ is plasma frequency.

$$\sigma^*=\sigma\left(\omega_{pe}\right)^{-1},\ g^*=g\left(L_D\omega_{pe}^2\right)^{-1},\ k^*=kL_D,\ h^*=h\left(L_D\right)^{-1},$$
$$\omega_q^{*2}=\hbar^2\left(4m_em_iL_D^4\omega_{pe}^2\right)^{-1},\ V_A^{*2}=V_A^2\left(L_D^2\omega_{pe}^2\right)^{-1},\ k_x=k\cos\theta,\ \Omega^*=\Omega\left(\omega_{pe}\right)^{-1}$$

$$\sigma^{*4}+\sigma^{*2}\left[V_A^{*2}k^{*2}\cos\theta+\frac{4\Omega^{*2}\left(h^{*2}+4n^2\pi^2\right)}{\left(h^{*2}+4n^2\pi^2+4k^{*2}h^{*2}\right)}-\frac{4g^*k^{*2}h^{*2}}{\left(h^{*2}+4n^2\pi^2+4k^{*2}h^{*2}\right)}\right]+$$
$$V_A^{*2}k^{*2}\cos\theta\left[V_A^{*2}k^{*2}\cos\theta-\frac{4g^*k^{*2}h^{*2}}{\left(h^{*2}+4n^2\pi^2+4k^{*2}h^{*2}\right)}\right]=0 \tag{16}$$

Now solving above dimensionless dispersion relation by taking arbitrary values of parameters and drawing figures between growth rate (σ^*) and wave number (k^*) in variation of angular velocity and magnetic field. We have delineated figure 1 for configuration $h^*=1$, $n=1$, $g^*=10$, $V_A^*=0.4$, $\theta=0$ and found that the growth rate decreases on raising the value of angular velocity, this shows the angular velocity has stabilizing effect on the growth rate of Rayleigh Taylor instability of stratified. It indicates that the angular velocity has also stabilizing effect on the Rayleigh Taylor instability of

IOP Conf. Series: Journal of Physics: Conf. Series **836** (2017) 012009 doi:10.1088/1742-6596/836/1/012009

stratified plasma. Furthermore figure 2 has been depicted for configuration h* =1, n=1, g* = 10, Ω* = 0.4, VA* = 0.1, 0.3, 0.5 θ =0 to carry out the effect of magnetic field. We observe that the growth rate decreases on increasing the magnetic field which indicates that the magnetic field has stabilizing effect on the growth rate of the Rayleigh Taylor instability of stratified plasma. Therefore angular velocity and magnetic field have been predicted stabilizing influence on the Rayleigh Taylor instability of stratified fluids.

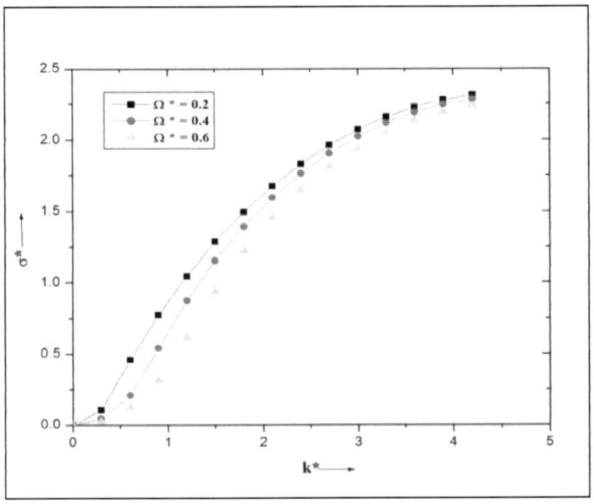

Figure 1. The growth rate (σ*) versus wave number (k*) in variation of angular velocity (Ω*).

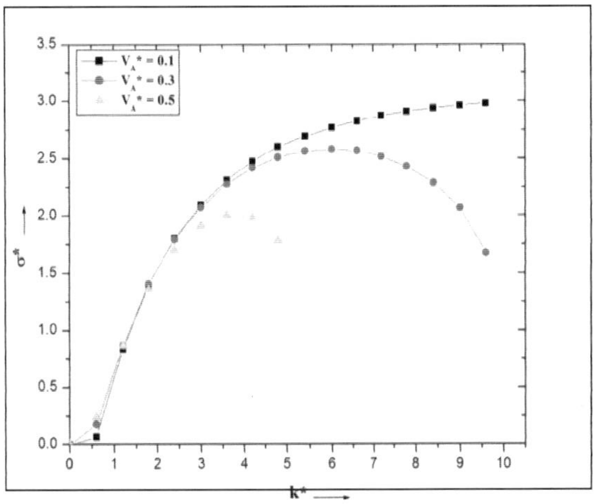

Figure 2. The growth rate (σ*) versus wave number (k*) in variation of magnetic field (V_A*).

References

[1] Ghasemizad A, Zarringhalam H and Gholamzadeh L 2009 *J. Plasma Fusion Res. SERIES* **8** 1234
[2] Hester J J *et al.* 1996 *Astrophys. J.* **456** 225
[3] Rayleigh L 1882 *Proc. London Math. Soc.* **14** 170
[4] Hoshoudy G A 2011 *Plasma Phys. Rep.* **9** 775
[5] Chandrasekhar S 1961 *Hydrodynamic and Hydromagnetic Stability* (Oxford: Clarendon press)
[6] Goldston R A and Rutherford P H 1997 *Introduction to Plasma Physics* (London: Institute of Physics)
[7] Chhajlani R K, Sanghvi R K and Purohit P 1984 *Z. Naturforsch.* **39** 939
[8] Sharma P K and Chhajlani R K 1998 *Phys. Plasmas* **5** 2203
[9] Sharma P K and Chhajlani R K 1998 *Z. Naturforsch.* **53** 937
[10] Cao J, Ren H, Wu Z W and Chu P K 2008 *Phys. Plasmas* **15** 012110
[11] Gupta R K and Singh M 2012 *Advances App. Sci. Res.* **3** 3253
[12] Hoshoudy G A 2009 *Phys. Plasmas* **16** 064501
[13] Hoshoudy G A 2014 *World J. Mech.* **4** 260

IOP Conf. Series: Journal of Physics: Conf. Series **836** (2017) 012010 doi:10.1088/1742-6596/836/1/012010

Theoretical investigation on the effect of nitrogen doping on the growth and field emission properties of the plasma-grown graphene sheet

N Gupta[a], S C Sharma, and R Gupta

Department of Applied Physics, Delhi Technological University, Delhi-110042, India

E-mail: [a)]nehagpt2105@gmail.com

Abstract. A theoretical model to study the impact of nitrogen-doping on the plasma-assisted growth and field emission properties of the graphene sheet (GS) has been developed. The model incorporates the charging rate of the GS, kinetics and energy balance of all plasma species i.e., electrons, positively charged ions and neutral atoms along with the nitrogen doping species, and growth rate of the GS. Numerical calculations on the effect of nitrogen doping on the thickness of the GS have been carried out for typical glow discharge parameters. It is found that the thickness of the GS decreases with nitrogen doping. The ramifications of nitrogen doping on the field enhancement factor of GS have likewise been examined on the premise of the above result. It is observed that the nitrogen doped GSexhibits better field emission as compared to undoped GS. Some of the results of the present investigation are in compliance with the experimental observations.

1. Introduction

Doping of graphene sheet (GS) using plasma enhanced chemical vapor deposition (PECVD) is considered as an effective method to deterministically control the growth of GS at low temperatures [1].Mehdipouret al.[2] and Sharma et al.[3] have developed the theoretical model to study the growth of carbon nano fibers and GS in the plasma environment respectively. Nitrogen is considered as one of the most popular candidate for doping of GS as it has similar radii as that of carbon and thus can replace carbon atoms in the graphene to induce n-type conduction in graphene [4]. Seoet al.[1] have reported that the thickness of the GS reduces upon addition of nitrogen to the $CH_4/Ar/H_2$ gas mixture. Moreover, the nitrogen doped GS exhibits high field enhancement factor [5]($\beta \approx \left(\lambda_{gr} / \tau_{gr}\right)$, where λ_{gr} is the height and τ_{gr} is the thickness of the GS) in comparison to undoped GS [6, 7].

2. Model

Following Sodha et al. [8], the present model considers the embryonic growth of GS in complex plasma consisting of electrons, positively charged ions and neutrals of CH_4 (denoted as A), neon (denoted as B), and nitrogen (denoted as C).Here, nitrogen is considered as the doping material.

The growth of GS in plasma environment can occur by many ways, namely, cluster formation, nucleation, coagulation, and growth of embryonic GS by condensation [9]. However, in the present investigation, we consider the embryonic growth of GS by condensation as we aim to explore the effect of nitrogen doping on the growth and field emission properties of the GS. Since the field enhancement factor β of carbon nanostructures is less influenced by the height of the carbon nanostructures [10], we study the influence of nitrogen doping on the thickness of the GS.

2.1. Growth rate equation of the mass of graphene sheet

$$\alpha_{gr}\rho_{gr}\frac{d\tau_{gr}}{dt} = \sum_{k}^{A,C} M_k \chi_k I_{kgr} + \sum_{k}^{A,C} M_{ik}\chi_{ik} I_{ikgr}, \tag{1}$$

where $\alpha_{gr}\left(=l_{gr}\times\lambda_{gr}\right)$ is the area, l_{gr} is the length of the GS, and ρ_{gr} is the surface mass density of the GS. M_k and M_{ik} are the masses of atom and ion k (k refers to A and C type of atom or ion), respectively. Equation (1) describes the increase in thickness of the GS due to atomic $\left(I_{kgr}\right)$ and ion collection current $\left(I_{ikgr}\right)$ of species A and C. χ_k and χ_{ik} are the sticking coefficients of atom and ion of type k , respectively.

2.2. Kinetic equation of positively charged ions and neutral atoms

$$\frac{d\eta_{ik}}{dt} = \varepsilon_k\eta_k - \delta_k\eta_e\eta_{ik} - \eta_{gr}I_{ikgr}, \tag{2}$$

$$\frac{d\eta_k}{dt} = \delta_k\eta_e\eta_{ik} - \varepsilon_k\eta_k - \eta_{gr}\left(1-\chi_{ik}\right)I_{ikgr} - \eta_{gr}\chi_k I_{kgr}, \tag{3}$$

where η_k and η_{ik} are the number densities of atom and ion of type k (k refers to A, B or C type of atom or ion) , respectively.

Equation (2) represents the kinetics of positively charged ions, (here, k refers to A, B or C type of ion) in the plasma due to ionization of neutral atoms (ε_k is the coefficient of ionization), recombination of electrons and ions (δ_k is the coefficient of recombination), and ion collection current at the GS surface.

Equation (3) describes the kinetics of neutral atoms in the plasma due to electron-ion recombination, ionization of neutral atoms, ion collection current at the GS surface, and the accretion of neutral atoms A and C at the surface of GS. However, the last term of equation 3 i.e., the accretion of neutral atoms would not be accounted for the kinetics of neutral atom B.

3. Results and discussions

We have carried out the calculations to study the effect of nitrogen doping on the growth and field emission properties of the GS for typically glow discharge parameters. In particular, we have estimated the time variation of thickness of GS for undoped and nitrogen doped GS, and then premeditated the repercussion of nitrogen doping on the field enhancement factor β of the GS.

The figure 1 illustrates the variation of thickness of undoped and nitrogen doped GS with time. It can be seen from figure 1 that the thickness of GS decreases upon nitrogen doping. This can be attributed to the fact that the nitrogen atom, being pentavalent produces an extra electron on the GS surface which in turn increases the ionization of neutral atoms and thereby decreases the neutral atoms available for accretion. Since, the neutral atoms available for accretion decreases, the thickness of GS decreases on nitrogen doping. Moreover, the etching effect by nitrogen is additionally considered as the basis for decrease of thickness of nitrogen doped GS. The results of figure 1 are in consistence with the experimental observations of Seo et al. [1].

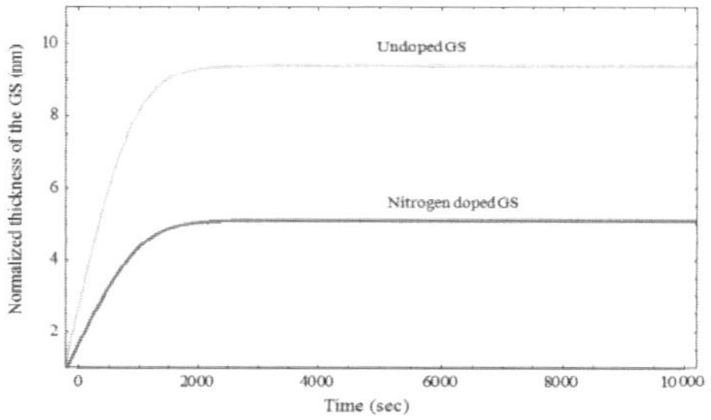

Figure 1. Variation of normalized thickness (in nm) of the undoped GS and nitrogen doped GS.

Using the above result, the field enhancement factor $\beta \approx (\lambda_{gr} / \tau_{gr})$ is calculated using the saturated value of thickness corresponding to undoped and nitrogen doped GS. Since height (λ_{gr}) has little influence on the field enhancement factor [10], it is assumed to be 1 μm for the calculation of β for undoped and nitrogen doped GS. It is found that the field enhancement factor is higher for nitrogen doped GS in comparison to undoped GS (cf. figure 2). The high field enhancement factor β of nitrogen doped GS is due to reduced thickness of GS upon nitrogen doping. The similar observation has been reported by Takeuchi et al. [6] and Palnitkar et al. [7].

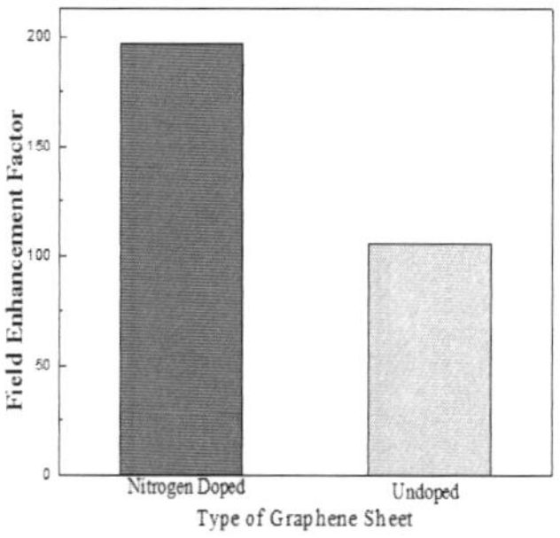

Figure 2. Field enhancement factor of undoped GS and nitrogen doped GS.

Acknowledgements
This work is partially supported by the Department of Science and Technology, Government of India.

References

[1] Seo D H, Kumar S, Rider A K, Han Z and Ostrikov K 2012 *Opt. Mat. Express* **2** 700
[2] Mehdipour H, Ostrikov K and Rider A E 2010 *Nanotechnology* **21** 455605
[3] Sharma S C and Gupta N 2015 *Phys. Plasmas* **22** 123517
[4] Rybin M, Pereaslavtsev A, Vasilieva T, Myasnikov V, Sokolov I, Pavlova A, Obraztsova E, Khomich A, Ralchenko V and Obraztsova E 2016 *Carbon* **96** 196
[5] Miller R, Lau Y Yand Booske J H 2007 *Appl. Phys. Lett.* **91** 074105
[6] Takeuchi W, Kondo H, Obayashi T, Hiramatsu M and Hori M 2011 *Appl. Phys. Lett.* **98** 123107
[7] Palnitkar U A, Kashid R V, More M A, Joag D S, Panchakarla L S and Rao C N R2010 *Appl. Phys. Lett.* **97** 063102
[8] Sodha M S, Misra S, Misra S K and Srivastava S 2010 *J. App. Phys.* **107** 103307
[9] Santolia I, Tewari A, Sharma S C and Sharma R 2014 *Phys. Plasmas* **21** 063508
[10] Xu Z, Bai X D and Wang E G 2006 *App. Phys. Lett.* **88** 133107

Effect of quantum correction and black body radiation on Jeans instability of porous medium

D L Sutar[1,a] and R K Pensia[2,b]

[1]Research Scholar, Mewar University Gangrar, Chittorgarh 312901, India
[2]Department of Physics, Govt. Girls College, Neemuch 458441, India

E-mail: [a)]devilalsutar833@gmail.com,[b)]rkpensia23456@gmail.com

Abstract. The Jeans instability of self-gravitating gaseous plasma is re-discussed in the framework of viscous, self-gravitating porous medium under the influence of quantum correction and black body radiation. The mathematical solution of the problem has been obtained through the normal mode analysis and the dispersion relation has been derived with the help of linearized perturbation equations. It is found that quantum corrections and black body radiation modify the fundamental Jeans criterion of gravitational instability and a new instability quantum corrected radiative instability is found. It is pointed here that the role of quantum correction is to stabilize the considered system by decreasing the value of critical Jeans wave-length. The stability of the system is discussed by applying Rought-Hurwitz criteria.

1. Introduction

There has been an accrued interest on astrophysics, motivated by applications in understanding the various collective process in gaseous plasma, which is ubiquitous in space, including diffuse and dense interstellar media, circumstellar shell, ionosphere, nova ejecta, star envelopes dark and dark interiors molecular clouds. The starting point for modern star cosmogony is that stars are formed and reach a state similar to that of the sum owing to the gravitational condensation of rarefied clouds of gas. It has been established as a fact that stars are formed as a result of the gravitational construction of fragments of molecular clouds, which are the sorts of collective behavior. The gravitational instability of an infinite homogeneous self-gravitating gaseous cloud was first discovered by Jeans [1]. The problem of instability of self-gravitating plasma in a magnetic field is of considerable astrophysical significance. A detailed analysis of the effect of magnetic field and rotation on the self-gravitational instability of an infinite homogeneous medium has been discussed by Chandrasekhar [2]. Chieze [3] has studied the formation of these clouds and suggested that the fragmentations occur near the gravitational instability threshold. These clouds are unstable against fragmentation when they approach gravitational instability in a fixed pressure environmental. The problem of an isothermal gas sphere under the action of external pressure has been investigated by Ebert [4], who pointed out that the disturbances of length scale approximately equal to the λ_j,Jeans length based on the central density were unstable to gravitational collapse. Since $M \propto \rho\lambda_j^3 \ \propto \rho^{\frac{-3}{2}}$, this considerably reduced the minimum unstable mass and demonstrated that an 'O' star cloud form in the center of an interstellar cloud. Hunter [5] has studied the growth of perturbations in a gravitationally contracting isothermal gas cloud and he points out that perturbations with the initial length scale of order less than λ_j, grew less rapidly relative to the background density than did perturbations of substantially large dimension. Recently Prajapati et al. [6] have discussed the problem of self-gravitational instability of rotating viscous hall plasma with arbitrary radiative heat-loss functions and electron inertia.In addition to this, there has been interesting in understanding the role of Black Body radiation in star formation and molecular cloud condensation process. An important point to be noted in this case, that the heat-loss process is a major cause for the condensation of large astrophysical compact objects. In this problem we have considered that for the case where the perturbations under heat-loss by black body radiation under the plasma correction, since radiative pressure in the inertial area of the star is usually small in comparison with the gaseous pressure, in interior of hot and large plasma cloud like H regions etc., where temperature is rather high while the density is low, it should be important to take into account the radiative processes. The radiative heat-loss functions considered earlier by Chandrasekhar [7].Many researchers [8-15]is studies of the self-gravitational instability of gaseous plasma, the role of quantum corrections and black body radiation has not been analyzed. It would, therefore, be of interest to examine the effect of black body radiation and quantum corrections. On the Jeans instability of self-gravitating viscous infinite homogeneous porous medium. We hope that the presence result will help to understand the structure formation in interstellar medium.

2. Equations of the problem

Let us consider an infinite homogeneous, self-gravitating viscous gaseous molecular clouds quantum plasma incorporating black body radiation. We introduced the quantum effects through the Bohm potential term in the momentum transfer equation. For simplicity of the problem, we ignore the additional convective instability, effects which may occur in such a system if the temperature T is a decreasing function of a coordinate, we shall assume T to be constant. The medium is taken optically thick and the black body radiation is assumed. The momentum transfer equation for magnetized quantum plasma is

$$\rho\frac{dv}{dt} = -\nabla p + \rho\nabla\phi + \rho\vartheta\left(\nabla^2 - \frac{1}{k_1}\right)v + \frac{\hbar^2\rho}{2m_e m_i}\nabla\left(\frac{\nabla^2\sqrt{\rho}}{\sqrt{\rho}}\right) \tag{1}$$

$$\varepsilon\frac{\partial\rho}{\partial t} + \nabla.(\rho v) = 0 \quad (2), \qquad \nabla^2\phi + 4\pi G\rho = 0 \quad (3), \qquad \frac{dT}{T} = (\Gamma - 1)\frac{d\rho}{\rho} \tag{4}$$

where $p = p_g + p_r$is the total pressure of the medium, $p_g = \rho RT$ is the gas pressure and$p_r = K_B T^4/3$ is the radiation pressure, K_B is the Boltzmann's constant. Also, the symbolsρ, γ,$\vec{v}$, ϕ, ϑ,G,T and Γ denotes the number density, adiabatic index, fluid velocity, gravitational potential, kinematic viscosity, universal gravitational constant, temperature and radiative heat function, respectively. $\hbar$Planck's constant divided by 2π,m_e and m_i are the electron and ion mass, respectively. Adiabatic changes in an enclosure containing matter and radiation are given by equation (4), (Chandrasekhar, 1957)

$$\Gamma = \left(1 + \frac{\Gamma_1 - b}{4-3b}\right), \Gamma_1 = \left(b + \frac{(4-3b)^2(\gamma-1)}{b+12(1-b)(\gamma-1)}\right), b = \left(\frac{p_g}{p_g+p_r}\right) \tag{5}$$

If the radiation is negligible then, $\Gamma = \Gamma_1 = \gamma$while in the case of $p_g \ll p_r$is$\Gamma = \Gamma_1 = 4/3$.

3. Linearized perturbation equations

In the linearization, we write the space and time dependent physical quantitiesρ, P, $\vec{v}$ and ϕ, in the form of the sum of the equilibrium and perturbed part as'

$$\rho = \rho_0 + \delta\rho, \quad p = p_0 + \delta p, \phi = \phi_0 + \delta\phi, \vec{v} = \vec{v}_o + \delta\vec{v} \tag{6}$$

The terms with subscript '0' denote the equilibrium part of the physical quantities. Perturbation in fluid velocity, fluid pressure, fluid density and gravitational potential are given as,$\delta\vec{v} = (0,0,v_z)\delta p\delta\rho$and $\delta\phi$, respectively.Using equation (6) in equations, (1) to (4) we write the linearized perturbation equations of infinite conducting quantum plasma, by removing '0' from subscript in the equilibrium quantities, for simplicity. Thus, we obtain linearized perturbation equations of the considered system as,

$$\frac{\partial \delta \vec{v}}{\partial t} + \frac{RT}{\rho}\frac{\partial \delta \rho}{\partial z} + R\left(1+4R_p\right)\frac{\partial \delta T}{\partial z} - \frac{\partial \delta \phi}{\partial z} - \rho\vartheta\left(\nabla^2 - \frac{1}{k_1}\right)\vec{v} - \frac{\hbar^2}{4m_e m_i}\frac{\partial}{\partial z}\frac{\left(\nabla^2 \delta\rho\right)}{\rho} = 0 \tag{7}$$

$$\varepsilon\frac{\partial \delta\rho}{\partial t} + \rho\frac{\partial \delta\vec{v}}{\partial z} = 0 \quad (8), \qquad \frac{\partial^2 \delta\phi}{\partial z^2} + 4\pi G\delta\rho = 0 \quad (9), \qquad \frac{\partial \delta T}{\partial t} - (\Gamma - 1)\frac{T}{\rho}\frac{\partial \delta\rho}{\partial t} = 0 \tag{10}$$

Where, $v \equiv v_z R_p = p_{r0}/p_{g0}$ denote the ratio of the radiation pressure and gas pressure, and Γ is assumed to be constant.

4. Dispersion relation

Let us assume the perturbation of all the quantities very as, $\exp\{-i\sigma t + ikz\}$ (11)

Where, σ is the frequency of harmonic perturbations and $k$is the wave number in z- direction. Combining equations (7) to (10), we get the dispersion relation

$$\omega^2 - \vartheta_k\omega + \frac{k^2RT}{\varepsilon}\left[1 + \left(1+4R_p\right)(\Gamma - 1)\right] - \frac{4\pi G\rho}{\epsilon} + \frac{\hbar^2 k^4}{4m_e m_i \varepsilon} = 0 \tag{12}$$

Where, $\omega = i\sigma$,the growth rate of instability, $\vartheta_k = \vartheta\left(k^2 + \frac{1}{k_1}\right)\vartheta$ is viscosity, k_1 is permeability. This dispersion relation (12) shows the combined influence of viscosity, black body radiation, porosity and quantum correction on the Jeans instability of optically thick quantum plasma. If we ignore the quantum correction and viscosity then this dispersion relation reduces to Vranjes and Cadez [10]. Thus, the dispersion relation is modified by the presence of viscosity and quantum correction. The stability of the system, represented by preceding equation, since all the coefficient of equation (12) is positive that the necessary condition for instability of the system is satisfied. To obtain the sufficient condition, the principal diagonal minors of Hurwitz matrix must be positive, which are shown below.

$$\Delta_1 = \vartheta_k > 0, \Delta_2 = \vartheta_k\frac{k^2RT}{\varepsilon}\left[1 + \left(1+4R_p\right)(\Gamma - 1)\right] + \frac{4\pi G\rho\vartheta_k}{\varepsilon} + \frac{\hbar^2 k^4\vartheta_k}{4m_e m_i \varepsilon} > 0,$$

$$\Delta_3 = \left(\frac{k^2RT}{\varepsilon}\left[1 + \left(1+4R_p\right)(\Gamma - 1)\right] + \frac{4\pi G\rho}{\varepsilon} + \frac{\hbar^2 k^4}{4m_e m_i \varepsilon}\right)\Delta_2 > 0,$$

We see that all $\Delta's$ are positive so we find that a Quantum Correction, Black body radiation, the viscous porous medium is stable the system. The condition of instability is obtained from the constant term of equation (12) and the condition is instability is given as,

$$\frac{k^2RT}{\varepsilon}\left[1 + \left(1+4R_p\right)(\Gamma - 1)\right] + \frac{\hbar^2 k^4}{4m_e m_i \varepsilon} - \frac{4\pi G\rho}{\varepsilon} < 0 \tag{13}$$

Thus the system is unstable if

$$\frac{\lambda_J}{\lambda_c} \geq \left[\frac{1}{\gamma\varepsilon}\left\{1 + \left(1+4R_p\right)(\Gamma - 1)\right\} + \frac{\hbar^2}{4\lambda_c^2 m_e m_i \varepsilon c_s^2}\right]^{-1/2} \tag{14}$$

where $\lambda_J = \sqrt{\pi c_s^2/\rho G}$is critical Jean's wavelength and $\lambda_c = 2\pi/k$, is the critical wavelength associated with black body radiations. The above equation (14) shows the quantum and black body corrected condition of relative instability. It is clear that this relative instability is affected by the presence of viscosity. In the absence of quantum correction, the above condition of relative instability is identical to Vranjes and Cadez [10]. Thus we have found the new condition of instability which is modified by quantum corrections term. It is also noted from relative instability condition (14) that the role of quantum corrections terms is to stabilize the considered system by increasing the value of critical Jean's wavelength.

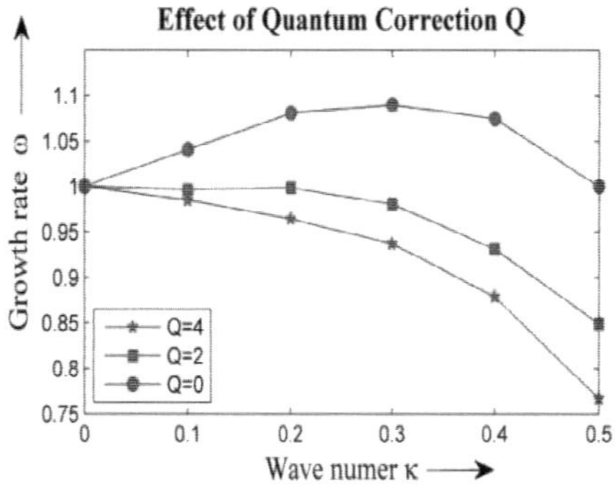

Figure1. Effect of quantum correction

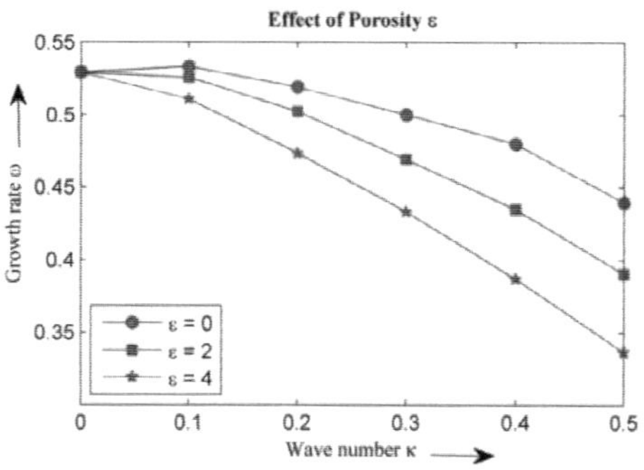

Figure 2. Effect of porosity

Now we plot the curves between normalized growth rates of Jeans instability versus normalized wave number. In figure 1and 2 we have plotted the real root of growth rate against wave number for various values of quantum correction and porosity. It is clear from this figure that for any wave number value the growth rate of instability decreases as increasing the value of quantum correction and porosity. Hence the quantum correction and porosity have a stabilizing influence on the system.

5. Conclusions

In the present paper, it is found that the viscosity has a dissipative effect but do not affect the Jeans expression. We have obtained a new modified relative instability which is affected by black body radiations, quantum corrections, and Porosity terms. It is pointed out that the black body radiation has a stabilizing influence on the considered problem which is further increased by quantum corrections terms.

References

[1] Jeans J H 1902 *Phil. Trans. Ray. Soc. London* **199** 1
[2] Chandrasekhar S 1961 *Hydrodynamics and Hydromagnetic Stability* (Oxford: Clarendon) p 588
[3] Chieze J P 1987 *Astron. Astrophys.* **171** 225
[4] Ebert R 1957 *Z. Astrophys.* **42** 263
[5] Hunter J H 1966 *Mon. Not. Roy. Astron. Soc.* **139** 239
[6] Prajapati R P, Pensia R K, Kaothekar S and Chhajlani R K 2010 *Astrophys. Space Sci.* **327** 139
[7] Chandrasekhar S 1957 *An Introduction to the Study of Stellar Structure* (New York: Dover Publication) p 55
[8] Vaghela D S and Shrivastava H S P 1994 *Czechoslovak J. Physics* **44** 905
[9] Radwan A E 2004 *Appl. Math. Comp.* **148** 331
[10] Vranjes J and Cadez V 1990 *Astrophys. Space Sci.* **164** 329
[11] Stiele H Lesch H and Heitsch F 2006 *Mon. Not. Roy. Astron. Soc.* **372** 862
[12] Sharma R C and Thakur K P 1982 *Astrophys. Space Sci.* **81** 95
[13] Masood W, Sallimullah M and Shah H A 2008 *Phys. Lett. A* **372** 6757
[14] Tsintsadze N L, Chaudhary R, Shah H A and Murtaza G 2008 *J. Plasma Phys.* **74** 847
[15] Patidar A K, Pensia R K and Shrivastava V 2012 *Canad. J. Phys.* **90** 1209

Jeans instability of a monatomic gas in the presence of thermal radiation

A Kumar[1,a], D L Sutar[2,b] and R K Pensia[3]

[1]Research Scholar, Faculty of Science, Pacific Academy of Higher Education and Research University, Udaipur 313003 India
[2]Research Scholar, Mewar University, Gangrar, Chittorgarh 312901, India
[3]Department of Physics, Govt. Girls College, Neemuch 458441, India

E-mail: [a)]akchoudhary23456@gmail.com,[b)]devilalsutar833@gmail.com

Abstract. The Jeans instability in magneto-radiative gas dynamics [MRGD] is discussed in the framework of magneto-gas dynamics [MGD] incorporating rotation and viscosity of the medium. The system of the present problem is assumed to adopt the Landau and Lifshitz model. The dispersion relation is obtained with the help of linearized perturbation equations. We find that the Jeans condition remains valid but the expression of the critical Jeans wave number is modified. Owing to the inclusion of radiative effect the sound speed changes which means that the critical wavelength depends on the amount of rate of radiation. We hope that our discussion on the present problem will help to improve our knowledge of interstellar medium [ISM].

1. Introduction

During few decade or so vigorous research activities have been launched to understand the various astrophysical problems. The simplest theory that describes the aggregation of masses in space the self-gravitational instability. The system comprising of particles that can aggregate together depends on the relative magnitude of the gravitational force to pressure force. Whenever the internal pressure of a gas is too weak to balance the self-gravitational force of a mass density perturbation, a collapse occurs. Such mechanism was first discovered by Jeans's [1]. It was shown that the disturbances would grow if their wavelength excelled a certain minimum wavelength λ_j, given by $\lambda_j = \left(\frac{\pi c^2}{G\rho} \right)^{1/2}$, where $\rho and c$ denote the density and sound velocity respectively and G is the gravitational constant. The Jean's criteria of instability are of central importance in understanding the process of formation of stars, planets, comets, asteroids, dark molecular clouds, commentary tails solar prominence and other astrophysical objects. A detailed contribution of the self-gravitational instability with different assumptions on the magnetic field and rotation has been given by Chandrasekhar [2]. In this connection, many investigators have discussed the gravitational instability of a homogeneous plasma considering the effects of various parameters [3-8].
On the other hand, the Coriolis force has been a subject of great interest for the last several decades. This interest was motivated by numerous mechanical applications in various disciplines, such as to understand the relation between angular momentum and rotational kinetic energy and how an inertial force can have

a significant effect on the moment of a body and still without doing any work. The magnetic force may be viewed as a kind of Coriolis force due to Thomas rotations, induced by successions of non-collinear Lorentz boosts. Maheshwari and Bhatia [9] have discussed the frictional effects with neutrals, in the presence of Hall currents, when the magnetic field is uniform and horizontal. Sankhla and Bhatia [10] have studied the combined influence of Hall current and Coriolis forces on the stability of a self-gravitating plasma. The characteristics of the magnetohydrodynamic wave in self-gravitating rotating plasma have been discussed by Hau and Chou [11]. Prajapati et al. [12] have investigated the problem of self-gravitational instability of rotating anisotropic heat conducting plasma. Recently, interest has been developed in understanding the role of thermal radiations in the star formation and molecular cloud condensation process in connection with thermal instability. Gomez-Pelaez and Moreno-Insertis [13] have investigated the problem of the thermal instability in a cooling and expanding medium including self-gravity and conduction in the neutral fluid dynamics. Shaikh et al. [14] have investigated the self-gravitational instability of thermally conducting partially ionized Hall plasma in the presence of a variable magnetic field. Recently, Prajapati et al. [15] have discussed the problem of self-gravitational instability of rotating viscous Hall plasma with arbitrary radiative heat-loss functions and electron inertia.

In addition to this magnetic fields can provide pressure support and inhibit the contraction and fragmentation of interstellar clouds. The magnetic field interacts directly only with the ions, electrons and charged grains in the gaseous medium. The collisions of the ions with the predominantly neutral gas in the clouds are responsible for the indirect coupling of the magnetic field to the bulk of the gaseous medium. The degree to which the magnetic pressure is important depends on the fields strength and the fractional abundance. The effects of simultaneous inclusion of viscosity, rotation, electron inertia and magnetic field on Jean's instability of a monatomic gaseous medium is not investigated yet.

In view of the importance of thermal radiation effects in astrophysical contexts, we have incorporated thermal radiation effects in the investigation of Jean's instability of an infinite homogeneous, viscous, rotating and magnetized monatomic gaseous medium in the presence of a finite electron inertia.

2. Linearized perturbation equations

We consider an infinite homogeneous rotating hot gaseous plasma in the presence of magnetic field B $(B_x, B_y, B_z)$and rotation assuming as $\Omega(0, \Omega_y, \Omega_z)$.The linearized perturbation equations of the problem are written as.

$$\rho\frac{\partial\vec{v}}{\partial t} + \frac{P}{\rho}\vec{\nabla}\delta\rho - \left(\frac{P}{T} + K_B a T^3\right)\vec{\nabla}\delta T - \rho\vec{\nabla}\delta U - \frac{1}{4\pi\mu}\left(\nabla\times\vec{b}\right)\times\vec{B} - \rho\vartheta\nabla^2\vec{v} - 2\rho\left(\vec{v}\times\vec{\Omega}\right) = 0 \quad (1)$$

$$\frac{\partial\delta\rho}{\partial t} + \rho\vec{\nabla}.\vec{v} = 0 \quad (2)$$

$$\frac{\partial\vec{b}}{\partial t} = \vec{\nabla}\times\left(\vec{v}\times\vec{B}\right) + \frac{c^2}{\omega_{pe}^2}\frac{\partial}{\partial t}\nabla^2\vec{b} \quad (3)$$

$$4\pi G\rho = 0 \quad (4)$$

$$\vec{\nabla}.\vec{b} = 0 \quad (5)$$

$$3K_B\left(\frac{\rho}{2m} + aT^3\right)\frac{\partial\delta T}{\partial t} - \frac{P}{\rho}\vec{\nabla}\delta\rho - K_B T\left(\frac{1}{m} + \frac{aT^3}{\rho_0}\right)\frac{\partial\delta\rho}{\partial t} = 0 \quad (6)$$

The parameters ρ, p, T, B, K_B, ϑdenote the density, pressure, temperature, magnetic field, Boltzmann constant, viscosity and t the time, $a = \frac{4\pi^2 k^3}{45h^3c^3}$, his Planck's constant c is the velocity of light, μ is the permeability of magnetic field, G, gravitational constant, m is the mass of particle,ω_{pe} electron plasma frequency. All perturbed quantity are given as $q = q + \delta q$, where v, $\delta\rho$, δT, δU, b, are the perturbed velocity, fluid density, magnetic field, temperature, gravitational potential, and magnetic field respectively.

3. Dispersion relation and discussions

We assume that all the perturbed quantity vary as $\exp(i\omega t + ik_z)$ (7)

where ω is the frequency of harmonic disturbances and k is the wave number in the z-direction. Using equations (1) – (7) we obtain the following dispersion relation.

$$\alpha_0\sigma^6 + \alpha_2\sigma^4 + \alpha_4\sigma^2 + \alpha_6 = 0 \tag{8}$$

We have assumed the following substitutions.

$$i\omega = \sigma,\ \Omega_j^2 = k^2c_R^2 - 4\pi G\rho\ ,\ c_R^2 = \frac{5K_BT}{3m}\left[1 + \frac{2a^2(\delta T^6)}{5n\{n+2a(\delta T^3)\}}\right],$$

$n = \frac{\rho}{m}$ is number density of the particles in the unperturbed state.

$$\alpha_0 = 16\pi^2\mu^2\rho^2$$

$$\alpha_2 = 16\pi^2\mu^2\rho^2\vartheta_k^2\Omega_j^2 + 4\,\pi\mu\rho[k^2\frac{(B^2 + B_Z^2)}{A_1} + 16\pi\mu\rho\vartheta_k\Omega^2]$$

$$\alpha_4 = 8\pi\mu\rho\left(\frac{k^2B_Z^2}{A_1} + 8\pi\mu\rho\vartheta_k\Omega_z^2\right)\Omega_j^2 + k^2[\frac{k^2B^2B_Z^2}{A_1^2} + 16\pi\mu\rho\frac{\left(B_y\Omega_z - B_z\Omega_y\right)^2}{A_1} + 16\pi\mu\rho\vartheta_k B_z^2\Omega_z^2]$$

$$\alpha_6 = k^4\vartheta_k^2B_Z^4\Omega_j^2$$

where, $A_1 = (\sigma f),\ and\ f = \left(1 + \frac{C^2k^2}{\omega_{pe}^2}\right)$

Equation (8) gives the combined influence of rotation, viscosity, electron inertia, magnetic field and thermal radiation on Jean's instability of the homogeneous gaseous plasma. Equation (8) does not allow a positive real or a complex root whose real part is positive and the system is stable because $B_z \neq 0$it follows that when $\Omega_j^2 < 0$then one of the root of equation (8) is positive, that means instability occurs with

$$k_j = \left(\frac{4\pi G\rho}{c_R^2}\right)^{1/2} \tag{9}$$

where, $k_j$is modified critical Jeans wave number and corresponding critical wavelength is given by

$$\lambda_j = \left(\frac{\pi c_R^2}{G\rho}\right)^{1/2} \tag{10}$$

The gaseous molecular cloud is unstable for all wavelengths of perturbation which are greater than Jeans wavelength and for all wave number which is less than Jeans wave number $(k < k_j)$.

If we take $B_z = 0$ then,$\alpha_6 = 0, \alpha_4 = 16\pi\mu\rho\Omega_z^2\left(4\pi\mu\rho\vartheta_k\Omega_j^2 + \frac{k^2B^2}{A_1}\right)$ and $V^2 = \frac{B^2}{4\pi\mu\rho}$is the Alfven velocity.

Since $\Omega_z \neq 0$ then from equation (8) we get modified critical Jean's wavelength as.

$$\lambda_{j1} = \left(\pi\left(c_R^2 + \frac{V^2}{f}\right)(G\rho)^{-1}\right)^{1/2} \tag{11}$$

Comparing equations (10) and (11) we see that in the transverse direction the magnetic field increase the critical wavelength, therefore is evident that magnetic field has a stabilizing influence on the gaseous molecular cloud whereas the electron inertia has a destabilizing influence on the system. If we take $\Omega_z = 0$ then modified critical Jeans wavelength is given as

$$\lambda_{j2} = \left(\pi\left(c_R^2 + \frac{V^2}{f}\right)\left(G\rho - \frac{\Omega^2}{\pi}\right)^{-1}\right)^{1/2} \tag{12}$$

Comparing equations (11) and (12) we see that the rotation increase the critical Jean's wavelength. Thus the rotation has a stabilizing influence on the gaseous plasma. Thus the stabilizing effects of magnetic field and rotation have been removed by finite electron inertia on Jean's criterion of rotating gaseous plasma in the presence of thermal radiation.

IOP Conf. Series: Journal of Physics: Conf. Series **836** (2017) 012012 doi:10.1088/1742-6596/836/1/012012

4. Conclusions

In the present paper, we have investigated the problem of Jean's instability of viscous rotating homogeneous monatomic gaseous plasma in the presence of thermal radiation, a finite electron inertia, and magnetic field. The equations of the problem are stated and the general dispersion relation is obtained, which is modified due to the presence of these parameters. The effect of the magnetic field is found to stabilize the system whereas the effect of finite electron inertia is found to destabilize the system. The rotation parameter increases the stabilizing effect of magnetic field.

References

[1] Jeans J H 1902 *Phil. Trans. Ray. Soc. London* **199** 1
[2] Chandrasekhar S 1961 *Hydrodynamics and Hydromagnetic Stability* (Oxford: Clarendon Press) p 588
[3] Lehnert B 1955 *J. Astrophys.* **121** 481
[4] Bhatia P K and Hazarika A B R 1955 *Phys. Scripta* **51** 775
[5] Lima J A S,Silva R and Santos J 2002 *Astron. Astrophys.* **396** 309
[6] Aggarwal M and Talwar S P 1969 *Mon. Not. Roy. Astron. Soc.* **146** 235
[7] Shaikh S, Khan A and Bhatia P K 2007 *Plasma Phys.* **47** 147
[8] Shaikh S, Khan A and Bhatia P K 2008 *Physics Lett. A.* **372** 1451
[9] Maheshwari S L and Bhatia P K1976 *Beitr. Plasma Phys.* **16** 251
[10] Sankhla V Dand Bhatia P K1978 *Indian J. Phys.* **52** 416
[11] Hau L N and Chou C K 2001 *Phys. Plasmas* **8** 1095
[12] Prajapati R P, Parihar A Kand Chhajlani R K 2008 *Phys. Plasmas* **15** 012107
[13] Gomez-Pelaez A J and Moreno-Insertis F 2002 *Astrophys. J.* **569** 766
[14] Shaikh S, Khan A and Bhatia P K 2007 *Astrophysics Space Science* **312** 35
[15] Prajapati R P, Pensia R K, Kaothekar S and Chhajlani R K 2010 *Astrophys. Space Sci.* **327** 139

IOP Conf. Series: Journal of Physics: Conf. Series **836** (2017) 012013 doi:10.1088/1742-6596/836/1/012013

Electron-Phonon coupling in magnetized semiconductor quantum plasmas

S Ghosh and Apurva Muley

School of Studies in Physics, Vikram University, Ujjain (456010), India

Abstract. Present paper deals with electron-phonon coupling in piezoelectric n-type magnetized semiconductor plasma under quantum regime. A quantum modified dispersion relation is derived for the evolution of desired electron-phonon coupling in semiconductor plasma using quantum hydrodynamic (QHD) model. The main ingredients of this study are the role of non-dimensional quantum parameter-*H* and externally applied magneto-static field. The presence of quantum parameter-*H* includes the contributions of Fermi degenerate pressure and quantum diffraction. It represents the ratio of plasmon energy to Fermi energy of the system, hence is a function of doping concentration n_0. An expression for gain coefficient of acoustic wave is obtained in terms of quantum parameter-*H* and magnetic field under the collision dominated limit. We present the effects of doping in medium and orientation of magnetic field on gain profile of acoustic wave. The results show that the presence of magnetic field and quantum effects through quantum parameter-*H* effectively modifies the gain per unit length of acoustic wave.

1. Introduction

Dynamics of acoustic phonons in semiconductor plasma has been discussed from many angles. Plasma physicists and device engineers have mainly emphasized on the interaction of acoustic wave with free carriers of semiconductor [1-4]. Electron-phonon coupling in semiconductor plasma medium subsequently generate an intense acoustic flux due to background lattice vibrations which finally give rise to a gain of acoustic wave when the electron drift becomes supersonic i.e. $\vartheta_0 > \vartheta_s$ (here ϑ_0 and ϑ_s are electron drift and sound velocities, respectively). Piezoelectric semiconductors are found to be natural candidates for converting mechanical energy to electric energy or vice versa due to coupling between electrons and acoustic phonons through piezoelectricity [5-6]. Thus phenomena of electron-phonon coupling in piezoelectric semiconductor plasma have become one of the extremely important areas of study in plasma physics.

In addition, the propagation of acoustic wave in magnetized plasma has been a topic of important research. It is well known that the plasmas are strongly influenced by magnetic field [7-8] because the plasma parameter may vary with the application of magnetic field. The study of behavior of different waves in plasmas in presence of magnetic field is among the oldest field in plasma physics. Therefore many physicists have published relevant papers about it based on various models [9-10].

Since semiconductors may be doped with external impurities which make it denser by introducing free charge carriers in excess. In such dense medium, quantum effects become unavoidable. Thus semiconductor quantum plasma has gained significant attention from plasma researchers for more than a decade due to their tremendous nano-scale applications [11-12]. Moreover semiconductor quantum plasma can host a variety of plasma waves in the presence of an ambient static uniform magnetic field. But in this context, very few reports have been found [13-14] in the literature. Therefore, in the present paper, we aim to investigate the electron-phonon coupling in semiconductor quantum plasma in presence of magnetic field and quantum corrections through quantum parameter-H which represents the ratio of plasmon energy to Fermi energy of the carriers under the field geometry defined in Steele and Vural [15].

2. Theoretical formulation

The coupling between electrons and acoustic phonons in piezoelectric semiconductor quantum plasma is governed by the macroscopic model of piezoelectric media and quantum hydrodynamic (QHD) model. Hence following Steele and Vural [15], a quantum modified dispersion relation, in terms of quantum parameter-H and magneto-static field via electron cyclotron frequency, is derived as

$$\left(\omega^2 - k^2\vartheta_s^2\right)\left(1 - \frac{\omega_p^2}{(\omega - k\vartheta_0)F_Q(\omega,k)}\right) = K^2k^2\vartheta_s^2 \tag{1}$$

where $F_Q(\omega,k) = \left(\omega - k\vartheta_0 - i\nu - \frac{\nu D_F k^2\left(1 + \frac{k^2\vartheta_F^2}{4\omega_p^2}H^2\right)}{(\omega - k\vartheta_0)}\right) + \left(\frac{\omega_c^2\sin^2\theta(\omega - k\vartheta_0 - i\nu)}{\omega_c^2\cos^2\theta - (\omega - k\vartheta_0 - i\nu)^2}\right)$

Here $H = \hbar\omega_p/2K_BT_F$ is the non-dimensional quantum parameter, measuring the relevance of quantum effects, and is proportional to quantum diffraction. $\vartheta_F\left(=(2K_BT_F/m)^{1/2}\right)$is Fermi velocity and $D_F = \vartheta_F^2/\nu$ is diffusion constant at Fermi temperature T_F. Other symbols have their usual meaning.

The first bracket of L.H.S. of the equation (1) represents an acoustic mode while second bracket is a quantum modified electro-kinetic mode. The term on R.H.S. represents the coupling between electron and phonon through $K^2 = \bar{e}^2/\varepsilon C$, the dimensionless electro-mechanical coupling constant, and $\vartheta_s = \sqrt{C/\rho}$ the acoustic velocity in the medium.

Now using the approximation $\frac{k\vartheta_s}{\omega} = 1 + i\alpha$ [15], where the gain per radian $\alpha << 1$, in the collision dominated regime $(\omega, k\vartheta_0 << \nu)$, equation (1) reduces to

$$\alpha = \frac{\frac{1}{2}K^2\gamma\,\omega_R/\omega\varphi}{\left(\left(\frac{\omega_R}{\omega\varphi}\right)^2\left(1 + \frac{\omega^2\left(1 + \frac{k^2\vartheta_F^2}{4\omega_p^2}H^2\right)}{\omega_R\omega_{DF}}\right)^2 + \gamma^2\right)} \tag{2}$$

Here $\omega_R = \omega_p^2/\nu$ is dielectric relaxation frequency, $\omega_{DF} = \vartheta_s^2/D_F$ is diffusion frequency at Fermi temperature, $\gamma = \frac{\vartheta_0}{\vartheta_s} - 1$ and $\varphi = \frac{1 + \omega_c^2/\nu^2}{1 + \omega_c^2\cos^2\theta/\nu^2}$. The above expression represents attenuation coefficient in terms of quantum parameter-H, and magnetic parameter φ. At $\alpha > 0$ i.e. when attenuation coefficient becomes positive, one obtains the amplification of acoustic wave.

3. Results and discussions

For the numerical estimation of our results, we consider n-InSb as our study medium. Using the typical parameters of n-InSb, we investigate the dependency of gain per unit length $(\alpha\omega/\vartheta_s)$ of acoustic wave on the doping concentration (n_0) and orientation angle (θ) of magnetic field and are presented in the following figures

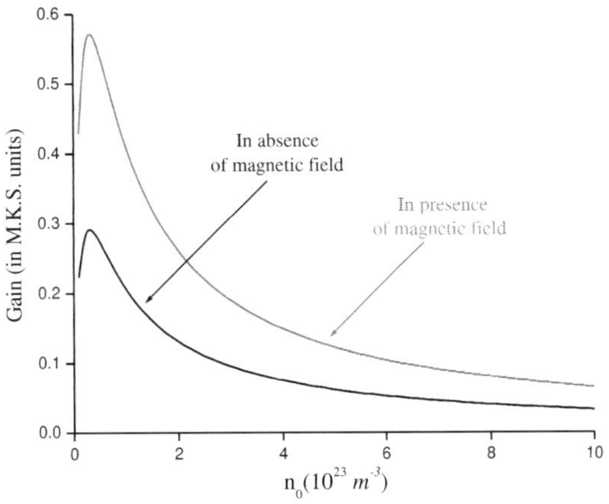

Figure 1. $\alpha\omega/\vartheta_s$ versus n_0

One may infer from figure 1 that $\alpha\omega/\vartheta_s$ varies with n_0 in identical manner both in presence and absence of magnetic field. It increases initially, attains maxima and then starts decreasing with n_0. The magnitude of $\alpha\omega/\vartheta_s$ is found to be high in presence of magnetic field. Hence, the presence of magnetic field becomes favourable for achieving higher gain.

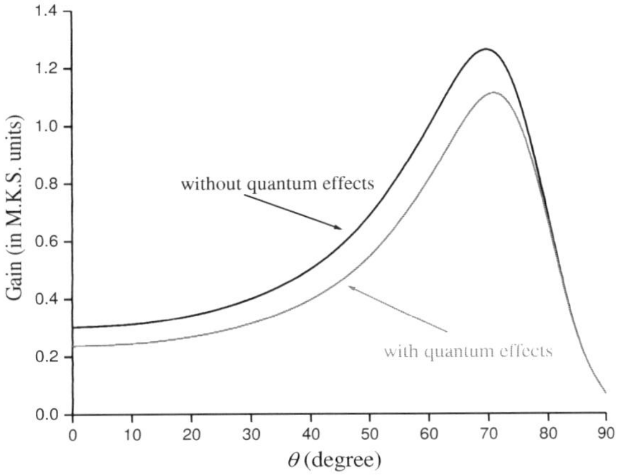

Figure 2. $\alpha\omega/\vartheta_s$ versus θ

The gain per unit length of acoustic wave as a function of orientation angle of magnetic field is depicted in figure 2 with quantum effect as parameter. Figure shows that on enhancing the value of θ, $\alpha\omega/\vartheta_s$ increases, attains maxima at $\theta \approx 72^\circ$ and then starts decreasing with θ. It is also observed from this figure that the magnitude of $\alpha\omega/\vartheta_s$ is found to be low as one includes the quantum corrections through quantum parameter-*H* and its effect on $\alpha\omega/\vartheta_s$ becomes vanished as θ approaches 84°.

4. Conclusions

The electron-phonon coupling in magnetised semiconductor plasma is studied under quantum regimes in which quantum effects are considered through non-dimensional quantum parameter-*H*. The results

IOP Conf. Series: Journal of Physics: Conf. Series **836** (2017) 012013 doi:10.1088/1742-6596/836/1/012013

of present study reveal that the quantum parameter-*H* and magnetic field significantly modify the dispersion relation and gain coefficient of acoustic wave. It is also found that higher gain can be obtained in presence of magnetic field and in absence of quantum parameter-*H*. Authors may conclude that this study may be useful in designing of many electronic devices such as acoustic wave amplifier, acousto-electric oscillators, acoustic filters etc.

References

[1] Mosekilde E 1974 *Phys. Rev. B* **9** 682
[2] Dey (Bose) M and Ghosh S 1990 *Phys. Stat. Sol. (b)* **157** 159
[3] Alok U, Kumar M, Singh B K and Kumar P 2012 *Bibechana* **8** 67
[4] Dubey P and Ghosh S 2016 *Acta Acust. United AC* **102** 436
[5] Büyükköse S, Mínguez A H, Vratzov B, Somaschini C,Geelhaar L, Riechert H, Wiel W G V and Santos P V 2014 *Nanotechnology* **25** 135204
[6] Bouhemadoua A,Khenatab R, Bin-Omranc S,Murtazad G and Al-Dourie Y 2015 *Opt. Mater.* **46** 122
[7] Dutta M, Ghosh S, Roychoudhury R, Khan M and Chakrabarti N 2013 *Phys. Plasmas* **20** 042301
[8] Li C, Wu Z, Yang W and Chu P K 2014 *Phys. Plasmas* **21** 072114
[9] Vyas A, Singh R K and Sharma R P 2016 *Phys. Plasmas* **23** 012107
[10] Amin M R 2015 *Phys. Rev. E* **92** 033106
[11] Rasheed A, Jamil M, Siddique M, Huda F and Jung Y D 2014 *Phys. Plasmas* **21** 062107
[12] Amin M R 2015 *Phys. Scr.* **90** 015601
[13] Ghosh S and Muley A 2015 *Int. J. Mod. Sci. Eng. Tech.* **2** 42
[14] Rasheed A, Jamil M, Areeb F, Siddique M and Salimullah M 2016 *J. Phys. D: Appl. Phys.* **49** 175109
[15] Steele M C and Vural B 1969 *Wave Interactions in Solid State Plasmas* (New York: Mc-Graw Hill) p 136

IOP Conf. Series: Journal of Physics: Conf. Series **836** (2017) 012014 doi:10.1088/1742-6596/836/1/012014

Effect of Coriolis force on the self-gravitational instability of dusty plasma in the presence of magnetic field

R K Pensia[1], D L Sutar[2,a], V Kumar[3] and A Kumar[3]

[1]Department of physics, Govt. P.G. Girls College, Neemuch 458441, India
[2]Research Scholar, Mewar University Gangrar, Chittorgarh 312901, India
[3]Department of Physics, Pacific University, Udaipur 313024, India

Email: [a)]devilalsutar833@gmail.com

Abstract. In view of the importance of Coriolis force in an astrophysical context, the problem of self-gravitational instability of dusty plasma in the presence of magnetic field is investigated. Equations of the problem are stated and the dispersion relation has been derived with the help of linearized perturbation equations. We find that the Jeans criterion of instability remains valid but the expression of the critical Jeans wave-number is modified. Mathematical calculations have been performed and some figures are plotted between the growth rate of instability and wave numbers. From the curves, it is found that dust sonic speed has a stabilizing effect.

1. Introduction

The importance of dusty plasma in understanding the problem of star formation has increased the attraction and attention of scientists in the recent past. The experimental observations from satellite and space studies have revealed the presence of charged dust grains in terrestrial atmosphere and also in some laboratory devices. The dust grains also exists in comets, planetary rings, asteroids and the magnetosphere, tokamaks MHD generators and plasma etching techniques are the examples of laboratory plasma in which dust particles present. The probable reason of anomalous scattering of radio waves in the ionosphere may be due to the dust grains in that region. The size of dust particles may range between .05μm to 10μm.When the grain average separation is greater in comparison to plasma Debye length, then we call it “dust in plasma”, otherwise “dusty plasma”. The role of dust grains is important for the process of formation of stars and planets. The collapsing of dusty plasma into the small clouds of mass less than 1 M_Θ has been discussed by Horedt [1], Alfvén [2] and Carlqvist [3].The analyses of dust related problems with the three-fluid model were studied by D’Angelo et al. [4] and D’Angeloet al. [5]. Many researchers [6-11] have analyzed in this field. In all the above studies the gravitational instability of the plasma system with rotating charged dust particles has not been investigated. Therefore, in the present work, we shall discuss the effect of charged dust grains on the gravitational instability of plasma system.

IOP Conf. Series: Journal of Physics: Conf. Series **836** (2017) 012014 doi:10.1088/1742-6596/836/1/012014

2. Basic equations and dispersion relation

To study the present problem, we consider a three component fluid model of dusty plasma. This model consists electrons, ions and charged dust particles subjected to an external field $\vec{B}_0$ .As a simplified model, we assume that dust grains have uniform mass, behave like a point charge. The self-gravitation for both ions and dust particles is considered by taking Poisson's equation for each. The momentum equation for each of dust, ion and electrons are

$$mnD\boldsymbol{v} = ZneE + \frac{Zne}{c}(\boldsymbol{v}\times\boldsymbol{B}) - \gamma T\nabla n + 2mn(\boldsymbol{v}\times\Omega) \quad (1)$$

$$m_i n_i D_i \boldsymbol{v}_i = n_i e\boldsymbol{E} + \frac{n_i e}{c}(\boldsymbol{v}_i\times\boldsymbol{B}) - \gamma_i T_i\nabla n_i + m_i n_i \nabla U_i \quad (2)$$

$$0 = -n_e e\boldsymbol{E} - \frac{n_e e}{c}(\boldsymbol{v}_e\times\boldsymbol{B}) - \gamma_e T_e \nabla n_e \quad (3)$$

where n_j, v_j, m_j, T_j and γ_j are, respectively, the total number density, velocity, mass, temperature and ratio of specific heat of the particle species j (j equals e for electrons and i for ions). The gravitational potential for ion and dust are denoted by U_i and U respectively. The terms n, v, m, T and γ are the number density, velocity, mass, temperature and ratio of specific heat for dust grains, respectively. The Coriolis force acting on the dust fluid is $2mn(v\times\boldsymbol{\Omega})>>2m_i n_i(v_i\times\boldsymbol{\Omega})$.The value of Z is taken positive or negative according to the positively or negatively charged dust particles. The numerical value of γ, γ_e and γ_i are to be appropriately chosen. For dust particles and ions, the continuity equations and Poisson equations are

$$\frac{\partial n}{\partial t} + [\nabla\cdot(n\boldsymbol{v})] = 0 \ (4), \quad \frac{\partial n_i}{\partial t} + [\nabla\cdot(n_i\boldsymbol{v}_i)] = 0 \ (5), \quad \nabla^2 U = -4\pi Gmn \ (6), \quad \nabla^2 U_i = -4\pi G m_i n_i \quad (7)$$

The electromagnetic fields $\boldsymbol{E}$ and $\boldsymbol{B}$, are given by Maxwell equations coupled to the equations (1)-(3).By assuming quasi-neutrality the displacement current for low frequencies the simplified form of these equations can be written as

$$\nabla\times\boldsymbol{B} = \frac{4\pi e}{c}(n_i\boldsymbol{v}_i + Zn\boldsymbol{v} - n_e\boldsymbol{v}_e) \ (8), \quad \nabla\times\boldsymbol{E} = -\frac{1}{c}\frac{\partial\boldsymbol{B}}{\partial t} \ (9), \quad n_e = n_i + Zn \quad (10)$$

The value of $\boldsymbol{v}_e$ from equation (8) is substituted in equation (3) to obtain an expression for $\boldsymbol{E}$ i.e.

$$\boldsymbol{E} = -\frac{n_i}{cn_e}\boldsymbol{v}_i\times\boldsymbol{B} - \frac{Zn}{cn_e}\boldsymbol{v}\times\boldsymbol{B} + \frac{1}{4\pi n_e e}(\nabla\times\boldsymbol{B})\times\boldsymbol{B} - \frac{\gamma_e T_e}{n_e e}\nabla n_e \quad (11)$$

By eliminating $\boldsymbol{E}$ from equation (11) and using equation (10), equation (1) and (2) can be rewritten as

$$mnD\boldsymbol{v} = \frac{Zn}{4\pi n_e}(\nabla\times\boldsymbol{B})\times\boldsymbol{B} + \frac{Zenn_i}{cn_e}(\boldsymbol{v}-\boldsymbol{v}_i)\times\boldsymbol{B} - Z\gamma_e T_e\frac{n}{n_e}\nabla n_e - \gamma T\nabla n + mn\nabla U + 2mn(\boldsymbol{v}\times\Omega) \quad (12)$$

$$m_i n_i D_i\boldsymbol{v}_i = \frac{n_i}{4\pi n_e}(\nabla\times\boldsymbol{B})\times\boldsymbol{B} - \frac{Ze}{c}\frac{n_i n}{n_e}(\boldsymbol{v}-\boldsymbol{v}_i)\times\boldsymbol{B} - \gamma_e T_e\frac{n_i}{n_e}\nabla n_e - \gamma_i T_i\nabla n_i + m_i n_i\nabla U_i, \quad (13)$$

After eliminating $\boldsymbol{E}$ between equation (1) and (9),we have

$$\frac{\partial\boldsymbol{B}}{\partial t} = \nabla\times(\boldsymbol{v}\times\boldsymbol{B}) + \frac{2mc}{Ze}\nabla\times(\boldsymbol{v}\times\Omega), \quad (14)$$

where we have assumed that for low-frequency wave $\omega <<ZeB_0/mc$,so the term $(mc/Ze)\nabla\times(D\boldsymbol{v})$ can be neglected in comparison to the first term. The set of equations (4), (7), (10) and (12)-(14) represents completely a self-gravitating system of magnetized rotating dusty plasma. In real, the dust particle is much heavier than ions so the ion inertia can also be neglected. With this approximation, we get by adding equations (12) and (13) and using equation (10).

$$mnD\boldsymbol{v} = \frac{1}{4\pi}(\nabla\times\boldsymbol{B})\times\boldsymbol{B} - (\gamma_e T_e + \gamma_i T_i)\nabla n_i - (Z\gamma_e T_e + \gamma T)\nabla n + mn\nabla U + 2mn(\boldsymbol{v}\times\Omega) + m_i n_i\nabla U_i \quad (15)$$

Equation (13) yields with the above approximation

$$\boldsymbol{v}_{i\perp} = \frac{\mathbf{1}}{\boldsymbol{B}^2}\boldsymbol{B}\times(\mathbf{v}\times\boldsymbol{B}) - \frac{\boldsymbol{c}}{n_e}\frac{1}{4\pi nB^2}\boldsymbol{B}\times[(\nabla\times\boldsymbol{B})\times\boldsymbol{B}] - \frac{cn_e m_i}{Zn_e B^2}\boldsymbol{B}\times\nabla\boldsymbol{U}_i + \frac{c\gamma_e T_e}{Zn_e B^2}\boldsymbol{B}\times\nabla\boldsymbol{n}_e + \frac{cn_e}{Zen_i}\frac{\gamma_i T_i}{nB^2}\boldsymbol{B}\times\nabla\boldsymbol{n}_e, \quad (16)$$

In equation (16) the last two terms appear due to the thermal effects. Moreover, the first term in the equation corresponds to $\boldsymbol{E}\times\boldsymbol{B}$ drift and is a dominant one; therefore, we neglect the thermal terms. We substitute the value of $\boldsymbol{v}_{i\perp}$ in equation (5) to get

$$\frac{\partial n_i}{\partial t} + \nabla_\perp \cdot \{ \frac{n_i}{B^2}\boldsymbol{B}\times(\boldsymbol{v}\times\boldsymbol{B}) - \frac{c}{Ze}\frac{n_i}{4\pi nB^2}\boldsymbol{B}\times[(\nabla\times\boldsymbol{B})\times\boldsymbol{B}] - \frac{cn_e}{ZneB^2}m_i n_i \boldsymbol{B}\times\nabla U_i \}, \quad (17)$$

3. Dispersion Relation

Equations (4), (14), (15) and (17) can be used as the basic equations for the considered self-gravitating dusty plasma system. These equations are linearized by assuming that perturbed quantity variesexp $i(k.x - \omega t)$.The perturbation state is given by

$$n = n_0 + \delta n,\ n_i = n_{i0} + \delta n_i, n_e = n_{e0} + \delta n_e, \boldsymbol{B} = \boldsymbol{B_0} + \boldsymbol{b}, \boldsymbol{v} = \boldsymbol{v_0} + \boldsymbol{v_1}, U = U_0 + \delta U, D = \frac{\partial}{\partial t} + \boldsymbol{v}.\nabla$$

The subscripts "0" and "1 denotes the equilibrium and the perturbed quantities, respectively. On solving these equations the general dispersion relation is obtained as,

$$\omega^2 = [V_A^2 + C_S^2 - \frac{4\pi G}{k^2}(\frac{m^2 n_0^2 + m_i^2 n_{i0}^2}{mn_0}) + \Omega B_0 - 2\Omega V_A]k^2 \ (18), \quad \boldsymbol{V_A} = \frac{\boldsymbol{B_0}}{(4\pi G m n_0)^{1/2}}, \quad C_S^2 = \frac{\gamma_e n_{e0} T_e + \gamma_i n_{i0} T_i + \gamma n_0 T}{mn_0}\boldsymbol{V_A}$$

represents the Alfvén velocity in terms of dust particles mass density and C_S represents the sound speed as the ratio of total particle pressure to the dust mass density. The dispersion relation (18) shows the combined influence of temperature, sound speed, Alfven speed in terms of the dust particle mass density and rotation parameter on the self- gravitational instability of a dusty plasma. The present results are also similar to that Rao[10] in the absence of self-gravitation and rotation in that case. Thus with these correlations, we find that the dispersion relation(18) is modified due to the combined effects of rotation, sound speed, Alfven velocity in terms of the dust particle mass, rotation and self-gravitation of the system.

The system is unstable for $k < k_{j1}$, where k_{j1} is modified Jeans critical wave number and it is given as

$$k_{j1} = [\frac{4\pi G}{(V_A^2 + C_S^2 - \Omega B_0 + 2\Omega V_A)}(\frac{m^2 n_0^2 + m_0^2 n_{i0}^2}{mn_0})]^{1/2} \quad (19)$$

Thus, we note that the Jeans criterion is modified due to the inclusion of rotation and dust particles in the system and the critical Jeans wave number depends on the rotation and mass density of dust along with the density of ions and electrons. In the absence of rotation, the relation (18)reduces to that of critical wave number already obtained by Chhajlani and Parihar [11].Thus we find the expression of the critical Jeans wave number which is modified due to the presence of rotation, sound speed and Alfven velocity in terms of the dust particle mass density. If we ignore the self-gravitation of the system then we obtain a fast Alfven wave of phase velocity $V_A^2 + C_S^2$. If fluid pressure dominates the magnetic pressure then wave turned into acoustic wave moving with velocity C_s . If the magnetic pressure is much greater in comparison to the fluid pressure then this turns as Alfven wave with velocity V_A.The Alfven velocity depends on the dust mass density. We write the dispersion relation (19) in non-dimensional form in terms of self-gravitation as

$$\omega^{*2} = V_A^{*2} + V_{sd}^{*2} + k^{*2} - (1+\delta) - \Omega^* V_A^* \quad (20)$$

where the various non-dimensional parameters are defined as

$$\omega^* = \frac{\omega}{(4\pi G\rho_d)^{1/2}}, V_S^{*2} = \frac{B_0^2 4\pi G\rho_d k^2}{4\pi\rho_d}, \delta = \frac{p_i^2}{\rho_d^2}, V_{sd}^{*2} = \frac{\gamma T/m}{4\pi G\rho_d}k^2, \Omega^* = \frac{\Omega}{(4\pi G\rho_d)^{1/2}}\ k^* = \frac{(\rho_e\gamma_e T_e/m_e) + (\rho_i\gamma_i T_i/m_i)}{(4\pi G\rho_d{}^2)}k^2$$

In figures 1and 2, we have depicted the non-dimensional growth versus nondimensional wave number for various arbitrary values of the dust sonic speed(V_{sd}^{*}),rotational(Ω^{*}) and the ratio of the density of dust particles (δ). Figure1 is plotted for the growth rate of an unstable mode(positive real root of (ω^{*}) against the wavenumber (k^{*}) with variation in the sound speed of dust (V_{sd}^{*}) parameter with other values as V_A^{*}=5, δ=10 and Ω^{*}=1. Hence the sonic speed of dust has a destabilizing influence on the growth rate of the instability.

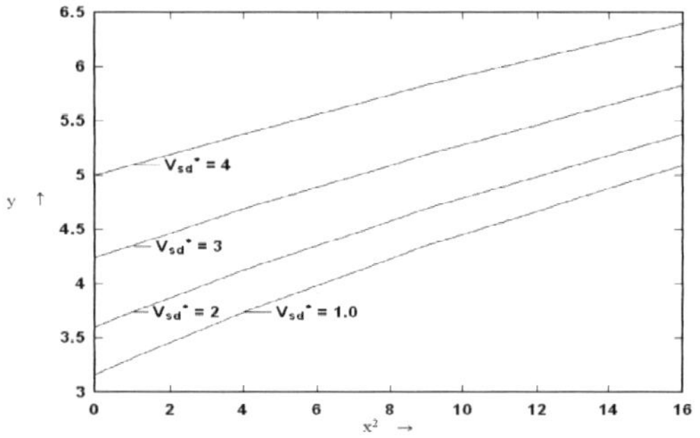

Figure 1

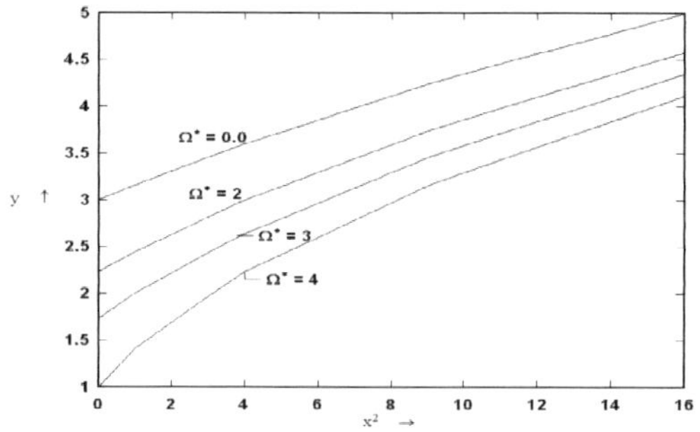

Figure 2

Figure 2 curve plotted for the growth rate of unstable mode against the wave number taking ω^{*} as y and k^{*} as x with variation in the Ω^{*} other values as V_{sd}*=4 ,δ=10 and V_A^{*}=2.We find that the growth rate of the instability decreases with increase in Ω^{*}. Hence the rotation parameter Ω^{*}has a stabilizing influence on the growth rate of instability of the system.

5. Conclusions

We have investigated the self-gravitational instability of magnetized dusty plasma under the effect of rotation. The general dispersion relation is obtained, which is modified due to the presence of these parameters. This dispersion relation is reduced for propagation perpendicular to the z-axis and propagation parallel to the z- axis. We find that the Jeans criterion of instability remains valid but the expression of the critical Jeans wave number is modified. We obtain a self-gravitating Alfven mode modified by the presence of rotation. We find that the condition of instability is affected by the presence of magnetic field and rotation. From the curves, it is found that rotation have a stabilizing effect but the sonic speed of dust has a destabilizing influence on the system.

Acknowledgements
The authors are grateful to professor N K Dabkra, Department of Physics, Vikram University Ujjain, for their constant encouragement.

References

[1] Horedt G P 1976 *Astrophys. Space Sci.***45** 353
[2] Alfvén H 1981 *Cosmic Plasma* (London: Reidel)
[3] Alfvén H and Carlqvist P 1978 *Astrophys. Space Sci.* **54** 279
[4] Angelo N D, Goeler S V and Ohe T 1965 *Phys. Fluids* **9** 1605
[5] Angelo N D and Merlino R L 1986 *IEEE Trans. Plasma Sci.* **514** 285
[6] Verheest F G 1967 *Physica* **34** 17
[7] James C R and Vermeulen F 1968 *Canadian J. Phys.* **46** 855
[8] RaoN N, Shukla P K and Yu M Y 1990 *Planet Space Sci.* **38** 543
[9] Rao N N 1993 *Planet Space Sci.* **41** 21
[10] Sharma R C and Sharma K C 1980 *Astrophys. Space Sci.* **71** 325
[11] Chhajlani R K and Parihar A K 1994 *Astrophys. J.* **422** 746

IOP Conf. Series: Journal of Physics: Conf. Series **836** (2017) 012015 doi:10.1088/1742-6596/836/1/012015

Theoretical investigation of the ultra-intense laser interaction with plasma mirrors in radiation pressure dominant regime

Krishna Kumar Soni[a], K P Maheshwari and N K Jaiman

Department of Pure & Applied Physics, University of Kota, India 324005

E-mail: [a)]sonikrishna1490@gmail.com

Abstract. At laser intensity in the range ~ 10^{22} -10^{23}W/cm^2, the radiation pressure starts to play a key role in the interaction of an intense electromagnetic wave with a dense plasma foil. Depending upon the incident laser intensity, polarization of the incident beam and also on the density of the thin plasma layer the mirror motion may be assumed to be uniform, accelerated, or oscillatory. A solid dense plasma slab, accelerated in the radiation pressure dominant (RPD) regime, can efficiently reflect a counter-propagating relativistically strong source pulse consisting of up-shifted frequency and high harmonics. In this RPD regime we present our numerical results for the frequency and brightness of the reflected radiation from a uniformly moving plasma mirror. Our numerical results show that for the appropriate laser and plasma parameters in the case $2\gamma < \left(n_e \lambda_s^3\right)^{1/6}$ there are approximately 8.03×10^{42} photons / (mm^2 - mrad^2 - sec.-0.1% bandwidth) in the energy range ~ 10keV. In the case when $2\gamma > \left(n_e \lambda_s^3\right)^{1/6}$ for the same parameters and $a_d = 300$, $\lambda_d = 0.8\mu m$,the brightness is found to be 3.27×10^{34} photons / (mm^2 - mrad^2 - sec. - 0.1% bandwidth) in the energy range ~100 keV.

1. Introduction

With the availability of a new range of laser intensities by the development of CPA (chirped pulse amplification) technique has revived interest in high harmonic generation (HHG) from plasmas [1]. The ultra-short and relativistically strong laser pulses provided by CPA turn a solid target almost immediately into overdense plasma [2]. Because of the huge radiation pressure associated with laser intensities ~ 10^{22} - 10^{23} W/cm^2, combined with ponderomotive force and Coulomb attraction of ions, the plasma electron fluid moves with relativistic velocity and acts as relativistically moving mirror [3]. Depending upon the incident laser intensity, polarization of the incident beam and also on the density of the thin plasma layer the mirror motion may be assumed to be uniform, accelerated, or oscillatory [4]. In the case if an accelerated dense thin plasma slab meets a counter-propagating intense laser pulse, the reflected laser pulse is compressed in the longitudinal direction, intensified, and its frequency is Doppler-upshifted by a factor of $4\gamma^2$[5]. This phenomenon is of great interest for developing compact sources of coherent radiation in the ultraviolet and X-ray range of photo energies [6]. X-ray sources have a broad range of applications from single-molecule imaging to medicine, oncological hadron therapy [7] and they are required in material science and for the investigation of fundamental science problems such as in nonlinear quantum electrodynamics (QED) [4].

In the present paper we consider analytically and numerically the interaction of a uniformly moving plasma mirror with a counter-propagating intense plane electromagnetic wave (source wave) at normal/oblique incidence. In this paper we give our numerical results of the brightness of a source. The numerical results of the spectral brightness of the reflected radiation from a uniformly moving plasma mirror as a function of the photon energy are obtained under the assumption $2\gamma < \left(n_e \lambda_s^3\right)^{1/6}$ and also in the opposite case when $2\gamma > \left(n_e \lambda_s^3\right)^{1/6}$.

2. Reflection of electromagnetic wave from an uniformly moving thin plasma foil

An electromagnetic wave incident on an infinitely thin plasma foil, representing a mirror moving along x-axis represented by the coordinate $X_M(t)$, satisfies the wave equation [4,8]

$$\frac{\partial^2 A}{\partial t^2} - c^2 \frac{\partial^2 A}{\partial x^2} + \frac{4\pi e^2 n_e l \,\delta\left(x - X_M(t)\right)}{m_e \gamma} A = 0, \tag{1}$$

where A is the vector potential, n_e is initial electron density, l is thickness of plasma slab, m_e is the mass of electron and $\gamma = \left[1 - \dot{X}_M^2 c^{-2}\right]^{-1/2}$. Introducing the dimensionless variables $\bar{x} = k x$, $\bar{t} = c k t$ and new variables $\xi = (\bar{x} - \bar{t})/2$, $\eta = (\bar{x} + \bar{t})/2$, equation (1) becomes

$$\frac{\partial^2 A}{\partial \eta \partial \xi} = 2 n_e l \frac{e^2}{m_e c^2}\left(\frac{2\pi}{k}\right)\frac{\delta\left(\psi(\xi,\eta)\right)}{\gamma(\xi,\eta)} A, \tag{2}$$

where $\psi(\xi,\eta) = \left(\eta + \xi - X_M(\eta - \xi)\right)$. Solution of equation (2) with proper boundary conditions at the plasma interface can be written in the form of two differential equations [9]

$$a_1'(\xi) = \chi\left(a_1(\xi) + a_0 e^{2i\eta_0(\xi)}\right) F\left(\xi, \eta_0(\xi)\right), \tag{3}$$

$$2 i a_0 e^{2i\eta_0(\xi)} - a_2'(\eta) = \frac{\chi}{F\left(\xi_0(\eta), \eta\right)} a_2(\eta). \tag{4}$$

Here a_0, $a_1(\xi)$ and $a_2(\eta)$ are the incident, reflected and transmitted wave amplitudes respectively, plasma density parameter $\chi = 2 n_e l r_e \lambda, \lambda = (2\pi/k), r_e$ is the classical radius of electron and $F = \left[1 + X_M'(\eta - \xi) \big/ 1 - X_M'(\eta - \xi)\right]^{1/2}$. Solution of the equation (3) for the uniformly moving mirror gives the reflected wave amplitude [8, 9]

$$a_1 = -a_0 \frac{\chi}{\chi + 2 i F_0} \exp\left(-2 i F_0^2 \xi\right). \tag{5}$$

The expression for the reflection coefficient is given by $R = \left|a_1/a_0\right|^2 = \chi^2/\chi^2 + 4F_0^2$. (6)

Similarly transmission coefficient can be calculated by the relation $R + T = 1$, where $n_e = 480\, n_{cr} = 5.4 \times 10^{23}\text{cm}^{-3} \times (1\mu m/\lambda)^2$, l=0.01 λ, r_e=2.8$\times 10^{-13}$ cm, λ=0.8 μm , $F_0 = \left(1 + \beta_M/1 - \beta_M\right)^{1/2}$. The frequency of a reflected wave depends on the incidence angle and the mirror velocity as

$$\omega_r = \omega_0 \frac{1 + 2\beta_M \cos\theta_0 + \beta_M^2}{1 - \beta_M^2}. \tag{7}$$

Depending on whether the wave and mirror are co-propagating $(\beta_M < 0)$ or counter-propagating $(\beta_M > 0)$ in the lab-frame, we have either the frequency downshift or frequency upshift. In the simplest configuration of normal incidence of the wave on the mirror (θ_0= 0), then $\omega_r = \omega_0\left(1 + \beta_M/1 - \beta_M\right) = \omega_0\left(1 + \beta_M\right)^2 \gamma^2$, where $\gamma = \left(1 - \beta_M^2\right)^{-1/2}$. In the ultra-relativistic limit $\gamma >> 1$, the reflected wave frequency is higher by a factor $\approx 4\gamma^2$.

3. Spectral brightness

A light beam from a finite source can be characterized by its beam divergence $\Delta\Omega$, source size (usually surface area A), spectral power density $P(\nu)$ (watts per hertz of bandwidth). From these parameters it is useful to determine the "spectral brightness β_ν " of the source, which is defined as the power flow per unit area, per unit band width, and steradian namely $\beta_v = P_v / A \ \Delta \ \Omega \ \Delta \ \nu$. If dS denotes the elemental surface area of the source, the power dP emitted by dS into the solid angle $d\Omega$ around a direction making the angle θ with respect to the surface, can be written as $dP = B \cos\theta \ dS \ d\Omega$, where B is the brightness.

For mirror velocities greater than some critical value $(2\gamma) > (n_e \lambda_s^3)^{1/6}$, the wavelength of the reflected light from the moving mirror in the frame of the mirror becomes shorter due to relativistic Doppler effect. In this case the scattering of light from the plasma mirror electrons is incoherent and corresponding brightness [4] is given by $B \approx a_d \ \varepsilon_s \ (\hbar\omega_r)^2 r_e \ \lambda_s^2 / 8\pi^4 \hbar^3 c^2 \lambda_d^3$, where ε_s is source pulse energy, a_d is the dimensionless vector potential of driver pulse, λ_s is the source pulse wavelength and λ_d is the wavelength of driver pulse. The coherent scattering occurs when the condition $(2\gamma) < (n_e \lambda_s^3)^{1/6}$ is satisfied [4]and corresponding brightness is given by $B \approx \varepsilon_s (\hbar\omega_r)^3 \lambda_s / 4\pi^5 \hbar^4 c^3$. In both the cases one can get bright high frequency radiation source.

4. Results and discussions

Figure 1 shows the variation of the frequency of the reflected wave (ω_r / ω_0) when the counter-propagating laser pulse from the source meets the plasma mirror at an angle θ for a given mirror velocity $\beta_M = 0.999$.

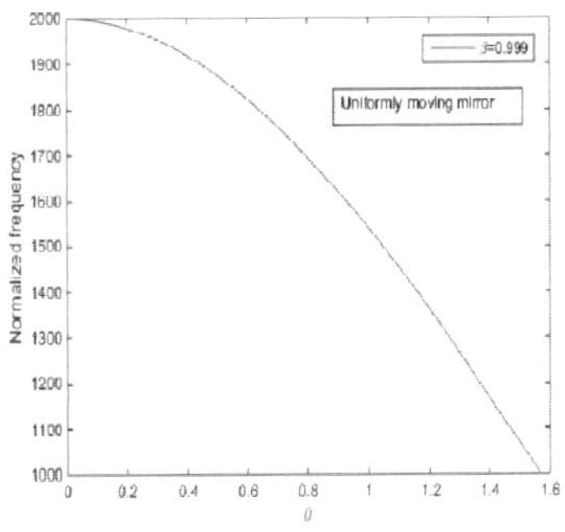

Figure 1. Figure showing the variation of normalized frequency of the reflected radiation versus incidence angle of the mirror.

Figure (2) shows the variation of brightness of the reflected radiation versus energy of the reflected photon in the case when $2\gamma < (n_e \ \lambda_s^3)^{1/6}$ for different values of source pulse energy ε_s. From this figure we see that in the coherent interaction for the parameter $\lambda_s = \lambda = 0.8\mu m$, maximum brightness corresponding to 10 keV photon energy is approximately 4.82×10^{42} photons / (mm^2 - mrad^2 - sec. - 0.1% bandwidth) for ε_s = 6 J, 6.43×10^{42} photons / (mm^2 - mrad^2 - sec. - 0.1% bandwidth) for ε_s = 8 J, 8.03×10^{42} photons / (mm^2 - mrad^2 - sec. - 0.1% bandwidth) for ε_s = 10 J,and 9.64×10^{42} photons / (mm^2 - mrad^2 - sec. - 0.1% bandwidth) for ε_s = 12 J. This energy of the reflected photon is corresponding to the energy of hard X-ray source. Figure (3) shows the variation of brightness of the reflected radiation versus energy of the reflected photon in the case when $2\gamma > (n_e \ \lambda_s^3)^{1/6}$ for different values of source pulse energy ε_s . In this case when the mirror is moving uniformly, the reflection of the photons are incoherent. From this figure we see that in the incoherent interaction using parameters $\lambda_s = \lambda_d = 0.8\mu m$, $a_d = 300$, and $r_e = 2.8 \times 10^{-13}$ cm maximum brightness corresponding to 100 keV photon energy is approximately 1.96×10^{34} photons / (mm^2 - mrad^2 - sec. - 0.1% bandwidth) for $\varepsilon_s = 6$ J, 2.62 $\times 10^{34}$ photons / (mm^2 - mrad^2 - sec. - 0.1% bandwidth) for $\varepsilon_s = 8$ J,3.27 $\times 10^{34}$ photons / (mm^2 - mrad^2 - sec. - 0.1% bandwidth) for $\varepsilon_s = 10$ J, and 3.93 $\times 10^{34}$ photons / (mm^2 - mrad^2 - sec. - 0.1% bandwidth) for $\varepsilon_s = 12$ J. This energy of the reflected photon is corresponding to the energy of gamma- ray.

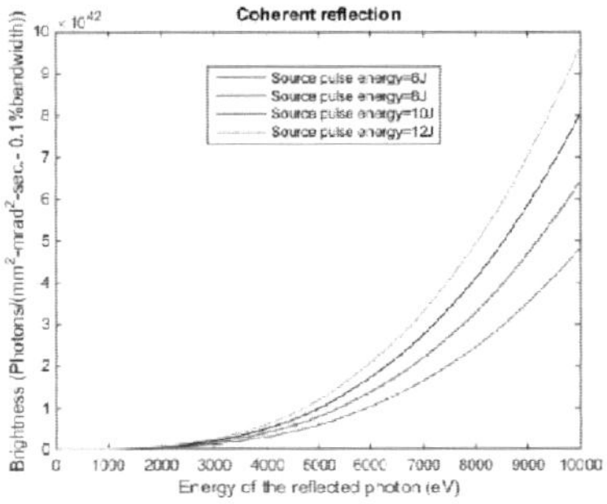

Figure 2. Variation of brightness of the reflected radiation versus energy of the reflected photon in the case $2\gamma \langle \left(n_e \lambda_s^3\right)^{1/6}$ when mirror is moving uniformly for different values of source pulse energy ε_s=6J, 8J, 10J, 12J

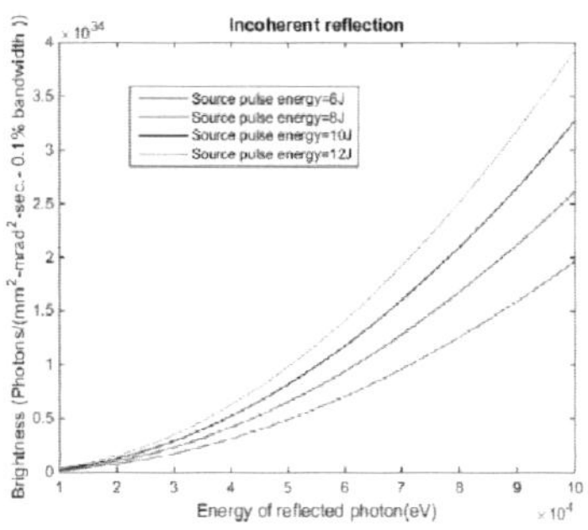

Figure 3. Variation of brightness of the reflected radiation versus energy of the reflected photon in the case $2\gamma \rangle \left(n_e \lambda_s^3\right)^{1/6}$ when mirror is moving uniformly for different values of source pulse energy ε_s=6J, 8J, 10J, 12J

5. Conclusions

A solid dense plasma slab, accelerated in the RPD regime, can efficiently reflect a counter-propagating relativistically strong source pulse consisting of up-shifted frequency and high harmonics. This reflected source pulse is chirped due to mirror acceleration. Our numerical results for brightness show that when the plasma mirror velocities are greater than some threshold then the distance between electrons in the slab becomes longer than the incident wavelength so the reflection from plasma slab is not coherent. In effect, one can develop a compact source of high-brightness X-rays and short gamma rays.

Acknowledgment

Financial support from the Department of Science & Technology, New Delhi, (Government of India) is thankfully acknowledged.

References

[1] Mourou G A, Barty C P J and Perry M D 1998*Phys. Today***51** 22
[2] BrabecT andKrausz F 2000*Rev. Mod. Phys.***72** 545
[3] Bulanov S V,Esirkepov T Zh and Tajima T 2003*Phys. Rev. Lett.***91**085001
[4] Bulanov S V, Esirkepov T Zh, Kando M,Pirozhkov A S and Rosanov N N 2013*Phys.Usp.* **56** 429
[5] Cherepenin V A and Kulagin V V 2004*Phys. Lett. A***321** 103
[6] Kulagin V V, Cherepenin V A, Hur M Sand Suk H 2007*Phys. Plasmas***14** 113101
[7] Bulanov S VandKhoroshkov V S 2002*Plasma Phys. Rep.***28** 453
[8] Jain V, Maheshwari K P,Jaiman N KandMalav H 2014*Las. Particle Beams* **32** 253
[9] Esirkepov T Zh, Bulanov S V,Zhidkov A G, Pirozhkov A Z andKando M 2009*Phys. Rev. Lett.***103** 025002

IOP Conf. Series: Journal of Physics: Conf. Series **836** (2017) 012016 doi:10.1088/1742-6596/836/1/012016

Up and downstream density scale asymmetries in Aditya tokamak scrape-off layer 3D simulations

Bibhu Prasad Sahoo[1,a], Devendra Sharma[1], Ratneshwar Jha[1] and Yuhe Feng[2]

[1]Institute for Plasma Research, HBNI, Bhat, Gandhinagar-382428, India
[2]Max-Planck-Institut fur Plasmaphysik, Wendelsteinstr. 1, Greifswald, D-17491, Germany

E-mail: [a)]bibhuprasad@ipr.res.in

Abstract. In magnetically confined tokamak plasmas the width of plasma Scrape-off Layer (SOL) allows direct estimation of anomalous diffusivity in limited and diverted versions of plasma configurations. The diffusivity coupled to edge and SOL plasma fluctuations is to be a decisive factor in stability of fusion grade plasmas and can be characterized for its dependence on 3D effects in moderate size tokamaks like Aditya. For Aditya tokamak configuration, 3D effects in SOL plasma transport are studied using 3D Monte-Carlo transport code combination EMC3-EIRENE to complement the limited probe measurements during experimental operation of the device. As an important 3D effects, variation from 1D estimates of up and downstream density scale lengths are recovered to generate appropriate corrections for estimates of plasma diffusivity and other transport propertied measured in Aditya tokamak plasma.

1. Introduction

The tokamaks are magnetic confinement devices to achieve improved plasma confinement and control in an ideally two dimensionally symmetric toroidal magnetic field configuration. Beacause of a toroidally localized, poloidal ring-like limiter, Strong 3-dimensional character of Scrape-off Layer plasma transport properties is expected in Aditya tokamak where the magnetized plasma experiments relevant to relatively larger fusion grade plasma device can be performed at moderate scales. In order to provide more sophisticated 3D global SOL plasma transport approximations, complementing the probe measurements, 3D Monte-Carlo coupled plasma-neutral transport code EMC3-EIRENE is implemented to the Aditya SOL plasma configuration as presented in this work [5-8]. As an important pointer to strength and character of edge and plasma turbulence that enable transition to high confinement regime, the SOL widths are strong indicator of plasma diffusivity and are readily available diagnostics to estimate its values. The SOL widths also decisive for intensity of heat fluxes incident on the targets and must be watched in to effectively optimizing the heat flux intensities in high power fusion grade plasma operations in large devices like ITER [3,4]. In rest of this paper we present, description of the model and application of the EMC3-EIRENE code to the Aditya SOL region followed by our results in terms of density gradient scale lengths of up and downstream SOL plasma as a function of anomalous plasma diffusivity. The comparison of the density scale lengths is done with simplistic expression of these lengths in a 2D symmetric SOL plasma and their discussion is followed by presenting important conclusions.

IOP Conf. Series: Journal of Physics: Conf. Series **836** (2017) 012016 doi:10.1088/1742-6596/836/1/012016

2. EMC3-EIRENE simulations of Aditya SOL plasma transport

Aditya is a moderate size tokamak having major and minor radii of 75 cm and 25 cm, respectively, with a maximum toroidal magnetic field of 0.72 T at the magnetic axis. In all discharges an outboard shift of plasma center from the magnetic axis and in outboard limited plasma discharges a slight shift of the plasma strike-point from the exact θ=0 (mid plane) location is generally noted depending on the orientation of the toroidal magnetic field B_t. A complex distribution of connection lengths length (L_c) is already identified in the SOL region of Aditya in previous studies using EMC3-EIRNE [7,8]. The presently adopted version of the EMC3-EIRENE code works by executing Monte-Carlo plasma transport realizations along a 3D spatial mesh constructed following the equilibrium magnetic configuration of Aditya [11,12]. Originally applied to stellarators, helical devices [13,14] and recently to many tokamaks the underlying model model of the EMC3-EIRENE includes the balance equations for density, momentum, and energies of electrons and ions. Coupling of plasma transport code with neutral transport code EIRENE [6], provides their sources. An iterative procedure ensures a self-consistent solutions for both plasma and neutral particle parameters.

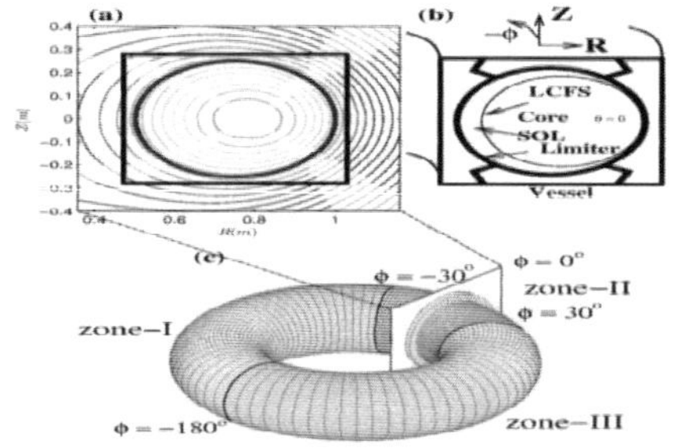

Figure 1. (a) The equilibrium magnetic flux surfaces for Aditya with a view of the poloidal ring limiter (thick circle centered at 75 cm) and the device vessel (thick square). (b) A schematic view describing the relative arrangement of the plasma core, LCFS, SOL and the experimentally explored ring limiter (located at $\Phi = 0^o$) in the device. The EMC3-EIRENE simulation grid. (c) for the ring limiter case.

As illustrated in figure 1(c), divided into 3 toroidal zones for optimum computation with zone-2 having higher resolution because of presence of ring limiter. Considering the relevant discharge conditions we use plasma current, I_p = 25 kA, plasma duration =25 ms, core electron temperature =100 eV, center chord averaged plasma density 5 x 10^{18} m^{-3}, SOL plasma density 5 x10^{17} m^{-3} where hydrogen is the working gas [6]. For the loop voltage V_L = 2 to 3 V with the plasma current I_p =25 kA, an input power about 60 kW is used into the SOL from the core boundary surface. The converged output provides plasma density, temperature and flow velocity (Mach number). A good statistical accuracy ensures that their gradients can also be estimated.

3. Up and downstream plasma parameter characteristics

For the case of $D_\perp$ = 1 m^2s^{-1}, in Fig. 2(a) and (b) the simulated plasma density profiles (2D surfaces) describing the effects of presence of ring limiter at the downstream toroidal location (Φ= 0°) is compared with that at the upstream location (Φ = -180°) where the sharp gradients of the profiles are largely smeared off. The radial profiles of density extracted from these results at various poloidal angles representing inboard and outboard radial extension across the SOL are plotted in figure 2(c) and (d), corresponding to upstream and downstream locations, respectively. A characteristic drop of a factor of about two in the density value from upstream location to downstream locations in the SOL region (r - r0 > 0) is recovered in each case validating the essential accuracy of the simulation results. With availability of 3D spatial distributions of plasma parameters on the mesh their gradients are obtainable for a range of D⊥ values such that the density gradient scale lengths and their dependence on the plasma diffusivity can be characterized. In order to extract the effects introduced purely by 3D attributes of the SOL due to the toroidally localized ring limiter and poloidally localized plasma strike-zone, these lengths can now be compared with simpler one or two dimensional versions of these characteristics. As a preliminary effort the comparison of radial density gradient length, recovered here as function of effective plasma diffusivity is made with the most essential model that approximates diffusivity and SOL width relationship given as follows [1,2],

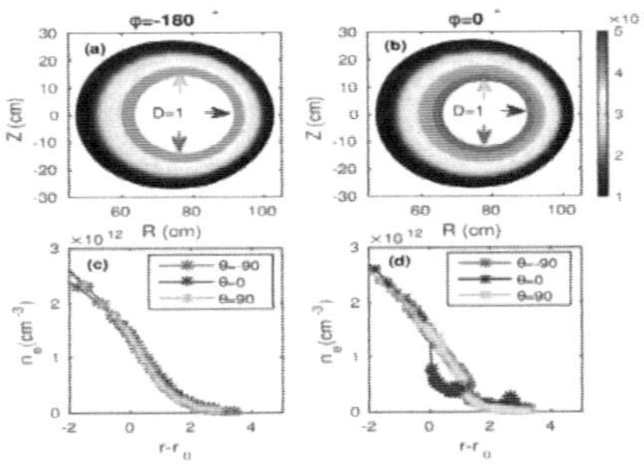

Figure 2. Simulated density profiles (a) upstream (Φ = -180°) and (b) downstream (Φ = 0°) and their radial variation at (c) upstream and (d) downstream at poloidal angles (θ= -90°) ,(θ= 0°) and (θ= 90°).

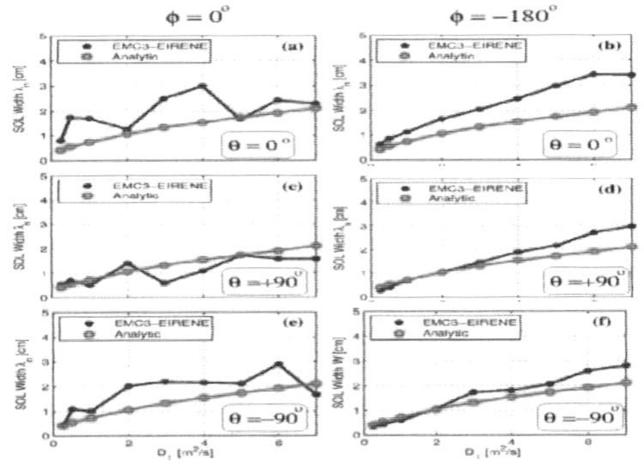

Figure 3. Dependence of SOL width λ_n on D⊥ from analytic model (1) (circles) compared to computed value of (stars).

Figure 3 obtained from the simulated density profiles and their gradients at (Left) downstream _ = 0 and (Right) upstream locations for poloidal angles (a),(b) θ = 0°, (c),(d) θ = 90° and (e),(f) θ= -90°. The marks on the density profiles at 25.5 cm indicate the location of the density data used for computing the SOL width λ_n . The data is derived from the location r = 25.5 cm, i.e., 0.5 cm radially outwards from the limiter inner radius r = 25 cm and the used diffusivity values are D⊥= 0.25, 0.5, 1, 2, 3, 4, 5, 6 and 7 $m^2 s^{-1}$.

$$\lambda_n = LD_\perp/Cs \tag{1}$$

where L is the magnetic connection length and c_s is the ion acoustic velocity. Below we calculate SOL widths for a range of diffusivity values from Model equation (1) at the upstream (Φ = -180°) and downstream (Φ = 0°) locations to compare them with the numerically obtained diffusive SOL width extracted from the EMC3-EIRENE data.

3.1. SOL width and gradient scales at upstream and downstream locations

The radial density profiles from up and downstream location at various poloidal angle values are used to characterize the relationship (1), between diffusivity and SOL width. The procedure adopted here involves determination of the plasma density (n) and its radial gradient in the SOL region from the simulation data and obtaining λ_n = n/(dn/dr) for a range of D⊥ values. For a selected radial location, r = 25.5 cm in the SOL (where r = 25 cm corresponds to the LCFS location), the λ_n values are obtained for the complete range of diffusivity value used for the simulations. For the upstream location these values of λ_n are presented in figure (3) (b), (d) and (f) corresponding to poloidal locations θ = 0°, θ= 90° and θ= -90°, respectively. The local connection length value on the respective flux tubes and average sound velocity along the flux tube are used on the other hand with the value of diffusivity in each case to obtain the alternate value of λ_n by substituting them in expression (1).

For very small diffusivities, e.g., D⊥ = 0.25 m^2s^{-1}, the λ_n values estimated by (1) shows good agreement with the simulated data as presented in figure 3(b), (d) and (f). However, the radial density decay length λ_n = n/(dn/dr) obtained from a local gradient shows a finite fluctuation with variation in D⊥ . The similar analysis is made using the density data at an upstream toroidal location Φ = -180° presented in figure 3 (b), (d) and (f). The upstream profiles are free from any non-monotonic variation with respect to D⊥.This simplification at upstream is attributed to smearing of the structures present in the plasma parameters due to stronger diffusive effect over larger distance at upstream. Note that the limiter is about a length $\pi(R + a)$ away from this toroidal location towards the downstream location $\Phi = 0^o$ where it intersects the LCFS at r = 25 cm at the mid plane. At the downstream location large density jump across the LCFS is present and gradients, as visible from figure 2(d) downstream profiles, change sharply along the radial direction. This variation is however relaxed at poloidal angles away from the outboard mid plane. On the other hand, very smooth change in gradient is present in the upstream profiles throughout the poloidal extent as seen from figure 2(c).

4. Summary and conclusions

The 3D simulations of plasma transport in the SOL of non axisymmetric toroidally discontinuous poloidal ring-limiter configuration of Aditya tokamak plasma are done. The simulations are optimized to capture 3D effects in the transport properties of the SOL plasma and generate a substitute for largely 2D simplistic models in use for interpretation of the limited probe measurements. The validation of the simulation results in terms of conventional up and downstream characteristics of the plasma owing along the individual flux tubes is done. With availability of complete 3D spatial plasma parameter distributions, the computation of local gradients allows numerical estimation of the density gradient scale length and corresponding correction in them arising from the 3D effects. The density gradient scale lengths are analysed and compared with respect to simplistic estimates of the SOL widths applicable to one or two dimensional plasma flow along the open field lines. The contrast between the SOL widths recovered at upstream and downstream locations could be characterized showing that the SOL widths at the downstream locations have a rather complex dependence on the diffusivity. In addition to deviation in the SOL widths derived from the radial plasma from the simplistic estimates because of presence of 3D effects, in presence of stronger fluctuations, or correspondingly at larger diffusivity values used in the simulation, the complexity of 3D effects observed enhancing. The origin of this enhancement is expect to be in the increased possibility of mutual interaction between spatially wider 3D helical plasma structures at larger diffusivities.

References

[1] Uehara K, Yamamoto Y G T, Suzuki N, Maeno M, Hirayama T, Shimada M, Konoshima S and Fujisawa N 1979 *Phys. Plasma* **21** 89
[2] Stangeby P C, Tagle A, Erents S K and Lowry C 1988 *Plasma Phys. Control. Fusion* **30** 1787
[3] Kobayashi M, Feng Y, Loarte A, Federici G, Strohmayer G, Shimada M, Sardei F, Reiter D and Sugihara M 2007 *Nucl. Fusion* **47** 61
[4] Stangeby P C 2010 *Nucl. Fusion* **50** 035013
[5] Feng Y, Sardei F and Kisslinger J 1999 *J. Nucl. Mater* **266-269** 812
[6] Reiter D 1984 *Technical Report Juelich-1947*
[7] Sahoo B P, Sharma D, Jha R and Feng Y 2015 *Nucl. Fusion* **55** 063042
[8] Sharma D, Jha R, Feng Y and Sardei F 2013 *J. Nucl. Mater.* **S554-S558** 438
[9] Sangwan D, Jha R, Brotankova J and Gopalkrishna M V 2013 *Phys. Plasma* **20** 062503
[10] Srinivasan R and Deshpande D 2007 *Technical Report*, IPR/**RR-393**/2007 (AUGUST, 2007)
[11] Johnson J L, Dallied H E and Greene J M 1979 *J. Comp. Phys.* **32** 212
[12] Feng Y, Sardei F and Kisslinger J 2006 *Nucl. Fusion* **46** 807
[13] Sharma D, Feng Y, Sardei F and Reiter D 2005 *Nucl. Fusion* **45** 825
[14] Kobayashi M, Feng Y, Sardei F and Reiter D 2007 *J. Nucl. Mat.* **363-365** 294

IOP Conf. Series: Journal of Physics: Conf. Series **836** (2017) 012017 doi:10.1088/1742-6596/836/1/012017

Analytical modeling to study the effect of hydrogen plasma on the growth of multi-walled carbon nanotubes

U Sharma[a], N Gupta, R Gupta[1]and S C Sharma

Department of Applied Physics, Delhi Technological University, Delhi-110042, India

Email: [a)]umangsharma2810@gmail.com

Abstract. An analytical model has been developed to study the effect of hydrogen plasma on the growth of Multi-walled Carbon Nanotubes (MWCNTs). The model incorporates the charging rate of the MWCNT, kinetics of plasma species, namely, electrons, positively charged ions, and neutral atoms, growth rate of MWCNT in the presence of CH_4/H_2 plasma. Numerical calculations have been carried out to study the influence of hydrogen number density on the dimensions of MWCNT for typical glow discharge plasma parameters. The impact of hydrogen on the hydrocarbon density has also been studied in the present investigation. The present work can serve as a major tool in better understanding of highly proficient field emitters based on MWCNTs.

1. Introduction

The carbon nanotube (CNT) is a novel carbon structure found since 1991 [1]. CNTs due to their extraordinary mechanical properties, have been getting much attention for a wide variety of applications. Maruyama et al. [2] prospered in producing high-purity single-walled CNTs (SWCNTs) with a high number density. Hata et al. [3] discovered a way to grow high-purity SWCNTs with a high growth rate by adding a small amount of H_2O vapor to the carbon source. They advocated that OH radicals worked as a weak oxidizer to eradicate impurities without damaging the CNT. Zhang et al. [4] argued the effects of H radicals in the gas phase on SWCNT growth. They suggested that the balance of carbon and hydrogen radicals were vital for CNT growth. Plasma enhanced chemical vapour deposition (PECVD) has appeared as a key growth technique to yield vertically-aligned nanotubes [5]. Researchers using plasma processing have examined the correlation between a plasma gas phase and deposition mechanisms. Ostrikov et al. [6] examined $CH_4/H_2/Ar$ plasma for synthesis of carbon nanostructures. They noted that H atoms affected catalyst activity and that cations were provided as precursors for the carbon nanostructure. Denysenko et al. [7] exhibited that C_xH_y ions increased with the supply of the carbon source and that H atoms also increased with the C_xH_y molecules formed in the $CH_4/H_2/Ar$ plasma. Hydrogen plays a dual role in the PECVD growth of CNT. It acts as an activator for the dissociation of hydrocarbon species into more active species and also as an etching reagent of CNT. In the present work, we analyze the effect of hydrogen on the growth of CNTs in plasma processing, analytically. In section 2, we develop a theoretical model incorporating charging rate of the CNT, kinetics of plasma species i.e., neutral atoms, growth rate of CNT in the presence of CH_4/H_2 plasma. Results and discussions are given in section 3. Finally, the conclusion is given in section 4.

2. Model

We consider plasma comprising electrons, positively charged ions of types A and B, neutral atoms of type A and B, and CNT grown in the existence of plasma. The positively charged ions are anticipated to be singly ionized. For simplicity of the problem, average radius of MWCNT has been considered.

IOP Conf. Series: Journal of Physics: Conf. Series **836** (2017) 012017 doi:10.1088/1742-6596/836/1/012017

Charge neutrality equation

$$\frac{dZ}{d\tau} = n_{iAct} + n_{iBct} - \gamma_e n_{ect}$$

where Z=the charge on the CNT, n_{ijct}= the ion collection current for A and B type ions and n_{ct}= the number density of the CNT. Growth rate equation of the mass of CNT

$$\frac{dm_{ct}}{d\tau} = (m_A \gamma_A n_{Act} + m_{iA} \gamma_{iA} n_{iAct})$$

where $m_{ct} = \frac{4}{3}\pi a^3 \rho_{ct}$is the mass of CNT and the terms are the gain in the mass density due to the collection of atomic and ionic species A. Energy balance equation for neutral atoms

$$\begin{aligned}
\frac{d}{d\tau}\left[\frac{3}{2}(n_A + n_B)k_B T_n\right] &= \left[\left(\frac{3}{2}k_B\right)(\alpha_A n_e n_{iA} + \alpha_B n_e n_{iB})(T_e + T_i)\right) + \left(\alpha_A n_e n_{iA} I_{pA} + \alpha_B n_e n_{iB} I_{pB}\right)\Big] \\
&+ \left(\frac{3}{2}k_B\right)\{n_e(v_{eA}\delta_{eA} + v_{eB}\delta_{eB})(T_e - T_n)[(v_{iAA}\delta_{iAA} + v_{iAB}\delta_{iAB})n_{iA} \\
&+ (v_{iBA}\delta_{iBA} + v_{iBB}\delta_{iBB})n_{iB}](T_i - T_n)\} + [\left(\frac{3}{2}k_B\right)n_{ct}[(1 - \gamma_{iA})n_{iAc} + n_{iBc}]T_{ct} \\
&- \left(\frac{3}{2}k_B\right)n_{ct}[n_{Ac}[\gamma_A T_n + \delta_{Act}(1 - \gamma_A)(T_n - T_{ct})] + n_{Bc}\delta_{Bct}(T_n - T_{ct})]] \\
&- \left(\frac{3}{2}k_B\right)(\beta_A n_A + \beta_B n_B)T_n - E_{diss}
\end{aligned}$$

where the first term is the power gained per unit volume by the neutral species due to the recombination of electrons and positively charged ions, the second term is the rate of power gained per unit volume by neutral atoms in elastic collisions with electrons and positively charged ions, and the third term is the energy gained per unit volume per second due to the formation of neutrals at the surface of the CNT due to accretion and elastic collisions with CNT. The fourth term refers to the thermal energy lost per unit volume per unit time by neutral atoms due to ionization. The last term is the energy dissipation rate per unit volume by neutral atoms into the surrounding atmosphere.

3. Results and discussions

We have solved the above mentioned equations along with others mentioned in the work of Sharma and Tewari [8] simultaneously for the charging of the CNT and for the kinetics and energy balance of electrons, ions, neutrals with appropriate boundary conditions, viz., at τ=0, n_{ct}=10^6 cm^{-3}, n_{iA0} = 0.58n_{e0}, n_{iB0}= 0.42 n_{e0}, n_{A0} = n_{B0}= 5.1 x 10^{11} cm^{-3}, n_{e0} = 1.1 x 10^{12} cm^{-3}, T_{e0}= 0.6 eV, T_{i0}= 2400 K, and T_{n0}= T_{ct}= 2000 K.

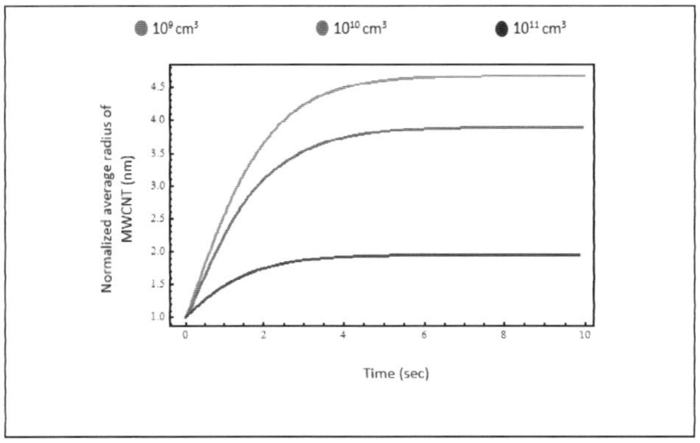

Figure 1. Time variation of normalized average radius of CNTs for different hydrogen ion density

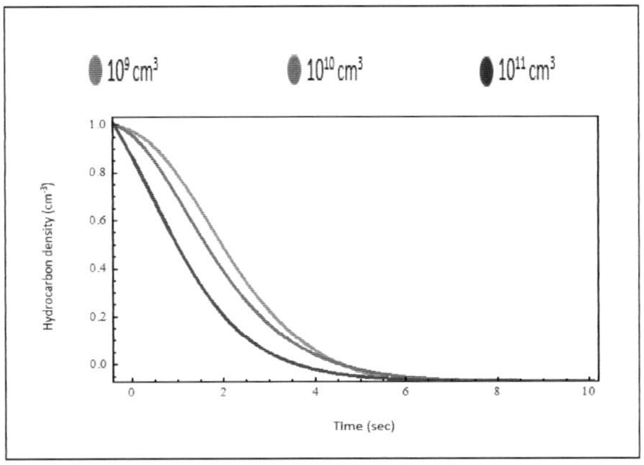

Figure 2. Time variation of hydrocarbon density of MWCNTs for different hydrogen ion density

Figure 1illustrates variation of the normalized radius of a CNT with time for different hydrogen ion number densities. It can be seen that the normalized radius of the CNT first increases with time and then attains a saturation value. The figure also shows the decrease of the normalized radius with the hydrogen ion density. This happens because for larger values of the hydrogen density, the hydrocarbon density decreases. Figure 2 illustrates variation of the hydrocarbon density with time for different hydrogen ion number densities. It can be seen that with the increase in hydrogen ion densities with time, the hydrocarbon density decreases which further affects the radius of MWCNT.

4. Conclusions

An analytical model for the growth kinetics of CNT in plasma has been developed. The approach is based on the charge neutrality and number density of electrons, positively charged ions, neutral atoms, and CNTs. It is analysed from our study that on increasing the hydrogen ion density, the hydrocarbon density and the average radius of MWCNT decreases. The above mentioned results can prove to be beneficial in the fabrication of MWCNTs.

References

[1] Iijima S 1991 *Nature* **354** 56
[2] Maruyama S, Kojima R, Miyauchi Y, Chiashi S and Kohno M 2002 *Chem. Phys. Lett.* **360** 229
[3] Hata K, Futaba DN, Mizuno K, Namai T, Yumura M and Iijima S2004 *Science* **306** 1362
[4] Zhang G, Mann D, Zhang L, Javey A, Li Y, Yenilmez E, Wang Q McVittie J P, Nishi Y, Gibbons J and Dai H 2005 *Proc. Nat. Acad. Sci. USA* **102** 16141
[5] Ostrikov K, Tzakadze Z,Rutkevych P P, Long J D, Xu S and Denysenko I 2005 *Contrib. Plasm. Phys.* **45** 7
[6] Ostrikov K 2005 *Rev. Mod. Phys.* **77** 489
[7] Denysenko I B, Xu S, Long J D, Rutkevych P P, Azarenkov N A and Ostrikov K 2004 *J. App. Phys.* **95** 2713
[8] Sharma S C and Tewari A 2011 *Phys. Plasmas* **18** 063503

Phase mismatched optical parametric generation in semiconductor magnetoplasma

Swati Dubey, S Ghosh and Kamal Jain[a]

School of Studies in Physics, Vikram University, Ujjain 456010 (M.P.) INDIA

E-mail: [a)] jainrupam1987@gmail.com

Abstract. Optical parametric generation involves the interaction of pump, signal, and idler waves satisfying law of conservation of energy. Phase mismatch parameter plays important role for the spatial distribution of the field along the medium. In this paper instead of exactly matching wave vector, a small mismatch is admitted with a degree of phase velocity mismatch between these waves. Hence the medium must possess certain finite coherence length. This wave mixing process is well explained by coupled mode theory and one dimensional hydrodynamic model. Based on this scheme, expressions for threshold pump field and transmitted intensity have been derived. It is observed that the threshold pump intensity and transmitted intensity can be manipulated by varying doping concentration and magnetic field under phase mismatched condition. A compound semiconductor crystal of n-InSb is assumed to be shined at 77 K by a 10.6μm CO_2 laser with photon energy well below band gap energy of the crystal, so that only free charge carrier influence the optical properties of the medium for the I.R. parametric generation in a semiconductor plasma medium. Favorable parameters were explored to incite the said process keeping in mind the cost effectiveness and conversion efficiency of the process.

1. Introduction

New generation of nonlinear optical crystals revived intensive research interest in the field of nonlinear optics and transformed parametric devices from evidence of principle demonstrations to variable sources of coherent light for practical applications. Now a day's operating domain of parametric devices has been extended from selected spectral and temporal regimes to new U.V. well into the mid I.R. and encompassed all time scale from continuous wave (CW) to ultrafast femtosecond regime. At the same time availability of high power laser sources with enhanced spectral and spatial coherence has enabled the development of parametric devices at record power levels and established them as an important class of practical coherent light sources.

Optical parametric frequency conversion based on second and third order nonlinearities are achieved by mixing coherent waves, e.g. light beams from lasers, in a nonlinear material. The strongest parametric processes, and consequently the first to be discovered, are those generated in nonlinear materials possessing a non-zero second order nonlinearity. There are a lot of challenges in harnessing large $\chi^{(2)}$ coefficient of compound semiconductors near their band gap. Most of the semiconductor materials are isotropic so that no birefringence phase-matching scenario are accessible [1]. Nonlinear optical processes require high intensity light beams to make their effects observable and useful. Although coherent light can distort the oscillating polarization, Fourier components of these distortions can create or enhance new fields that then radiate from the oscillating nonlinear polarization [2].

In nonlinear optical interactions, because of the dispersive nature of crystal, the fundamental frequency and the generated frequency travel at different speeds in the material, so there is a phase mismatch between the interacting waves. Achieving phase matching in semiconductors with significant dispersion and large $\chi^{(2)}$ is quite difficult. However main concern is to get larger conversion efficiency in case of optical parametric generation process [3-4].

There have been many efforts to achieve phase matching through poling, birefringence of the medium. But it seems that uptill now no studies were undertaken to study the parametric generation process which is phase mismatched. Hence present paper describes a way to identify and quantify the various implications of the optical phase mismatching on IR generation in the presence of external magnetic field. Parametric process pumped by 10.6μm CO_2 laser where bulk $\chi^{(2)}$ nonlinearity is considered as phase mismatched for optical parametric generation process as a consequence of three wave mixing. The process gets initiated by a single intense pump field at frequency ω_0 at the input to a nonlinear crystal. This field provided by CO_2 laser in turn mixes through the nonlinear susceptibility with a signal field at $\omega_1 = \omega_0\text{-}\omega_a$. The idler field so generated in turn mixes back with the pump to produce additional signal and the generated signal mixed with the pump to produce more idler. The process continues in this way until power is gradually transferred from the strong pump at the initially weak signal and idler fields through the nonlinear interaction mediated by $\chi^{(2)}$. This wave mixing process is well analyzed by using coupled mode theory and expressions for threshold pump field, transmitted intensity have been derived.

2. Theoretical Formulations

This section deals with the theoretical formulation of induced nonlinear polarization and transmitted intensity in doped semiconductor magnetoplasma medium. The model used in the analysis is the well-known hydrodynamic model of one component plasma. A pump electric field $E_0 \exp(k_0x - \omega_0 t)$ is assumed to be incident on a semiconductor sample, parallel to the wave vector k (along x axis) and external magnetic field is applied along z axis. In the presence of a strong input pump, the signal and idler fields in the parametric interaction can experience growth as they propagate through the nonlinear crystal. Generation of signal and idler are also hoped to be strongly dependent functions of the phase mismatch parameter $\Delta k = k_0 - k_1 - k_a$ where k_0, k_a and k_1 are wave vectors of pump, acoustic and signal modes respectively. This corresponds to the more realistic case as exact phase matching is very difficult to achieve experimentally. Following basic equations are used for analysis:-

$$\frac{\partial \vartheta_0}{\partial t} + \vartheta_0\left(\frac{\partial \vartheta_0}{\partial x}\right) + \upsilon\vartheta_0 = \frac{e}{m}\left(E_0 + \vartheta_0 \times B_0\right) \tag{1}$$

$$\frac{\partial \vartheta_1}{\partial t} + \vartheta_0\left(\frac{\partial \vartheta_1}{\partial x}\right) + \vartheta_1\left(\frac{\partial \vartheta_0}{\partial x}\right) + \upsilon\vartheta_1 = \frac{e}{m}\left(E_1 + \vartheta_1 \times B_0\right) \tag{2}$$

$$\frac{\partial n_1}{\partial t} + n_0\frac{\partial \vartheta_1}{\partial x} + n_1\frac{\partial \vartheta_0}{\partial x} + \vartheta_0\frac{\partial n_1}{\partial x} = 0 \tag{3}$$

$$\rho\frac{\partial^2 u}{\partial t^2} + 2\gamma_s\rho\left(\frac{\partial u}{\partial t}\right) + \beta\frac{\partial E_s}{\partial x} = C\frac{\partial^2 u}{\partial x^2} \tag{4}$$

$$\frac{\partial E_s}{\partial x} + \frac{\beta}{\varepsilon}\frac{\partial^2 u}{\partial x^2} = \frac{n_1 e}{\varepsilon} \tag{5}$$

Meanings of all the symbols are given in [6]. Following standard approach of the Neogi and Ghosh [5] and utilizing equations (1)-(5) a simplified expression for the density perturbation is obtained as

$$\frac{\partial^2 n_1}{\partial t^2} + \upsilon \frac{\partial n_1}{\partial t} + \overline{\omega_p}^2 n_1 + n_0 e\beta \frac{\partial^2 u}{\partial x^2}\left(\frac{\upsilon^2}{\omega_c^2 + \upsilon^2}\right) = -i(k_0 + k_1)n_1 \overline{E} \tag{6}$$

$$\overline{E} = \frac{e}{m} E_0 + \omega_c \vartheta_{0y}, \qquad \vartheta_{0y} = \frac{-e\omega_c E_0}{m(\omega_c^2 - \omega_0^2)} \qquad \text{and} \qquad \overline{\omega_p^2} = \omega_p^2\left(\frac{\upsilon^2}{\omega_c^2 - \upsilon^2}\right) \tag{7}$$

Induced nonlinear polarization may be obtained as

$$P_1(\omega_1) = \frac{ieAk\varepsilon\varpi_p^2 E_1^* E_0}{m\omega_1\left(\frac{c}{\rho}k_a^2 - \omega_a^2 + 2i\gamma_s\omega_a\right)}\left(\frac{\upsilon^2}{\omega_c^2 + \upsilon^2}\right)\left[(\varpi_p^2 - \omega_a^2 + i\upsilon\omega_a) - \frac{(k_0 + k_1)^2 \overline{|E|^2}}{(\varpi_p^2 - \omega_1^2 - i\omega_1\upsilon)}\right]^{-1} = \varepsilon_0 \chi^{(2)} E_1^* E_0 \tag{8}$$

Threshold pump field required for the onset of I.R. generation can be obtained by setting imaginary component of the second order susceptibility $\chi^{(2)}$equals to zero

$$E_{0th} = \frac{m}{e(k_0 + k_1)} \frac{\omega_c^2 - \omega_0^2}{\omega_0^2} \sqrt{\frac{(\varpi_p^2 - \omega_1^2)^2 \omega_a + \upsilon^2 \omega_1^2 \omega_a}{\omega_1}} \tag{9}$$

If the sample length is 10 to 10^2 orders greater than the pump wavelength [6], one can easily use the expression for induced polarization (P_1) deduced for an infinite medium, to express the transmitted intensity (I_T) in a crystal length L given by

$$I_T = \frac{1}{2}\eta\varepsilon_0 c|E_T|^2 \text{ where } E_T = -\frac{ik_1 L}{\varepsilon} P_1(\omega_1) \tag{10}$$

Equations (8), (9) and (10) may be used to determine induced polarization, transmitted electric field amplitude and transmitted intensity respectively.

3. Results and discussions

This section deals with the numerical appreciation of the theoretical formulations for different parameters in previous section. Data of a compound semiconductor crystal of InSb pumped by 10.6μm CO_2 laser at 77K is used [7] for the calculations.

Figure 1 illustrates the variation of E_{0th} with the magnetic field B_0. It is clear from this graph that threshold field linearly decreases with increasing magnetic field at constant carrier concentration $n_0=2\times10^{23}$ m^{-3} .This observation fulfills the prime requirement of lower threshold field at higher external magnetic field which is beneficial for a cost effective I.R. generation.

Figure 2 displays the variation of transmitted intensity I_T with respect to external magnetic field B_0. It is found that transmitted intensity first decreases with increasing external magnetic field till 6.7T and beyond this magnitude, transmitted intensity increases. Resonance between modified plasma frequency and idler wave frequency (ω_a) causes minimum transmission at 6.7T. Beyond this magnetic field modified plasma frequency becomes greater than ω_a and hence the transmitted intensity also increases with increasing magnetic field.

Figure 3 illustrates the variation of transmitted intensity I_T and carrier density n_0 at various cell lengths. Maximum transmitted intensity is envisaged for higher cell length and at higher carrier densities.

Phase mismatching condition is found to increase threshold pump requirements by a factor of 10 as compared to earlier study[6] done with phase matching in highly doped regime. It can be concluded that highly doped semiconductor plasma with high magnetic field may be helpful for a cost effective I.R. generation with improved conversion efficiency.

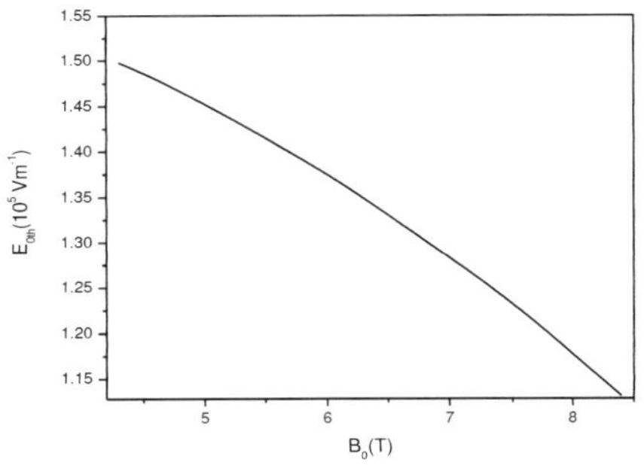

Figure 1. Variation of threshold field E_{0th} with magnetic field B_0 at $k = 5.5\times10^8\ m^{-1}$ & $n_0 = 2\times10^{23}m^{-3}$.

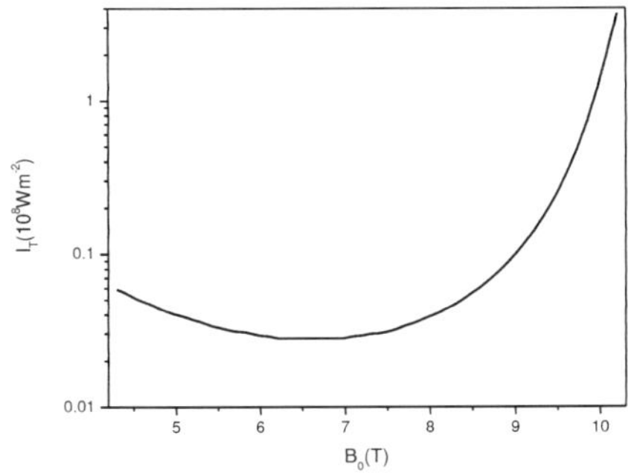

Figure 2. Variation of Transmitted Intensity I_T with external magnetic Field B_0 at $L=5\times10^{-5}m$, $k= 5.5\times10^8m^{-1}$, $E_0=3\times10^5\ Vm^{-1}$ & $n_0= 2\times10^{23}m^{-3}$.

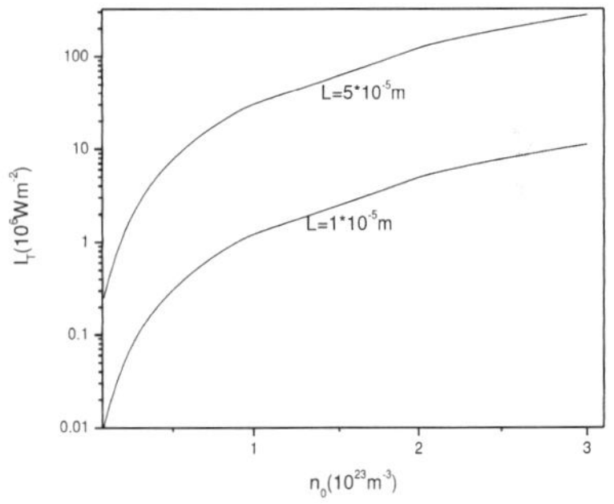

Figure 3. Variation of Transmitted Intensity I_T with carrier density n_0 at magnetic field $B_0=10T$, $k= 5.5\times10^8m^{-1}$, $E_0=3\times10^5Vm^{-1}$ & $n_0= 2\times10^{23}m^{-3}$.

Acknowledgment
One of the authors (KJ) is thankful to UGC, New Delhi, India for the award of Maulana Azad fellowship for Minority.

References
[1] Rosencher E and Vinter B 2002 *Optoelectronics* (New York: Cambridge University Press)
[2] Denman C A, Hillman P D, Moore G T, Telle J M, Preston J E, Drummond J D and Fugate R Q 2005 *Proceedings of SPIE* **5707** 46
[3] Averchi A, Faccio D, Berlasso R, Kolesik M, Moloney J V, Couairon A and Di Trapani 2008 *Phys. Rev.***77** 021802
[4] Danielius R, Piskarskas A, Stabinis A, Banfi G P, Tripani P Di and Righini R 1995 *J. Opt. Soc. Am. B* **10** 2222
[5] Neogi A and Ghosh S 1989 *Phys. Stat. Sol. (b)* **152** 691
[6] Simoda K 1982 *Introduction to Laser Physics* (Berlin: Springer-Verlag) p160
[7] Dubey S and Ghosh S 2009 *New J. Phys.* **11** 093030

IOP Conf. Series: Journal of Physics: Conf. Series **836** (2017) 012019 doi:10.1088/1742-6596/836/1/012019

Parametric interactions of acoustic waves in semiconductor quantum plasmas with strain dependent dielectric constants

N Yadav, S Ghosh and A Agrawal[a]

School of Studies in Physics, Vikram University, Ujjain (456010), India

Email: [a)]anjaliagrawal7675@gmail.com

Abstract. Using quantum hydrodynamic model (QHD) of semiconductor plasma for a one-component we present an analytical investigation on parametric interaction of a laser radiation in an unmagnetised material with a strain-dependent dielectric constant. The nonlinear current density and third order *susceptibility* are analyzed in different wave number regions in presence and absence of quantum effect. We present the qualitative behavior of threshold pump intensity with respect to wave number in presence and absence of quantum effect. The numeric estimates are made for n-$BaTiO_3$ crystals at 77k duly irradiated by pulsed 10.6μm CO_2 laser. It is found that the quantum correction through Fermi temperature and Bohm potential terms modifies the threshold characteristics.

1. Introduction

The field of quantum plasma physics is vibrant and evolving rapidly. Since last few years, many investigations in the field of quantum plasma are carried out. This has been characterized by high plasma particle densities and low temperatures, in contrast to classical plasma which has high temperatures and low particle number densities. Quantum plasmas are common in different environments, e.g. in super dense astrophysical bodies, in intense laser-solid plasma experiments, quantum dots, and in ultra-small electronic devices, quantum diodes, ultra cold plasma [1-8].

Quantum phenomena in semiconductor devices are increasingly important as the characteristic lengths of the modern devices are of the order of nanometer only. In fact, there are devices, like resonant tunneling diodes, whose behavior is essentially based on quantum effects. The quantum effects become important in plasmas, when the de-Broglie wavelength associated with particles is equal to or greater than the average inter particle distance [9]. In quantum plasmas, the electron are degenerate and obey the Fermi-Dirac distribution function, while non-degenerate strongly correlated ions are coupled with electrons via the electromagnetic fields. Here the quantum mechanics comes into the picture due to (1) Overlapping of electron wave functions owing to the Heisenberg uncertainty principle, leading to electron tunneling through the quantum Bohm potential and (2) electron exchange and electron correlations because of the electron one half spin effect [10]. The interaction between electrons and lattice vibrations is one of the fundamental interaction processes in solids. This interaction gives useful information regarding band structure of the semiconductor. The electron phonon interaction can lead to amplification of acoustic waves by the application of dc electric field that has been commercially exploited for the fabrication of delay lines, acoustic-electric amplifiers, acoustoelectric oscillators etc.

Usually two kinds of acoustic effects are studied. The first kind, namely, the sound attenuation and the change of sound velocity due to interaction with electrons, at small amplitudes are linear in the amplitude of the acoustic wave .The simplest phenomenon nonlinear in the sound amplitude is the so-called acoustoelectric effect, which is due to a drag of charge carriers by an acoustic wave[11]. Since the acoustic wave is 10^5 time slower than the electromagnetic waves, it enables one to design and

IOP Conf. Series: Journal of Physics: Conf. Series **836** (2017) 012019 doi:10.1088/1742-6596/836/1/012019

fabricate very sophisticated signal processing devices which are orders of magnitude smaller in size than its electromagnetic counterpart and perform the same functions[12].

The third order nonlinear optical susceptibility χ is, in general, complex quantity and is capable of describing the interference between various resonant and non-resonant processes. The present analysis is based on quantum hydrodynamic model (QHD) model for the electron dynamics. The QHD model for plasmas has been developed by Manfredi and Hass[13]. The QHD is an extension of classical fluid model used for plasma physics. The QHD model used in the paper includes two different quantum effects: (i) quantum diffraction, and (ii) quantum statistics. Quantum diffraction is due to the terms proportional to $\hbar^2$ in equations of motion and continuity in the QHD model. The QHD model consists of a set of equations describing the transport of charge, momentum and energy in a charged particle system interacting through a self consistent electrostatic potential. Mathematically the QHD model generalizes the fluid model for plasmas with the inclusion of a quantum correction term, i.e. Bohm potential. This extra term can appropriately describe negative differential resistivity in resonant tunneling diodes. The advantage of the QHD model is that they are able to describe directly the dynamics of physical observable and simulate the main characters of quantum transport model for charged particle systems [14]. Ghosh and Yadav [15] discussed the amplitude modulation and demodulation in strain dependent diffusive semiconductors. Quantum effect on parametric amplification characteristics in piezoelectric semiconductors was reported by Ghosh and his coworkers [16].

It appears from the available literature that in most of the previous reported works in field of parametric interaction of acoustic waves in quantum plasma SDDC effect has not taken into account. Thus, motivated by the above we have focused our attention on the modifications occurred by including the SDDC effect and we have analytically investigated the quantum effect on parametric process in material with high dielectric constant. The process is characterized by the effective third order susceptibility induced due to nonlinear current density in the high dielectric constant semiconductor plasma medium.

The paper is organized in the following way, the basic equations used under QHD model is given in section 2. In section 3 authors present numerical appreciation of the results obtained followed by physical discussion and final conclusion.

2. Theoretical formulation

In order to determine the third order susceptibility a spatially uniform pump field, $E_0 \exp\left[i\left(k_0 x - \omega_0 t\right)\right]$ is applied along the direction of wave propagation $\vec{k} = k\hat{x}$. The basic equations used in present analysis are as follows.

$$\frac{\partial V_0}{\partial t} + \nu V_0 = \frac{e}{m} E_0, \tag{1}$$

$$\frac{\partial V_1}{\partial t} + \nu V_1 + \left(V_0 \frac{\partial}{\partial x}\right) V_1 = \frac{e}{m} E_1 - \frac{1}{m n_0} \frac{\partial P_f}{\partial x} + \frac{\hbar^2}{4 m^2 n_0} \frac{\partial^2 n_1}{\partial x^2}, \tag{2}$$

$$\frac{\partial n_1}{\partial t} + V_0 \frac{\partial n_1}{\partial x} + n_0 \frac{\partial V_1}{\partial x} = 0, \tag{3}$$

$$\frac{\partial E_1}{\partial x} = \frac{e n_1}{\varepsilon} + \frac{\left(\varepsilon g E_0\right)}{\varepsilon} \frac{\partial^2 u^*}{\partial x^2}, \tag{4}$$

$$\rho \frac{\partial^2 u}{\partial t^2} = C \frac{\partial^2 u}{\partial x^2} - \left(\varepsilon g E_0\right) \frac{\partial E_1^*}{\partial x}. \tag{5}$$

where the symbols have usual meaning. Following standard approach and usingequation (1) to (5)

$$\frac{\partial^2 n_1}{\partial t^2} + \nu \frac{\partial n_1}{\partial t} + n_1 \bar{\omega}_p^2 - \frac{n_0 e k^2 \left(\varepsilon g E_0\right)}{m\varepsilon} u^* = -\bar{E} \frac{\partial n_1}{\partial x}, \tag{6}$$

The symbol $V_f'^2 = V_f\left(1 + \Gamma_e\right)$, $\Gamma_e = \hbar^2 k^2 / 8 m k_B T_f$, $\bar{E} = \left(e/m\right) E_0$, and $\bar{\omega}_p^2 = \omega_p^2 + k^2 V_f'^2$, where $\omega_p = \left(n_0 e^2 / m\varepsilon\right)^{1/2}$ is the plasma frequency. In the derivation of equation (6) we have neglected the Doppler shift under the assumption that $\omega_0 \geq \nu \geq k V_0$.

The coupled equations (7) and (8) are obtained from equation (6) under rotating wave approximation(RWA)

$$\frac{\partial^2 n_{1f}}{\partial t^2}+\nu\frac{\partial n_{1f}}{\partial t}+n_{1f}\overline{\omega}_p^2-\frac{n_0 \varepsilon k^2(\varepsilon g E_0)}{m\varepsilon}u^*=ik\overline{E}n_{1s}^*, \tag{7}$$

$$\frac{\partial^2 n_{1s}}{\partial t^2}+\nu\frac{\partial n_{1s}}{\partial t}+n_{1s}\overline{\omega}_p^2=ik\overline{E}n_{1f}^*. \tag{8}$$

Subscripts s and f stand for slow and fast components respectively. Asterisk (*) represents complex conjugate of the quantities. Using equations (7) and (8), we obtain

$$n_{1s}^*=-\frac{n_0 k^2(\varepsilon g E_0)^2 E_1}{E_0\rho\varepsilon(\omega_a^2-k^2V_a^2)}\left[1-\frac{(\delta_2^2+i\omega_a\nu)(\delta_1^2-i\omega_1\nu)}{k^2\overline{E}^2}\right]^{-1}, \tag{9}$$

where $\delta_1^2=\overline{\omega}_p^2-\omega_1^2$ and $\delta_2^2=\overline{\omega}_p^2-\omega_a^2$. The induced polarization as the time integral of current density we obtain

$$P=\frac{k^2(\varepsilon g E_0)^2 E_1\omega_p^2}{i\omega_a\rho(\omega_a^2-k^2V_a^2)(\nu-i\omega_0)}\left(1-\frac{(\delta_1^2-i\omega_1\nu)(\delta_2^2+i\omega_a\nu)}{k^2\overline{E}^2}\right)^{-1}=\varepsilon_0\chi^3E_0^2E_1. \tag{10}$$

This leads to the third order nonlinear susceptibility as

$$\chi^3=\frac{-k^2\varepsilon^2g^2\omega_p^2}{i\omega_a\omega_0\varepsilon_0\rho(\omega_a^2-k^2V_a^2)}\cdot\left[\frac{k^2\overline{E}^2\{(\delta_1^2\delta_2^2+\omega_1\omega_a\nu^2-k^2\overline{E}^2)-i(\omega_a\nu\delta_1^2-\omega_1\nu\delta_2^2)\}}{(\delta_1^2\delta_2^2+\omega_1\omega_a\nu^2-k^2\overline{E}^2)^2-(\omega_a\nu\delta_1^2-\omega_1\nu\delta_2^2)^2}\right]. \tag{11}$$

Now rationalizing equation (11) one obtains the real and imaginary parts of the complex third order susceptibility as

$$[\chi^3]_r=\frac{-k^4\varepsilon^2g^2\omega_p^2\overline{E}^2}{i\omega_a\omega_0\varepsilon_0\rho(\omega_a^2-k^2V_a^2)}\cdot\left[\frac{(\delta_1^2\delta_2^2+\omega_1\omega_a\nu^2-k^2\overline{E}^2)}{(\delta_1^2\delta_2^2+\omega_1\omega_a\nu^2-k^2\overline{E}^2)^2-(\omega_a\nu\delta_1^2-\omega_1\nu\delta_2^2)^2}\right], \tag{12}$$

$$[\chi^3]_i=\frac{k^4\varepsilon^2g^2\omega_p^2\overline{E}^2}{i\omega_a\omega_0\varepsilon_0\rho(\omega_a^2-k^2V_a^2)}\cdot\left[\frac{(\omega_a\nu\delta_1^2-\omega_1\nu\delta_2^2)}{(\delta_1^2\delta_2^2+\omega_1\omega_a\nu^2-k^2\overline{E}^2)^2-(\omega_a\nu\delta_1^2-\omega_1\nu\delta_2^2)^2}\right]. \tag{13}$$

The formula given in equations (12) and (13) reveals that the crystal susceptibility is influenced by quantum mechanical correction through δ_1, δ_2 and strain dependent dielectric constant. Here the damping of acoustic wave arises due to its acoustoelectric coupling with electron plasma waves. To compensate for the damping losses of the acoustic wave in the SDDC medium, one should apply a pump of a certain minimum amplitude called the threshold pump amplitude. The threshold pump amplitude E_{0th} may be obtained by setting the $[\chi^3]_r$ equation (12) equal to zero as

$$E_{0th}=\frac{m}{ek}(\delta_1^2\delta_2^2+\omega_1\omega_a\nu^2)^{0.5}. \tag{14}$$

3. Results and discussions

For analytical investigation of the parametric interaction processes we consider the irradiation of semiconductor sample $BaTiO_3$ by CO_2 laser at 77K. The parameters are $m=0.0145m_0$, $\varepsilon_s=2000$, $V_a=3\times10^3 ms^{-1}$, $\nu=5\times10^{11}s^{-1}$, $\omega_0=1.78\times10^{14}\sec^{-1}$, $\omega_a=2\times10^{12}\sec^{-1}$ $\rho=4\times10^3 kgm^{-3}$.

It is inferred from figures 1, that the $[\chi^3]_r$ increases with an increase of k in the positive group velocity dispersion regime $\omega_a>kV_a$e. $[\chi^3]_r$ is a positive quantity and increases with k. A slight increase in k beyond which point cause a sharp fall in $[\chi^3]_r$ making it vanish at $kV_a=\omega_a$. After this resonance condition, $[\chi^3]_r$ decreases sharply and then again starts increasing rapidly and saturates at larger acoustic wave number k values. It is worth mentioning that $[\chi^3]_r$ can be both positive and negative for $5\times10^8<k>7\times10^8$ .It can be observed that $[\chi^3]_r$ exhibits the usual dispersive characteristics of a medium and as increase in n_0, increases the negative and positive value of the $[\chi^3]_r$ without changing the resonant condition. One can infer from figure 1 doping level, and wavelength regime can enable one to achieve enhanced parametric dispersion.

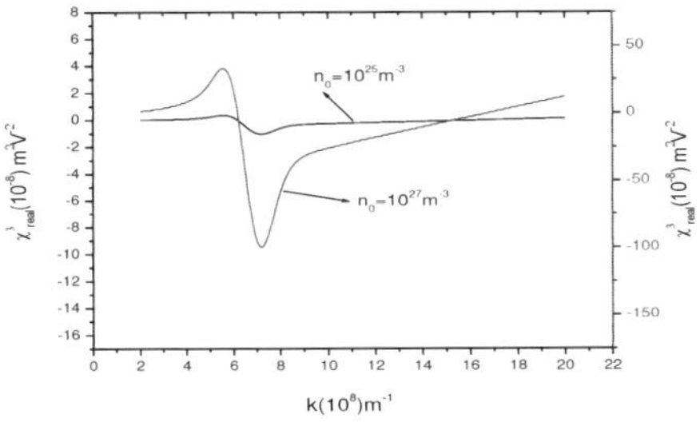

Figure1. Variation of real part of susceptibility $\left[\chi^3\right]_r$ with wave number k with carrier density n_0 as a parameter

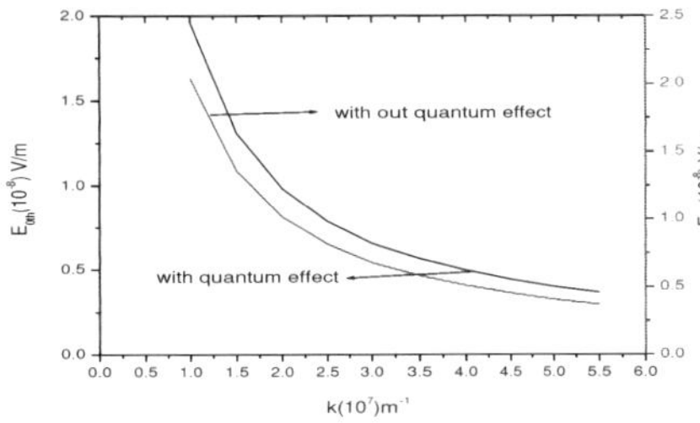

Figure 2. Variation of E_{0th} with k at $n_0 = 4\times10^{26} m^{-3}$ with and without quantum effect

Figure 2 displays the variation of threshold pump field viz E_{0th} on wave vector k using the material parameters with and without quantum effect.It is shown that as the wave vector increases the threshold electric field decreases for both the curves. While in term of pattern it is found that for lower values of k, E_{0th} decreases sharply while in case of the higher values of wave vector E_{0th} decreases with lower decapitation rate and becomes independent of it.Threshold value is also found to be influenced by quantum effect through the term δ_1. In high doping regime when $\omega_p^2 > k^2 V_f'^2$ and $\omega_{bp}^2 > \omega_1^2$ plasma mode dispersion dominates of quantum effect increases the value of E_{0th}.

4. References

[1] Bahaa F M and Rehab A 2013 *J. Mod. Phys.* **4** 327
[2] Jung Y D 2001 *Phys. Plasmas* **8** 3842
[3] Andreev A V 2000 *JETP Lett.* **72** 238
[4] Marklund M and Shukla P K 2006 *Rev. Mod. Phys.* **78** 591
[5] Markowich P A,Ringhofer C A and Schmeiser C 1990 *Semiconductor Equations* (Springer:Verlag New York)
[6] Shpatakovskaya G V 2006 *J. Exp. Theo. Phys.* **102** 466
[7] Wei L and Wang Y N 2007 *Phys. Rev. B* **75** 193407
[8] Ang L K 2004 *IEEE Trans. Plasma Sci.* **32** 410
[9] Chang D E, Sorensen A S, Hemmer P R and Lukin M D 2006 *Phys. Rev. Lett.* **97** 053002
[10] Shukla P K, Eliasson B and Stenflo L 2012 *Phys. Plasmas* **19** 07302
[11] Ghosh S and Khare P 2005 *Eur. Phys. J. D* **35** 521
[12] Galperin Yu M, Jin A and Shklovskii B I 1991 *Phys. Rev. B* **44** 5497
[13] Manfredi G and Hass F 2001 Phys. Rev. B **64** 07316
[14] Sharma A, Yadav N and Ghosh S 2013 *Int. J. Sci. Res. Pub.* **3** 2250
[15] Ghosh S and Yadav N 2007 *Acta Phys. Polo. A* **112** 29
[16] GhoshS, Dubey S and Vanshpal R 2010 *Phys. Lett. A* **375** 43

IOP Conf. Series: Journal of Physics: Conf. Series **836** (2017) 012020 doi:10.1088/1742-6596/836/1/012020

Gravitational instability of an anisotropic and viscoelastic plasma

P K Sharma[a], Anita Tiwari, Nusrat Khan and Shraddha Argal[b]

University Institute of Technology, Barkatullah University, Bhopal (M.P.) India 462026

Email: [a)]pks_buit30@yahoo.com, [b)]shraddhaargal@gmail.com

Abstract. The effect of pressure anisotropy is studied on the growth rate of gravitational instabilities in a viscoelastic medium. The problem is constructed with generalized hydrodynamic fluid model and Chew-Goldberger–Low fluid model for anisotropic pressure then a general dispersion relation for the viscoelastic medium is obtained using the normal mode analysis. The general dispersion relation is reduced for propagation along the magnetic field and propagation perpendicular to the magnetic field. These two modes are discussed for the classical or hydrodynamic and kinetic limits and conditions for jeans instability are obtained. We found that condition of Jeans instability is modified for viscoelastic medium under kinetic limit and depends on compressional viscoelastic mode. Numerical analysis for longitudinal mode for kinetic regime shows that the velocity of compressional viscoelastic mode has a stabilizing effect on the growth rate of Jeans instability. In the transverse mode, the Alfven velocity for kinetic regime has a stabilizing influence on the Jeans instability.

1. Introduction

To know the process of formation of astrophysical objects like stars, planets, comets, asteroids, the study of gravitational instability is of central importance. Jeans [1,2], Chandrasekhar [3] and Gliddon [4] studied the gravitational instability. Stellar matters exhibit both viscous and elastic characteristics together and behave like viscoelastic fluids. To study the gravitational instability of viscoelastic medium the generalized hydrodynamic model or GH model [5]-[6] is a more appropriate model. In this model normal viscosity coefficient is similar to a viscoelastic operator or Frenkel term [5]. The problem of gravitational instability has been extensively investigated for magnetized, isotropic and viscoelastic plasma by many investigators such that Kaw and Sen [6], Banerjee et al. [7], [8]. Janaki et al. [9], [10], Prajapati et al. [11], Prajapati and Chhajlani [12] and Sharma et al. [13] discussed the gravitational instability of strongly coupled plasma with GH model.

Chew et al. [14] obtained a closed set of hydro-magnetic equations by expanding the distribution function in the inverse power of ionic charge to mass ratio and by assuming that the particle correlation are due to a strong magnetic field rather than collisions. In view of the importance of the anisotropic pressure plasma various authors [15-17] have discussed instability problems using CGL equations. Although, MHD theory is widely applied in fusion plasma it gives a great support in illuminating the study of various instabilities. But in modern toroidal magnetic confinement devices, the plasma contains significant fast populations originated from neutral beam injection and ion cyclotron resonance heating, suggesting strong pressure anisotropy [17], [18]. Pressure anisotropy is not covered by the isotropic MHD theory, so the CGL approximations are used in the present work.

To study the gravitational instability of viscoelastic medium the GH is used. None of the investigators so far have studied the gravitational instability of anisotropic pressure for a viscoelastic medium under kinetic limit and hydrodynamic or classical limit using GH fluid model and CGL model. Our aim is to understand the physical behavior of linear waves occurring in anisotropic pressure viscoelastic magnetized plasma.

2. Linearized perturbation equations.

Consider a self-gravitating, viscoelastic fluid of uniform mass density ρ with pressure anisotropy ***P***. The medium is assumed to be embedded in a strong magnetic field ***B*** (0, 0, *B*). The scalar pressure is replaced by a pressure tensor $\boldsymbol{P} = p_{\perp}\boldsymbol{I} + (p_{II} - p_{\perp})\boldsymbol{nn}$, where $p_{\perp}$ and p_{II} are the components of pressure perpendicular and parallel to the direction of the magnetic field and ***I*** is the identity matrix. If the perturbations in the fluid pressure, density, fluid velocity, magnetic field and gravitational potential are p_1, ρ_1, $(\boldsymbol{u} = \partial\xi/\partial t)$, B_1, ϕ respectively the linearized equations for viscoelastic fluid are given by

$$\left(1+\tau\frac{\partial}{\partial t}\right)\left[\rho\frac{\partial \boldsymbol{u}}{\partial t} + \nabla\cdot\boldsymbol{\delta P} + \rho\nabla\phi_1 - \frac{1}{4\pi}(\nabla\times\boldsymbol{B}_1)\times\boldsymbol{B}\right] = \eta\nabla^2\boldsymbol{u} + \left(\chi+\frac{\eta}{3}\right)\nabla(\nabla.\boldsymbol{u}). \tag{1}$$

$$\frac{\partial\rho_1}{\partial t} + (\nabla.\boldsymbol{u})\rho = 0. \tag{2}$$

$$\nabla^2\varphi_1 = 4\pi G\rho_1. \tag{3}$$

$$\frac{\partial \boldsymbol{B_1}}{\partial t} = \nabla\times(\boldsymbol{u}\times\boldsymbol{B}). \tag{4}$$

$$\frac{\delta p_{II}}{p_{II}} + \frac{2B_1}{B} = \frac{3\rho_1}{\rho},\ \text{and}\ \frac{\delta p_{\perp}}{p_{\perp}} + \frac{B_1}{B} = \frac{\rho_1}{\rho}. \tag{5}$$

where symbols ϕ, τ, χ and η denote gravitational potential viscoelastic relaxation time, bulk viscosity and shear viscosity coefficient of the viscoelastic plasma. The term $(1+\tau\partial/\partial t)$ is called the Frenkel term. We considered that the medium is perturbed by Lagrangian displacement $\boldsymbol{\xi} = (\xi_x, \xi_y, \xi_z)$ and the perturbed quantities of the form $\exp i(\boldsymbol{k}.\boldsymbol{r} - \omega t)$ where $k = (k_x, k_y, k_z)$ is wave number and ω is the frequency of the harmonic disturbances. From the linearized equations we obtained the following equation.

$$\left[(1-i\omega\tau)\left(-\rho\omega^2 + 3k_{II}^2 p_{II} - 4\pi G\rho^2\frac{k_{II}^2}{k^2}\right) - k_{II}^2\left\{i\omega\left(\chi+\frac{\eta}{3}\right) + i\omega\eta\right\} - i\omega\eta k_{\perp}^2\right]\left[(1-i\omega\tau)\left\{-\rho\omega^2 + 2k_{\perp}^2\left(p_{\perp} + \frac{B^2}{8\pi}\right) + k_{II}^2\left(p_{\perp} - p_{II} + \frac{B^2}{4\pi}\right) - 4\pi G\rho^2\frac{k_{\perp}^2}{k^2}\right\} - k_{\perp}^2 i\omega\left\{\left(\chi+\frac{\eta}{3}\right)+\eta\right\} - i\omega\eta k_{II}^2\right] = \left[(1-i\omega\tau)\left(\frac{4\pi G\rho^2}{k^2} - p_{\perp}\right) + i\omega k_{II}\left(\chi+\frac{\eta}{3}\right)\right]^2 k_{II}^2 k_{\perp}^2, \tag{6}$$

where $\xi_{II} = \xi_z$, $\xi_{\perp}^2 = \xi_x^2 + \xi_y^2$, $k_{\perp}^2 = k_x^2 + k_y^2$, $k_{II} = k_z$, $k^2 = k_x^2 + k_y^2 + k_z^2$. Equation (6) is the general dispersion relation for anisotropic pressure viscoelastic magnetized plasma. On ignoring the viscoelastic effect for $\omega\tau << 1$ equation (6) reduces to the dispersion relation obtained by the MHD and CGL models. This can be easily verified by the other previous results, obtained by Gliddon [4], Bhatiya [15] and Kalra et al. [16].

3. Discussion of General Dispersion Relation.

We discussed the dispersion relation (6) by reducing it into two different modes.

3.1 Mode 1: Propagation along magnetic field.

For waves propagated in the direction of the magnetic field i.e., $k_{\perp} = 0$ the equation (6) reduces to

$$\left[(1-i\omega\tau)(\rho\omega^2-3k_{II}^2p_{11}+4\pi G\rho^2)+i\omega k_{II}^2\left(\chi+\frac{4\eta}{3}\right)\right]\left[(1-i\omega\tau)\left\{\rho\omega^2+k_{II}^2\left(p_{II}-p_{\perp}-\frac{B^2}{4\pi}\right)\right\}+i\omega\eta k_{II}^2\right]=0.(7)$$

3.1.1. Classical limit $\omega\tau<<1$. By taking $\omega\tau<<1$ in the first factor of equation (7) we obtained $k_{II}<(4\pi G\rho/3p_{II}/\rho)^{1/2}$. which is identical to condition of Jeans instability obtained by Gliddon [4]. This criterion is same that has been obtained by Jeans [1] and Bhatiya [15].

3.1.2. Kinetic limit. For $\omega\tau>>1$ we obtained the condition for Jeans instability from equation(7), which is $k_{II}<[4\pi G\rho/3p_{II}/\rho+v^2]^{1/2}$ where $v^2=(\chi+4\eta/3)\div\rho\tau$, v is the velocity of compressional viscoelastic mode. It is clear that the condition of Jeans instability is modified for kinetic limit and depends on compressional viscoelastic mode. From first factor of equation(7) we obtained dimensionless equation

$$\sigma^{*2}+3k_{II}^{*2}-1+k_{II}^{*2}v^{*2}=0\ . \quad (8)$$

Where $\sigma=-i\omega$, $\sigma^*=\sigma(4\pi G\rho)^{-1/2}$, $k_{II}^*=\left(\frac{p_{II}}{4\pi G\rho^2}\right)^{1/2}k_{II}$, $\eta^*=\frac{(4\pi G\rho)^{1/2}\eta}{p_{II}}$, $A^*=\frac{p_{\perp}+B^2/4\pi}{p_{II}}$, $\tau^*=\tau(4\pi G\rho)^{1/2}$

The normalized growth rate as a function of normalized wave number is plotted for various values of v^*(0.1, 0.3 and 0.6) in equation (8). We found from figure (1) that the velocity of compressional viscoelastic mode has stabilizing effect on the growth rate of Jeans instability.

3.2 Mode 2: Transverse mode of propagation ($k_{II}=0$).

In this mode the wave propagation is perpendicular to the magnetic field and we obtain the reduced dispersion relation by putting $k_{II}=0$ in equation (6).

$$(1+\tau\sigma)\left\{\rho\sigma^2+2k_{\perp}^2\left(p_{\perp}+\frac{B^2}{8\pi}\right)-4\pi G\rho^2\right\}+k_{\perp}^2\sigma\left(\chi+\frac{4\eta}{3}\right)=0\ . \quad (9)$$

Equation (9) shows a gravitating, shear viscous mode which depends upon magnetic field, pressure, shear and bulk viscosity coefficients. This generally leads to the compressional viscoelastic mode. This equation is discussed for classical and kinetic limits.

3.2.1 Classical limit $\omega\tau<<1$. From the constant term of equation (9) in classical regime the condition of instability is $(2k_{\perp}^2/\rho)(p_{\perp}+(B^2/8\pi))<4\pi G\rho$ which is modified due to pressure anisotropy and depends on magnetic field, pressure and density of the medium and is similar to the criterion obtained by Gliddon [4] and Sharma and Chhajlani [19] neglecting quantum effect in their case. Jean's criterion does not depend on the viscoelastic effect.

3.2.2 Kinetic limit. For $\omega\tau>>1$ from equation(9) we obtained the condition of instability $k<4\pi G\rho/(2p_{\perp}/\rho+B^2/4\pi\rho)+v^2$ to study the effects of Alfven velocity due to the magnetization and compressional velocity on the growth rate of Jeans instability, equation(9) is solved for $\omega\tau>>1$ numerically with the parameters

$$\chi^*=\frac{\chi(4\pi G\rho)^{1/2}}{p_{\perp}},\quad \eta^*=\frac{\eta(4\pi G\rho)^{1/2}}{p_{\perp}},\quad k_{\perp}^*=\left(\frac{p_{\perp}}{4\pi G\rho^2}\right)^{1/2}k_{\perp},\quad V_{\alpha}^{*2}=\frac{V_{\alpha}^2\rho}{p_{\perp}}\ ,\quad v^*=v(\rho/p_{\perp})^{1/2}$$

where $V_{\alpha}=B(4\pi\rho)^{-1/2}$ is Alfven velocity, we obtained the equation for kinetic regime in dimensionless form as

$$\sigma^{*2}+2k_{\perp}^{*2}+k_{\perp}^{*2}\left(V_{\alpha}^{*2}+v^{*2}\right)-1=0\ . \quad (10)$$

The growth rate of Jeans instability versus wave number for various values of Alfven velocity (V_{α}^* = 0.1, 0.3 and 0.5) with fixed value v^* = 0.2has been shown in figure2. The growth rate decreases with increase in value of Alfven velocity. Therefore, in the perpendicular direction of propagation the

Alfven velocity has a stabilizing influence on the Jeans instability.

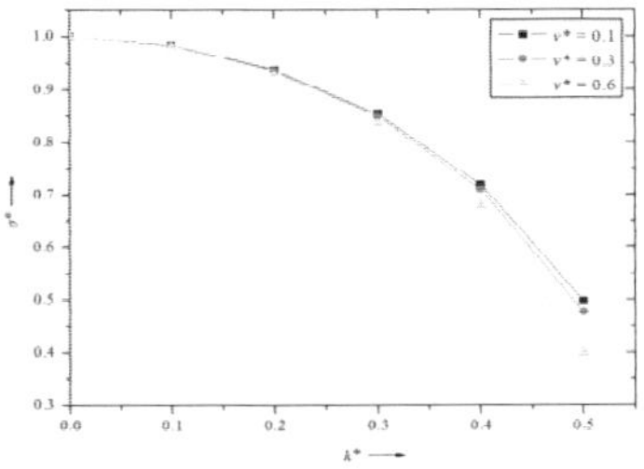

Figure 1.The growth rate versus wave number for different values of Alfven velocityfor perpendicular propagation (kinetic limit). (kinetic limit).

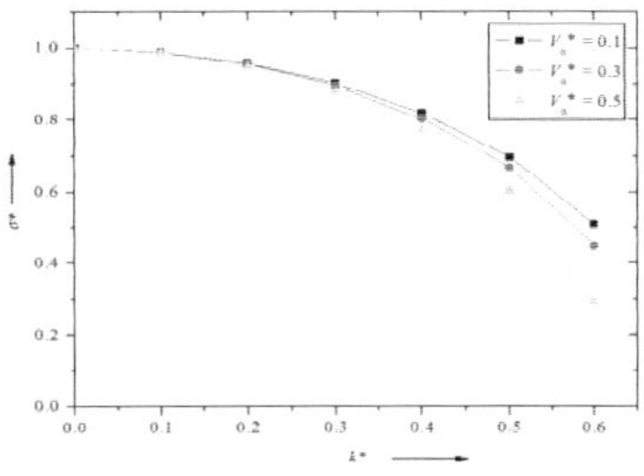

Figure 2. The growth rate versus wave number for different values of compre-ssional velocityfor parallel propagation

4. Conclusions

A dispersion relation for viscoelastic anisotropic magnetized fluid has been derived employing the normal mode analysis using GH model and CGL fluid models.The dispersion relationis discussed for two modes.For waves propagated in the direction of the magnetic field, in the hydrodynamic limit viscoelastic parameters have no effect on the condition of Jeans instability. In kinetic regime the condition of Jeans instability modified and depends on compressional viscoelastic mode. Numerical analysis shows that the velocity of compressional viscoelastic mode has a stabilizing effect on the growth rate of Jeans instability.In transverse mode of propagation for classical regime the Jean's criterion is modified due to pressure anisotropy and it depends on magnetic field, pressure and density. Under kinetic regime for transverse mode of propagation the critical Jeans wave number depends on relaxation time, shear and bulk viscosity or on compressional velocity. We numerically observed that Alfven velocity has a stabilizing influence on the growth rate of the Jeans instability.

References.

[1] Jeans J H 1902 *Phil. Trans. Roy. Soc. London* **A 199** 312
[2] Jeans J H 1929 *Astronomy and Cosmogony* (Cambridge UK: Cambridge University Press)
[3] Chandrasekhar S 1961 *Hydrodynamics and Hydromagnetic Stability* (Oxford: Clarendon Press)
[4] Gliddon E C 1966 *Astrophys. J.* **145**583
[5] Frenkel Y I 1946 *Kinetic Theory of Liquids* (Oxford: Clarendon Press)
[6] Kaw P K and Sen A 1998 *Phys. Plasmas* **5** 3552
[7] Banerjee D, Janaki M S, Chakrabarti N and Chaudhuri M 2010 *New J. Phys.* **12** 123031
[8] Banerjee D, Janaki MSand Chakrabarti N 2010 *Phys. Plasmas* **17** 113708
[9] Janaki M S, Chakrabarti N and Banerjee D 2011*Phys. Plasmas* **18** 092114
[10] Janaki M S, Chakrabarti N and Banerjee D 2011 *Phys. Plasmas* **18** 012901
[11] Prajapati R P, Sharma P K, Sanghvi R K and Chhajlani R K 2012 *J. Phys. Conf. Ser.* **365** 012040
[12] Prajapati R P and Chhajlani R K 2013 *Astrophys. Space Sci.* **344** 371
[13] Sharma P K, Argal S,Tiwari A and Prajapati R P 2015 *Z. Naturforsch.* **70** 39
[14] Chew G F, Goldberger M L and Low F E 1956 *Proc. Roy. Soc. London A* **236** 112
[15] Bhatia PK 1968 *Z. Astrophysik* **69** 363
[16] Kalra G L., Hosking R J and Talwar S P 1970 *Astrophys. Space Sci.* **9** 34
[17] Prajapati R P, Parihar A K and Chhajlani R K 2008 *Phys. Plasmas* **15** 012107
[18] Fasoli A *et al.* 2007 *Nucl. Fusion* **47** S264
[19] Sharma P and Chhajlani R K 2013 *Phys. Plasmas* **20** 092101

Amplitude modulation and demodulation of an electromagnetic wave in magnetized ion-implanted semiconductor plasmas having SDDC

N Yadav[1], S Ghosh[1] and P S Malviya[2,a]

[1]School of Studies in Physics, Vikram University, Ujjain 465010, India
[2]Department of Physics, Govt. J.N.S. Post Graduate College, Shujalpur 465333, India

E-mail: [a)]psm_sehore@rediffmail.com.

Abstract. In communication processes, amplitude modulation is very helpful to save power using a single band transmission. Using the hydrodynamical description of semiconductor plasma analytical investigations are made for the amplitude modulation as well as demodulation of an electromagnetic wave in magnetized ion implanted semiconductor plasmas having strain dependent dielectric constants. Analysis is made under different wave number regimes over a wide range of cyclotron frequencies without and with colloids. Numerical estimations are made for n-doped $BaTiO_3$ crystal irradiated by pump wave frequency $1.78\times10^{13}s^{-1}$. It has been found that ion implantation of negatively charged colloids modifies nearly $\approx 10^5$ of magnitude of amplitude modulation and demodulation processes. Ion implantation plays a key role in development of optoelectronics.

1. Introduction

The problem of modulation in semiconductor plasmas has been studied in number of works [1,2]. The modulation of microwave while propagating through a piezoelectrically active semiconductor media duly irradiated by an acoustic wave was first predicted by Mathur and Sagoo [3]. The periodic variation of the propagation parameter leads to the modulation of an electromagnetic wave passing through plasma. This periodic variation in the propagation parameters, caused by the time varying change in the carrier density of the plasma, may be due to a modulated magnetic field, the modulation of power in an rf discharge or the propagation of an acoustic wave. The phenomena of modulation of an electromagnetic wave by an acoustic wave are of special interest in the problem of communication of different types. This is due to fact that the scattering of light from sound or low frequency electromagnetic waves affords a convenient means of controlling the frequency, intensity and direction of an optical beam [4].

This type of modulation makes possible a large number of applications involving the transmission, display and processing of information, the fabrication of some optoelectronic devices, such as acousto-optic modulators, these are based on the interaction of an acoustic wave or low frequency electromagnetic wave with the incident laser beam. Acousto-optic interaction in dielectrics and semiconductors is playing an increasing role in optical modulation and beam steering. However, the integrated process becomes a serious limitation due to high power requirements. The most direct approach to this problem is to tailor new material with more desirable acousto-optic properties. The study of modulation can be made with respect to amplitude, frequency and phase. Amplitude modulation is one of the oldest forms of modulation. One of the important problems in communication systems is that of developing an effective method of modulation as well as demodulation of the waves[5]. The strain dependent dielectric constant (SDDC) effect for high dielectric materials is not

IOP Conf. Series: Journal of Physics: Conf. Series **836** (2017) 012021 doi:10.1088/1742-6596/836/1/012021

fully understood yet, it shows broadened peaks and considerable deviation from the Curie Weiss type law near critical strains which are attributed to complete phonon softening [6-7].

Ion implantation is the widely used doping techniques in preparation of doped semiconductors. The implanted ions in a material can modify the magnetic coercivity and nonlinear optical properties. A small quantity of ions may alter the electrical conductivity of the host material by several orders of magnitude. This process is responsible for implanted metal ions being neutralized during slowing down processes and somehow agglomerating to form colloids. The study of colloid formation of metal ions by ion implanted techniques is carried out in a number of laboratory experiments [8].

Charged grain with size range of micrometer or less are exist in many laboratory, industrial and natural plasmas. The grains are often highly negatively charged since fast-moving thermal plasma electrons easily collide inelastically with them and become attached. The grains can be useful models for studying the structural dynamics of microscopic matter. Crystal like grain structures and their physical characteristics, such as melting and fracturing have been predicted and observed in experiments and numerical simulation [9-10].

In the present paper, we investigate the effects of negatively charged colloids on the amplitude modulation and demodulation of an intense electromagnetic wave in ion implanted semiconductor crystal. The effects of charged colloids in ion implanted semiconductor plasma medium add new dimensions to the analysis presented in n-doped SDDC magnetoactive semiconductors. The intense pump beam electrostrictively generates an acoustic wave within the semiconductor medium that induces an interaction between the free carriers (through electron and colloid plasma wave) and acoustic phonons (through material vibration). This interaction induces a strong space field that modulates the pump beam. Thus, the applied optical and generated acoustic waves in an acoustooptic modulator can produce amplitude modulation and demodulation effects. It is found that ion implantation of negatively charged colloids modifies the properties of material.

2. Theoretical formulation

We have considered the hydrodynamic description of homogeneous, semiconductor plasma of infinite extent. The Lattice vibration equation:

$$\frac{\partial^2 u}{\partial t^2}=\frac{c}{\rho}\frac{\partial^2 u}{\partial x^2}-\frac{\varepsilon g E_0}{\rho}\frac{\partial E_1^*}{\partial x} \tag{1}$$

If ε_s is dielectric constant in absence of any strain, then $g=\left(\varepsilon_s/3\right)$ is the coupling constant.

$$\frac{\partial E_1}{\partial x}=\frac{n_{1j}Z_j e}{\varepsilon}+gE_0\frac{\partial^2 u^*}{\partial x^2} \tag{2}$$

where $j=e,d$ (electrons, colloids). Here $Z_d=\frac{q_d}{e}$ is the ratio of negative charges q_d resided over the colloidal grains to charge e on the electrons and $Z_e=-1$ is assumed for further analysis.

$$\frac{\partial v_{1j}}{\partial t}+\upsilon_j v_{1j}+\left(v_{0j}\frac{\partial}{\partial x}\right)v_{1j}=\frac{Z_j e}{m_j}\left(E_l+v_l\times B_0\right)-\frac{k_B T}{m_j n_0}.\nabla n_1 \tag{3}$$

The subscript l stands for $0,+$ and $-$ modes and $j=e,d$ (e- electrons and d-colloids). Momentum transfer equation (4) may be used to obtain the oscillatory carrier fluid velocity in presence of pump electric field (E_0) and field of side band modes ($E_\pm$).

The total transverse current density in the medium is given by

$$J_{total}=-Z_j e.\left[\sum_l n_0.v_l+\sum_l n_1.v_0.\exp\left[i\left(k_a.x-\omega_a.t\right)\right]\right] \tag{4}$$

where $n_1 v_0\exp\left[i\left(k_a x-\omega_l t\right)\right]$ represents the current generated due to the interaction of the pump wave with acoustic wave. In order to obtain the modulation indices of modulated sideband modes. Using equations (5), (6) and (7) in the general wave equation

$$\frac{\partial^2 E_{total}}{\partial x^2}-\mu.\varepsilon\frac{\partial^2 E_{total}}{\partial t^2}-\mu.\frac{\partial J_{total}}{\partial t^2}=0 \tag{5}$$

and neglecting $\exp(\pm ik_a x)$ in comparison to 1, we obtain the following expression for modulation indices;

$$\frac{E_\pm}{E_0}=\frac{\omega_0\mu Z_j e\rho u v_a^2 A^2(\omega_0+i\upsilon_j)}{m_j\beta[(\omega_{cj}^2+\upsilon_j^2-\omega_0^2)-i(2.\omega_0.\upsilon_j)]}\cdot\frac{k_a}{(k_a\pm 2k)}\left[1+\frac{V_{Thj}^2}{2}\cdot\frac{k_a^2}{\omega_{pj}^2}\right] \quad (6)$$

in which $\beta = gE_0$. By rationalization of above equation, one can obtain the real part of modulational indices as

$$\frac{E_\pm}{E_0}=\frac{\omega_0^2\mu Z_j e\rho u k_a v_a^2 A^2}{m_j\beta(k_a\pm 2k)}\frac{(\omega_{cj}^2-\omega_0^2-\upsilon_j^2)}{\left[(\omega_{cj}^2+\upsilon_j^2-\omega_0^2)^2+4\omega_0^2\upsilon_j^2\right]}\cdot\left[1+\frac{V_{Thj}^2}{2}\cdot\frac{k_a^2}{\omega_{pj}^2}\right] \quad (7)$$

when $j=e$ (i.e. e-electrons) and $Z_e=-1$

$$\frac{E_\pm}{E_0}=-\frac{\omega_0^2\mu e\rho u k_a v_a^2 A^2}{m_e\beta(k_a\pm 2k)}\frac{(\omega_{ce}^2-\omega_0^2-\upsilon_e^2)}{\left[(\omega_{ce}^2+\upsilon_e^2-\omega_0^2)^2+4\omega_0^2\upsilon_e^2\right]}\cdot\left[1+\frac{V_{The}^2}{2}\cdot\frac{k_a^2}{\omega_{pe}^2}\right] \quad (8)$$

where $\omega_{ce}=e|B_0|/m_e$, $V_{The}=\left(2k_BT_e/m_e\right)^{1/2}$ and $\omega_{pe}=\left(n_0e^2/m_e\varepsilon\right)^{1/2}$ are the cyclotron frequency, the thermal speed and the plasma frequency respectively for electron.

when $j=d$ (i.e. d-negatively charged colloids) and Z_d. The colloid thermal speed is usually far smaller than the electron thermal speed:

$V_{Thd}=\left(m_e.T_d/m_d.T_e\right)^{1/2}.V_{The}$ and Debye length($j=d$) $\lambda_D=\left(\varepsilon.k_BT_d/n_{0d}Z_d^2e^2\right)^{1/2}$

$$\frac{E_\pm}{E_0}=-\frac{\omega_0^2\mu Z_d e\rho u k_a v_a^2 A^2}{m_d\beta(k_a\pm 2k)}\frac{(\omega_{cd}^2-\omega_0^2-\upsilon_d^2)}{\left[(\omega_{cd}^2+\upsilon_d^2-\omega_0^2)^2+4\omega_0^2\upsilon_d^2\right]}\cdot\left[1+k_a^2.\lambda_D^2\right] \quad (9)$$

where $\omega_{cd}=Z_de|B_0|/m_d$, $V_{Thd}=(2k_BT_d/m_d)^{1/2}$ and $\omega_{pd}=(n_{0d}Z_d^2e^2/m_d\varepsilon)^{1/2}$ are the cyclotron frequency, the thermal speed and the plasma frequency respectively for negatively charged colloid.

3. Results and discussions

The numerical calculations are performed for the n-type doped semiconductor sample ($BaTiO_3$) at 77 K duly irradiated by 10.6 μm CO_2 laser. The following material parameters are taken as representative values. $m_e=0.0145m_0$ (m_0 being the free electron rest mass), $\varepsilon_s=2000$, $m_d=1.67\times10^{-27}kg.$, $\rho=4\times10^3 kg.m^{-3}$, $n_{0e}=10^{25}m^{-3}$, $T_e=T_d=77K$, $n_{0d}=2\times10^{20}m^{-3}$ $\upsilon_e=5\times10^{11}s^{-1}$ $\upsilon_d=1.405\times10^9s^{-1}$, $\omega_0=1.78\times10^{13}s^{-1}$, $\omega_a=1.6\times10^{13}s^{-1}$, $v_a=2.5\times10^3 m.s^{-1}$.

Case: 1 when $k_a > 2k$

The variation of $\left(E_+/E_0\right)$ and $\left(E_-/E_0\right)$ with the applied magnetic field (ω_c) are depicted in figure 1 and 2 for both without and with colloids. It may be seen from these figures that at ($\omega_c\approx\omega_0$) the modulation indices become zero. When one applies weak magnetic field so that cyclotron frequency ω_c for both case without and with colloids becomes smaller than the carrier frequency ω_0, then both the modes are in phase with pump wave, which exhibits modulation process. But at a particular value of magnetic field when ($\omega_c\approx\omega_0$) modulation indices for both the modes become zero and complete absorption of wave takes place on neglecting the collision term in equations (8) and (9) for without and with colloids. On further increasing the cyclotron frequency ($\omega_c>\omega_0$) both side band go out of phase. These out of phase side bands then interact with the pump wave to produce demodulated acoustic wave. We conclude that demodulation process can be observed in the regime ($k_a>2k$)with($\omega_c>\omega_0$) for both without and with colloids. The variation of $\left(E_+/E_0\right)$ and $\left(E_-/E_0\right)$ with the applied magnetic field (ω_c) are depicted in figure 3 and 4 for both without and with colloids. From both the figures for same wave number regime the behavior of modulation indices for plus and minus modes are opposite in nature. In the same wave number regime the amplitude of the plus mode is positive under the condition ($\omega_c < \omega_0$) for both without and with colloids. Hence under this regime of cyclotron frequency, the amplitude of plus mode is in phase with the pump wave. This side band then interacts with the wave to produce modulated wave. However, when the carrier frequency become nearly equal to the cyclotron frequency, complete absorption of waves takes place in equations (8) and (9) separately without and with colloids. In the range defined as $\omega_c > \omega_0$, the modulation indices of plus mode is negative.

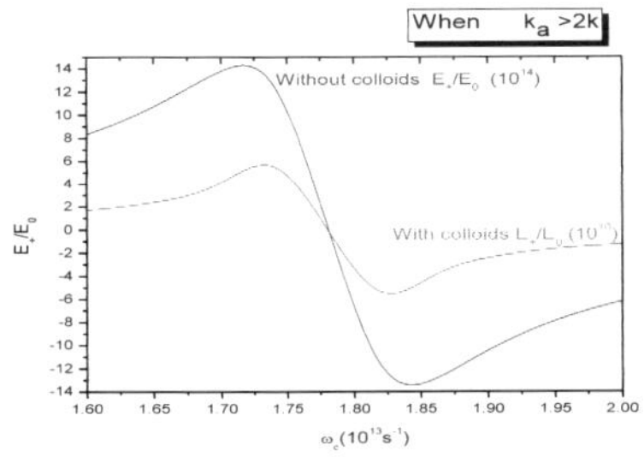

Figure 1. Variation of modulation index of plus mode ($k_a > 2k$) for without and with colloids respectively with magnetic field.

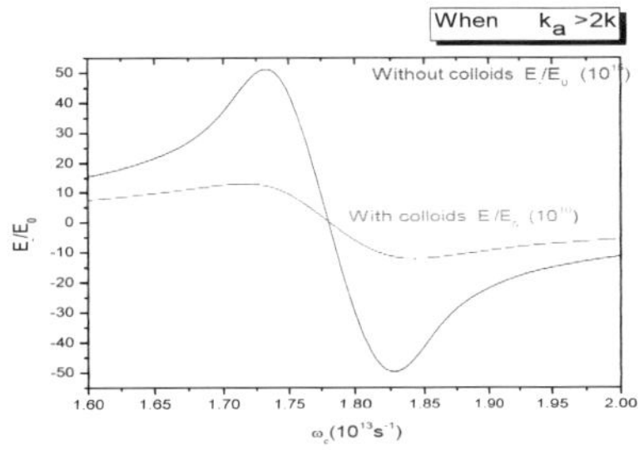

Figure 2. Variation of modulation index of minus mode ($k_a > 2k$) for without and with colloids respectively with magnetic field.

Case: 2 when $k_a < 2k$

This out of phase side band waves then interact with the pump to produce a demodulated wave which exhibits demodulation process. From figure 4 we may infer that in this wave number regime the amplitude of the minus mode is negative when $\omega_c < \omega_0$ and remain out of phase with the pump wave. This side band then interacts with the wave to produce demodulated acoustic wave for both without and with colloids [11]. Hence ,it can be seen easily that presence of negatively charged colloids modifies nearly $\approx 10^5$ of magnitude of amplitude modulation and demodulation processes.

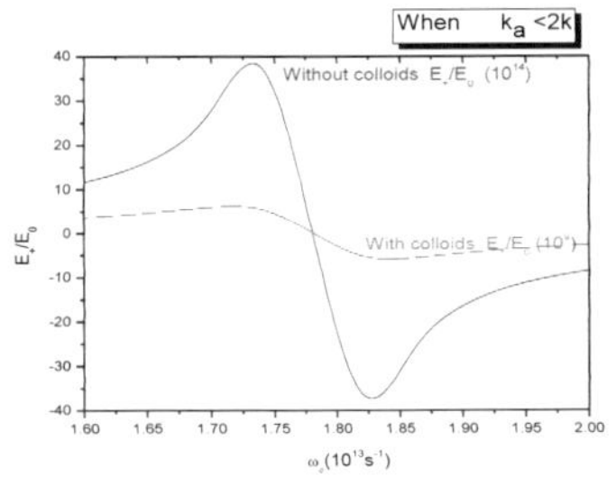

Figure 3. Variation of modulation index of plus mode ($k_a < 2k$) for without and with colloidsrespectively with magnetic field.

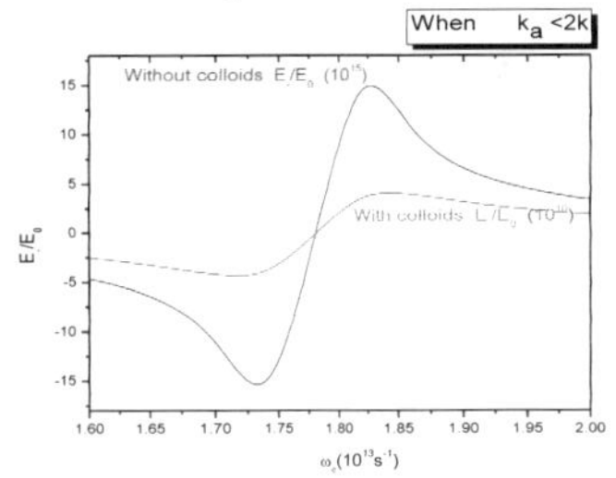

Figure 4. Variation of modulation index of minus mode ($k_a < 2k$) for without and with colloids respectively with magnetic field

References

[1] Neogi A 1995 *J. Appl. Phys.* **77** 191
[2] Sharma G and Ghosh S2000 *Eur. Phys. J. D* **11** 301
[3] Mathur S S and Sagoo M S 1973 *Canadian J. Phys.* **51**2 459
[4] Muller R 1988 Phys. *Stat. Solidi B* **150** 587
[5] Nimje N, Dubey S and Ghosh S 2012 *Opt. Laser Tech.* **44** 744
[6] Zhang X,Wang J B, Li X B, Zhong L, Lou X J and Zhou Y C 2011 *J. App. Phys.* **109** 126102
[7] Ong P V and Jaichan L 2012 *J. App. Phys.* **112** 014109
[8] Zeba I, Uzma Ch, Jamil M, Salimullah M and Shukla P K 2010 *Phys. Plasmas* **17** 032105
[9] Wang Y,YuM Y, Zhao J T, Wu L and Liu S 2014 *Phys. Scr.* **89** 125601
[10] Barros K and Luijten E 2014 *Phys. Rev. Lett.* **113** 017801
[11] Ghosh S and Yadav N 2007 *Acta Phys. Polo. A* **112** 29

IOP Conf. Series: Journal of Physics: Conf. Series **836** (2017) 012022 doi:10.1088/1742-6596/836/1/012022

Attosecond pulse generation from relativistic plasma mirror

B S Sharma[1,a], R C Dhabhai[1], A Sharma[2] and N K Jaiman[3]

[1] Department of Physics, Govt College Kota, Kota-324001, India.
[2] Department of Electrical Engineering and Photonics, IIT Kanpur-208016.
[3]Department of Pure and Applied Physics, University of Kota, Kota-324010, India.

E-mail: [a)]bs_phy@yahoo.com

Abstract. When an intense relativistic short laser pulse incident on an optically polished surface, it generates a high density plasma that acts as a relativistic plasma mirror. As this mirror reflects the intense laser field, its space time characterization changes due to high nonlinear response of the field. This lead to the phase variation both temporally and spatially that could affect the generation of high harmonics of the laser and attosecond pulses. The present theoretical study address the issues of the intensity dependent space-time characteristics on the generation of attosecond pulses from relativistic plasma mirror.

1. Introduction

The interaction of ultrashort-ultraintense laser ($I \geq 10^{20} W/m^2$) pulse with an optically reflected metal surface generates a dense plasma that acts as a plasma mirror(PM). These mirrors specularly reflect the main part of the laser pulse and can be used as an active optical elements to manipulate the spatial and temporal properties of the high harmonics. The modification in the temporal contrast lead to generate an intense attosecond extreme ultraviolet(XUV) or X-ray pulses of energy in the range 1-10 J through nonlinear harmonic up-conversion of the laser pulse. However, pressure exerted by the laser pulse deformed the PM surface non-uniformly This results in the rotation of the PM that effect the spatial and temporal contrast of the reflected laser field and the high harmonics.

PMs are routinely used at moderate light intensities ($10^{14} - 10^{16} W/cm^2$) as ultrafast optical switches, to enhance the temporal contrast of the femtosecond lasers. For intensities $I \geq 10^{16} W/cm^2$, nonlinear response of the PMs to the laser field results in sub-cycle temporal modulations of the reflected field , associated to the high harmonic generation in its spectrum (HHG).These harmonics generated through various mechanism are associated in the time domain to the attosecond pulses.For lasers with intensity $\geq 10^{18} W/cm^2$, the key HHG results in the relativistic oscillating mirror where the laser driven oscillation of the plasma surface induces a periodic Doppler effect on the reflected laser field [1, 2, 3, 4, 5, 8],which can result in harmonic orders of several thousands. In these high intensity applications, laser field exerts such a high pressure on the plasma ($\cong 5 Gbar for I \approx 10^{19} W/cm^2$) that it induces a significant motion of the PM surface, even during a femtosecond laser pulse. This leads spatial variation of intensity on the target and hence the deformation in the surface of the PM.

Our present work is based on the spatiotemporal coupling (STC) to the generation of isolated attosecond pulse. We consider deformation in the relativistic plasma mirror surface in the form of an elliptical curvature-which can affect the spatial and spectral properties of the reflected beam.

IOP Conf. Series: Journal of Physics: Conf. Series **836** (2017) 012022 doi:10.1088/1742-6596/836/1/012022

The intensity dependence of the harmonic spectrum and the attosecond pulse are described in section 2. Conclusions are presented in section 3.

2. Intensity Dependence

It is observed that the wave train of the attoseond pulses can be observed when intensity of the laser beam exceeds to $10^{18}W/cm^2$. It is further concluded that the rotation effect in relativistic oscillating plasma mirror change the denting mechanism of the reflected laser field and phase coherence in the attosecond pulses. The rotation effect of plasma mirror due to $E \times B$ changes the phase parameter of the harmonics and increases the value of f_h. This leads high repetition rate for attosecond pulses with increased intensity.

When an electromagnetic wave reflected from an oscillating mirror, its frequency spectrum extended to high frequency range and the wave breaks-up in the short waves. In relativistic oscillating mirror, harmonics of much higher frequency are generated. The reflected wave's electric field from the oscillating mirror in a reflection time $t' = t - x(t)/c$ is given as

$$\mathbf{E_r} = -\frac{\mathbf{1}}{\mathbf{c}}\frac{\partial \mathbf{A_L}(\mathbf{t}', \mathbf{x}')}{\partial \mathbf{t}'} \tag{1}$$

where x' and t' are the position and time of the reflected waves in observer's frame. The oscillating mirror model implies that the tangential components of the vector potential are zero at the mirror surface. The component of reflected electric field from the oscillating plasma surface will be parallel to incident electromagnetic wave. As a result of it,if the oscillating mirror moves with $\gamma_L \gg 1$ towards the laser pulse with oscillating frequency ω_{osc} and electric field $E_{||}$, and duration τ_L, the reflected electric field of the nth harmonics will be given as

$$E_r \propto n\gamma_L^2 E_l \tag{2}$$

and the pulse duration will be

$$\tau_L' \propto \frac{\tau_L}{n\gamma_L^2} \tag{3}$$

The Fourier spectrum of the electric field of reflected beam at position x' and time t' is

$$E_r(\Omega_r) = \frac{mc\omega}{e\sqrt{2\pi}} \int_{-\infty}^{\infty} \mathbf{a_L}(\mathbf{t}', \mathbf{x}')$$
$$\times \exp -\iota\omega_L t' - \iota\omega_L(x'/c) \exp(-\iota\phi_r) \tag{4}$$

where

$$t' - x(t')/c = t \tag{5}$$

and ϕ_r is the phase of reflected wave and is given by

$$\phi_r = \int_0^u \Omega_r(u)du = \omega_L[2t'(u) - u] \tag{6}$$

where $u = t' - x'(t')/c$ and spatial position x' and time t' and t'(u) can be obtained from the equation

$$t'(u) = u + x'(t'(u))/c \tag{7}$$

Differentiating above equation, we obtain

$$\phi_r' = \omega_L \frac{1 + \beta'(u)}{1 - \beta'(u)} \tag{8}$$

where $\beta'(u) = dx'(u)/cdt'$ is the mirror velocity normalized by c. Using both power law and the exponential decay parts, and properties of Array's function to analyze the spectrum modulation due to the $E \times B$ effect, we use Eq.(4-6) in Eq(2) to obtain

$$E_r = \left(\frac{\omega_L}{\Omega'_r n}\right)^{5/2} \exp\left(\frac{-16\sqrt{5}\Omega'_r}{5\sqrt{n^3}\omega_L\omega_p^{3/2}}\right) \times Re\frac{\exp(\iota\Omega'_r t - \iota\psi'_r)}{10\sqrt{2}/(5\sqrt{n^3}\omega_p) + \iota\omega_L} \quad (9)$$

where $\Omega'_r = \Omega_r + \Omega_{rot}$.
The amplitude of these reflected pulses decreases fast when Ω_r grows. However, the pulse duration does not depend on Ω_r. Since the fundamental frequency grows as Ω_r, the pulses obtained with an above cut-off filter are filled with electric field oscillations. We use Eq.(9) to obtain the intensity of nth harmonics as

$$I_{rn} \propto \left(\frac{\omega_L}{\Omega'_r n}\right)^{5} \exp\left(\frac{-32\sqrt{5}\Omega_r'^{5/2}}{5\sqrt{n^3}\omega_L\omega_p^{3/2}}\right) \times \frac{\left(n^3\omega_p^2 - \omega_L^2\right)}{8\omega_p^2} \quad (10)$$

The intensity of the reflected pulses decreases with higher harmonics and plasma frequency. However the pulse repetition rate increases when Ω_r grows. It is observed that the intensity of the attosecond pulses depend on the harmonic phases. if v_r is the ultra relativistic velocity of the reflected electric field of a particular harmonics at time $t'(u)$, than the maximum relativistic factor will be given as

$$\gamma_{max} = \frac{1}{\sqrt{1 - \frac{v_r(t'(u))^2}{c^2}}} \quad (11)$$

Consequently, for the surface γ factor during a relativistic spike, the highest harmonic will be generated over the time period

$$\Delta t \approx \frac{1}{\omega_L\gamma_{max}^3} \approx \frac{1}{\omega_L n_{cr}^{3/2}} \quad (12)$$

where n_{cr} is the critical density of plasma surface.
For this duration the reflected fields move with ultra-relativistic velocity in the direction of the emitted radiation. The intensity variation over this time interval for n^{th} harmonic can be written as

$$I_n \propto \left(\frac{n_{cr}}{n^3\Omega'_r}\right)^{5} \exp\left(\frac{-32\sqrt{5}\Omega_r'^{5/2}}{5\sqrt{n^3}\omega_L\omega_p^{3/2}}\right) \times \frac{\left(n^3\omega_p^2 - \omega_L^2\right)}{8\omega_p^2} \quad (13)$$

Eq.(13) shows a theoretical result of temporal structure of the intensity of attosecond pulse trains produced on plasma mirrors. The harmonic spectra associated with a train of attosecond pulses

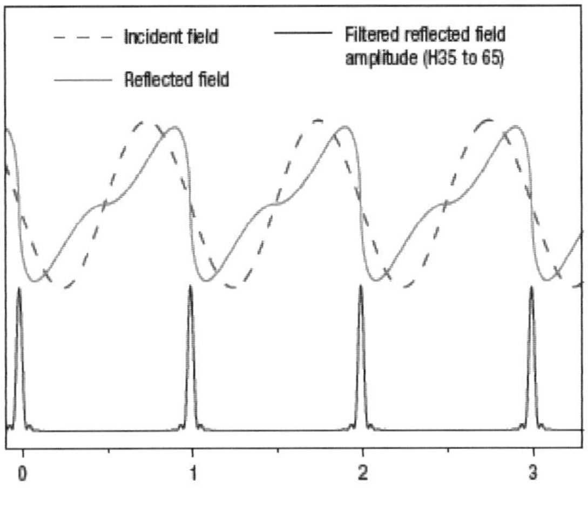

Figure 1. Harmonic spectra from plasma mirror

for different intensities is shown in the figure(1) The harmonic divergence θ_n of nth harmonics is given

$$\theta_n^2 = \left(\frac{\lambda_n}{\pi w_n}\right)^2 + \left(\frac{\lambda_n}{\pi w_n}\right)^2 (n\phi_L)^2$$

$$\theta_n = \theta_n^0\sqrt{1+(n\phi_L)^2} \tag{14}$$

where $\theta_n^0 = \lambda_n/\pi w_n$ is the harmonic divergence for the source size w_n, in the absence of PM curvature and rotational effect due to $E \times B$.

3. Conclusions

We have presented a simple analytical model for the generation of the attosecond pulse from the relativistic oscillating plasma mirror with $E \times B$ effect that leads rotation in the oscillating plasma mirror. The $E \times B$ effect changes the harmonic divergence which could change the pattern of extended ultra-broadband isolated attosecond pulse spectrum and repetition rate. We have also addressed the temporal characterization of the reflected electric field from plasma mirror, with temporal resolution going down to the attosecond range. It is further observed that the number of harmonics in the reflected laser field increases with intensity of the incident laser beam and electron plasma density.

References

[1] Thaury C.and F.Quéére, J. Phys. B, **43**,213001(2010).
[2] Licheters R., Meyer-ter-Vehn J., Pukhov.A, Phys. Rev. E **3**,3425(1996).
[3] Baeva T., Gordienko S., Pukhov A., Phys. Rev. **74**,046404(2006).
[4] Gonoskov,A.A., Korzhimanvo A.V.,Kim A.V.,Marklund M. and Sergeev A.M, Phys. Rev. E **84**, 046403(2011).
[5] Dromey B. *et al* . Nat. Phys. **2**,456(2006).
[6] Lehnert B., Dynamics of charged particles (New York: Interscience Publishers, (1964).
[7] Vincenti H. , Monchoc S. , Kahaly S. , Bonnaud G., Martin Ph. and Quéré F., Nat. Communication, (2014).
[8] Chini Michae, Zhao Kun and Chang Zenghu Nat. Photonics, **8**,178-186(2014).

IOP Conf. Series: Journal of Physics: Conf. Series **836** (2017) 012023 doi:10.1088/1742-6596/836/1/012023

Novel modes in electron-electron two stream interactions in semiconductor plasma embedded with a nanoparticle

S Ghosh and Priya Dubey

School of Studies in Physics, Vikram University, Ujjain (456010), India

Abstract. The temporal behaviour of electron-electron two stream instability in two valley semiconductor plasma embedded with a nanoparticle has been investigated. The analytical results obtained for nanoparticle embedded n-GaAs at 300K revealed the possibility of six electrostatic modes of propagation, out of which two modes are found to be induced by the nanoparticle present in the medium. We found that the temporal behaviour of four pre-existing modes does not depend on the signature of nanoparticle present within the medium whereas nanoparticle induced two novel modes are found to be evanescent and stable by nature. Both the new modes are found to be propagating with equal phase velocities but in opposite directions.

1. Introduction

The development of solid state devices operating at higher frequency region gained tremendous attention due to their possible applications in radio astronomy, industry and defence [1]. Here extremely small feature sizes are required, but to achieve required small sizes, carrier transit time always imposes limitations and these limitations could not be overcome by using conventional approaches. Modern workers proposed that to overcome this limitation, one has to explore the properties of nanoparticles embedded in existing plasma media. The presence of nanoparticle in the medium, due to their inherent properties, supposed to induce significant changes in bulk material. Recently, in few theoretical and experimental papers [2-7], we found the reports on the modifications in the thermal properties of bulk semiconductors when embedded with a metallic nanoparticle.

When dc electric field is applied on a semiconductor plasma medium doped with nanoparticle, the conduction electrons of the semiconductor medium acquire drift velocity, whereas the electrons of the cloud of nanoparticle get displaced from their equilibrium position. These two velocities get imposed to give an average velocity of electron. Hydrodynamically, it is the vector sum of these two velocities. On the other side, due to displacement in electron cloud of nanoparticle, a restoring force is also resulted. Due to this restoring force, electrons of the cloud start collective oscillations identical to plasma oscillations of free electrons of semiconductor medium. This effect is equivalent to making the spring constant stiffer, and therefore, leads to a higher plasma frequency. These two phenomena, resulted due to the presence of nanoparticle in semiconductor plasma medium, may become the cause of modifications in the dispersion and amplification characteristics of wave propagation.

Presently, using these unique properties of nanostructures, the present authors [8-10] have investigated the wave phenomena in semiconductor plasma doped with a nanoparticle. Ghosh and Dubey [8] have investigated the convective instability of longitudinal electrokinetic wave in semiconductor plasma consisting of a nanoparticle. They found that, for collisionless or nearly collisionless plasma, nanoparticle modifies only the dispersion characteristics of the modes, whereas in collision dominated region, gain characteristics of both the modes get modified. Ghosh and Dubey [9] studied the temporal instability of longitudinal electrokinetic wave in a nanoparticle impinged semiconductor plasma medium and found that two new modes are induced within the medium due to the inclusion of a nanoparticle. They have also explored the carrier drift and wave number dependency of spectral characteristics of all four propagating modes and reported modification due to impinged nanoparticle. The effect of a nanoparticle within the piezoelectric semiconductor on electro-acoustic wave spectrum is studied in detail and recently reported by Dubey and Ghosh [10]. They found significant modification in gain per radian of acoustic wave in terms of velocity ratio, wave frequency and free electron density of semiconductor medium. In presence of nanoparticle the resonant frequency is found to be higher than that found in absence of it, but attenuation to amplification cross-over point is found unaffected. Motivated by these, in present study, we have investigated the electron-electron two stream instability in two valley semiconductor (n-GaAs) and hoped that it may attribute to the fundamental knowledge being the very first attempt in this direction.

It is well known fact that the observation of electron-electron two stream instability in a solid depends on the existence of a double-humped electron distribution which could be approximated by a two-stream model. In semiconductors, with more than one conduction band minimum, like GaAs, it is possible to establish a double humped electron distribution [11]. Instabilities in n-GaAs using the peculiarities of its band structure have been proposed by a large number of workers [e.g., 12]. In the strong signal limit, it leads to Gunn effect [13] whereas in the small-signal limit, it leads to amplification of space charge or electrostatic waves. Such an amplification can be described in terms of two-stream instabilities [e.g., 12] and with proper feedback can lead to oscillations [14].

In present analytical investigation, we assumed that, the transferred electron mechanism creates two electron streams. Accordingly, at an applied static electric field greater than the threshold, the central and satellite valleys in the conduction band would be populated by significant number of electrons and these electrons would acquire different drift velocities because of different effective masses. Thereby, we get two-stream system of electrons in the medium.

2. Theoretical formulation

Assuming a one dimensional hydrodynamic model of a semiconductor plasma and small-signal approximation for a periodic perturbation proportional to $\exp[i(\omega t - kz)]$, we obtain the conduction current density for free electrons in two valley semiconductor like n-GaAs subjected to a static electric field E_{dc} in the direction of wave propagation as

$$J_z = -i\varepsilon\omega\left[\frac{\omega_{p1}^2}{\left[(\omega - k\vartheta_{01})(\omega - k\vartheta_{01} - i\nu_1) - \vartheta_{\theta1}{}^2 k^2\right]} + \frac{\omega_{p2}^2}{\left[(\omega - k\vartheta_{02})(\omega - k\vartheta_{02} - i\nu_2) - \vartheta_{\theta2}{}^2 k^2\right]}\right]E_{1z} \tag{1}$$

Subscripts 1 and 2 stand for central and satellite valleys respectively. Here, $\omega_{pj} = \sqrt{e^2 n_j / m_j \varepsilon}$ with $\varepsilon = \varepsilon_0 \varepsilon_l$. m_j , n_j, ν_j, $\vartheta_{\theta j}$ and ϑ_{0j} represent effective mass, number densities, collision frequencies, thermal velocities and drift velocities of electrons of jth-valley respectively.

For current density J_{np} of the electron cloud of the impinged nanoparticle, we will use equation (3) of Ref. [8] which is as follows

$$J_{np} = -i\varepsilon\omega\frac{\omega_{pn}^2}{\left(\omega^2 - \dfrac{\omega_{pn}^2}{3}\right)}E_{1z} \tag{2}$$

We obtain the resultant current density $J = J_z + J_{np}$ as

$$J = -i\varepsilon\omega\left[\frac{\omega_{p1}^2}{\left[(\omega - k\vartheta_{01})(\omega - k\vartheta_{01} - i\nu_1) - \vartheta_{\theta1}{}^2 k^2\right]} + \frac{\omega_{p2}^2}{\left[(\omega - k\vartheta_{02})(\omega - k\vartheta_{02} - i\nu_2) - \vartheta_{\theta2}{}^2 k^2\right]} + \frac{\omega_{pn}^2}{\left(\omega^2 - \dfrac{\omega_{pn}^2}{3}\right)}\right]E_{1z} \tag{3}$$

Using equations (3) in wave equation for present field geometry $(\omega\varepsilon_0\varepsilon_L E_{1z} = iJ)$, we may derive the dispersion relation as-

$$\varepsilon(\omega,k)=\left[1-\frac{\omega_{p1}^2}{\left[(\omega-k\vartheta_{01})(\omega-k\vartheta_{01}-i\nu_1)-\vartheta_{\theta1}{}^2k^2\right]}+\frac{\omega_{p2}^2}{\left[(\omega-k\vartheta_{02})(\omega-k\vartheta_{02}-i\nu_2)-\vartheta_{\theta2}{}^2k^2\right]}+\frac{\omega_{pn}^2}{\left(\omega^2-\frac{\omega_{pn}^2}{3}\right)}\right]=0 \quad (4)$$

Above relation (equation (4)) represents modified dispersion relation in presence of a nanoparticle within the two valley semiconductor plasma medium. In absence of nanoparticle and by neglecting the effect of diffusion, this dispersion relation reduces to equation (1) of Guha and Sen [16]. The modification induced due to plasma frequency of electron cloud present within the nanoparticle is evident through fourth term in square bracket.

Now, to examine the temporal behaviour of electrostatic modes in the medium under study, we rewrite [equation (4)] in form of a polynomial in terms of complex angular frequency ω as

$$A_6\omega^6+A_5\omega^5+A_4\omega^4+A_3\omega^3+A_2\omega^2+A_1\omega+A_0=0 \quad (5)$$

where

$$A_0=\left(-\tfrac{4}{3}\omega_{pn}^2\right)\left(k^2\vartheta_{01}^2+ik\vartheta_{01}\nu_1-\vartheta_{\theta1}^2k^2\right)\left(k^2\vartheta_{02}^2+ik\vartheta_{02}\nu_2-\vartheta_{\theta2}^2k^2\right)+$$
$$+\left(\frac{\omega_{pn}^2}{3}\right)\left[\omega_{p1}^2\left(k^2\vartheta_{02}^2+ik\vartheta_{02}\nu_2-\vartheta_{\theta2}^2k^2\right)+\omega_{p2}^2\left(k^2\vartheta_{01}^2+ik\vartheta_{01}\nu_1-\vartheta_{\theta1}^2k^2\right)\right]$$

$$A_1=\left(\tfrac{4}{3}\omega_{pn}^2\right)\left\{\left(2k\vartheta_{01}+i\nu_1\right)\left(k^2\vartheta_{02}^2+ik\vartheta_{02}\nu_2-\vartheta_{\theta2}^2k^2\right)+\left(2k\vartheta_{02}+i\nu_2\right)\left(k^2\vartheta_{01}^2+ik\vartheta_{01}\nu_1-\vartheta_{\theta1}^2k^2\right)\right\}-$$
$$-\frac{\omega_{pn}^2}{3}\left\{\omega_{p1}^2\left(2k\vartheta_{02}+i\nu_2\right)+\omega_{p2}^2\left(2k\vartheta_{01}+i\nu_1\right)\right\}$$

$$A_2=\left(k^2\vartheta_{01}^2+ik\vartheta_{01}\nu_1-\vartheta_{\theta1}^2k^2-\omega_{p1}^2\right)\left(k^2\vartheta_{02}^2+ik\vartheta_{02}\nu_2-\vartheta_{\theta2}^2k^2-\frac{\omega_{pn}^2}{3}\right)-$$
$$-\left(\tfrac{4}{3}\omega_{pn}^2\right)\left\{\left(2k\vartheta_{01}+i\nu_1\right)\left(2k\vartheta_{02}+i\nu_2\right)+\left(k^2\vartheta_{02}^2+ik\vartheta_{02}\nu_2-\vartheta_{\theta2}^2k^2\right)\right\}-$$
$$-\left\{\omega_{p2}^2\left(k^2\vartheta_{01}^2+ik\vartheta_{01}\nu_1-\vartheta_{\theta1}^2k^2-\frac{\omega_{pn}^2}{3}\right)+\omega_{pn}^2\left(k^2\vartheta_{01}^2+ik\vartheta_{01}\nu_1-\vartheta_{\theta1}^2k^2\right)\right\}$$

$$A_3=\left(\tfrac{4}{3}\omega_{pn}^2\right)\left\{\left(2k\vartheta_{01}+i\nu_1\right)+\left(2k\vartheta_{02}+i\nu_2\right)\right\}-\left(2k\vartheta_{01}+i\nu_1\right)\left(k^2\vartheta_{02}^2+ik\vartheta_{02}\nu_2-\vartheta_{\theta2}^2k^2-\omega_{p2}^2\right)-$$
$$-\left(2k\vartheta_{02}+i\nu_2\right)\left(k^2\vartheta_{01}^2+ik\vartheta_{01}\nu_1-\vartheta_{\theta1}^2k^2-\omega_{p1}^2\right)$$

$$A_4=\left(2k\vartheta_{01}+i\nu_1\right)\left(2k\vartheta_{02}+i\nu_2\right)+\left(k^2\vartheta_{01}^2+ik\vartheta_{01}\nu_1-\vartheta_{\theta1}^2k^2\right)+\left(k^2\vartheta_{02}^2+ik\vartheta_{02}\nu_2-\vartheta_{\theta2}^2k^2\right)-$$
$$-\left(\omega_{p1}^2+\omega_{p2}^2+\tfrac{4}{3}\omega_{pn}^2\right)$$

$$A_5=-\left(2k\vartheta_{01}+i\nu_1\right)-\left(2k\vartheta_{02}+i\nu_2\right)\text{and }A_6=1$$

This polynomial infers that, for a particular k, there exist six possible electrostatic modes of propagation in a two valley semiconductor plasma embedded with a single nanoparticle.

To appreciate the effect of embedded nanoparticle in the plasma medium, we write the dispersion relation (equation (4)) in polynomial form in terms of angular frequency ω in absence of nanoparticle $(\omega_{pn}\to 0)$ and obtain

$$B_4\omega^4+B_3\omega^3+B_2\omega^2+B_1\omega+B_0=0 \quad (6)$$

where,

$$B_0=\left(k^2\vartheta_{01}^2+ik\vartheta_{01}\nu_1-\vartheta_{\theta1}^2k^2-\omega_{p1}^2\right)\left(k^2\vartheta_{02}^2+ik\vartheta_{02}\nu_2-\vartheta_{\theta2}^2k^2\right)-\omega_{p2}^2\left(k^2\vartheta_{01}^2+ik\vartheta_{01}\nu_1-\vartheta_{\theta1}^2k^2\right)$$
$$B_1=-\left(2k\vartheta_{01}+i\nu_1\right)\left(k^2\vartheta_{02}^2+ik\vartheta_{02}\nu_2-\vartheta_{\theta2}^2k^2-\omega_{p2}^2\right)-\left(2k\vartheta_{02}+i\nu_2\right)\left(k^2\vartheta_{01}^2+ik\vartheta_{01}\nu_1-\vartheta_{\theta1}^2k^2-\omega_{p1}^2\right)$$

$$B_2 = (2k\vartheta_{01} + i\nu_1)(2k\vartheta_{02} + i\nu_2) + (k^2\vartheta_{01}^2 + ik\vartheta_{01}\nu_1 - \vartheta_{\theta 1}^2 k^2) + (k^2\vartheta_{02}^2 + ik\vartheta_{02}\nu_2 - \vartheta_{\theta 2}^2 k^2) - (\omega_{p1}^2 + \omega_{p2}^2)$$

$$B_3 = -(2k\vartheta_{01} + i\nu_1) - (2k\vartheta_{02} + i\nu_2) \text{ and } B_4 = 1$$

One may infer from equation (6) that, in absence of nanoparticle, only four modes of propagation of electrostatic wave is possible. Thus, one may conclusively interpret from equations (5) and (6) that the presence of a single nanoparticle induces two novel modes of propagation through plasma frequency of electron cloud of nanoparticle impinged in the two valley semiconductor.

3. Results and discussions

We have numerically solved the dispersion relation (equation (5)) for real positive $k \cong 10^6 m^{-1}$ in n-GaAs. The physical parameters of n-GaAs are obtained from references [17, 18]. We found that the dispersion and amplification characteristics of all the four pre-existing electrostatic modes in two valley semiconductor (n-GaAs) plasma are unaffected by the presence of nanoparticle within the medium; hence of no interest. The two newly induced electrostatic modes are found to be propagating in opposite directions to each other i.e., one is forward and other is backward propagating mode. Both transmit through the medium with equal phase velocity $\approx 2ms^{-1}$ which is independent of the magnitude of applied electric field E_{dc}. Due to slow phase velocities, they may be used as diagnostic tool by material scientists working with many valley semiconductors. While studying amplification characteristic of these two novel modes, we found that their gain coefficient is always zero for the range of electric field $E_{dc}(10^5 - 10^6 Vm^{-1})$ considered here for the study. Hence, these two modes may be termed as stable propagating electrostatic modes or evanescent modes of propagation in the range of parameter under study.

References

[1] Mustafa F and Hashim A M 2010 *Prog. Electromagn. Res.* **104** 403
[2] Kim W, Zide J, Gossard A, Klenov D, Stemmer S, Shakouri A and Majumdar A 2006 *Phys. Rev. Lett.* **96** 045901
[3] Zide J M, Klenov D O, Stemmer S, and Gossard A C 2005 *Appl. Phys. Lett.* **87** 112102
[4] Kawasaki J K, Timm R, Delaney K T, Lundgren E, Mikkelsen A and Palmstrom C J 2011 *Phys. Rev. Lett.* **107** 036806
[5] Schultz B D and Palmstrom C J 2006 *Phys. Rev. B* **73** 241407
[6] Lung F and Marinescu D C 2011 J. *Phys.: Condens. Matter* **23** 365802
[7] Choudhary K K 2013 *J. Nanopart. Res.* **15** 1362
[8] Ghosh S and DubeyPriya 2014 *Int. J. Phys. Math. Sci.* **4** 20
[9] Ghosh S and DubeyPriya 2014 *Int. J. Phys. Math. Sci.* **4** 44
[10] Dubey P and Ghosh S 2016 *Acta Acust. United AC* **102** 436
[11] Kakihana S and Buchmiller L 1967 *J. Appl. Phys.* **38** 5376
[12] Perlman B S 1973 *RCA Rev.* **34** 457
[13] Gunn J B 1963 *Solid State Commun.* **1** 86
[14] Steele M C and Vural B 1969 *Wave Interactions in Solid State Plasmas* (New York: Mc-Graw Hill)
[15] Jain S and Parashar J 2011 *Int. J. Nanosci. Nanotechnol.* **2** 147
[16] Guha S and Sen P K 1977 *Phys. Stat. Sol. (a)* **43** 89
[17] Ehrenreich H 1960 *Phys. Rev.* **120** 1951
[18] Rowland M C 1969 *The preparation and properties of Gallium Arsenide, in Gallium Arsenide lasers* (New York: Wiley-Interscience Publishers)

IOP Conf. Series: Journal of Physics: Conf. Series **836** (2017) 012024 doi:10.1088/1742-6596/836/1/012024

Theoretical study to investigate the impact of plasma parameters on the catalyst nanoparticle growth

R Gupta[a], S C Sharma and N Gupta

Department of Applied Physics, Delhi Technological University, Delhi-110042, India

Email: [a)]ravigpt0709@gmail.com

Abstract. The plasma kinetics based model is adopted to elucidate the effect of plasma parameters on the nucleation and growth mechanism of catalyst nanoparticle. The present model considers the plasma processing of thin catalyst film, power equalization at the film surface, flux and kinetics of plasma species (electrons, ions, and neutrals). In our investigation, it is found that catalyst nanoparticle diameter decreases with increase in ion number density in plasma. Moreover, it is also found that catalyst film thickness significantly affect the catalyst nanoparticle size i.e., catalyst nanoparticle diameter increases with catalyst film thickness. In addition, it is observed that the substrate temperature increases during the plasma processing and finally achieve saturation. Our theoretical results are in good agreement with the experimental results.

1. Introduction

The catalyst nanoparticles are the fundamental site for the nucleation of nanostructures and significantly influence the structure, shape and size of the nanostructure [1]. These catalyst nanoparticles can be synthesized using the plasma technique, i.e., plasma sputtering and etching of the thin film[2]or by thermal annealing of the thin film deposited on substrate surface. During the plasmatreatment of thin film, plasma parameters extensively affect the catalyst nanoparticle span (diameter). Chang et al. [3] experimentally observed that the catalyst nanoparticle size obtaining from thin catalyst film decreases with increases in plasma power.

So far only thermodynamics based theoretical models are available to depict the size of the catalyst nanoparticle obtained from the thin catalyst film. In the present paper, in section 2, we have devised the plasma based model to elucidate the growth mechanism of catalyst nanoparticle resulting from the etching and sputtering of the thin film in the reactive plasma, i.e. plasma treatment of the thin catalyst film. The results of the model are discussed in the section 3 and finally conclusion is given in section 4.

2. Model

The model acknowledges the kinetics and energy fluxes of various plasma species (electrons, positively charged ions and neutrals), power equalization at the catalyst film-substrate surface, and heat radiations. The present model accounts the reactive plasma consisting electrons, ions and neutrals of hydrogen. It is assumed that, the spherical catalyst nanoparticles are formed as a result of sputtering and etching of the catalyst film during the plasma treatment.

IOP Conf. Series: Journal of Physics: Conf. Series **836** (2017) 012024 doi:10.1088/1742-6596/836/1/012024

2.1. Kinetic equation of electrons, positively charged ions, and neutrals in the reactive plasma

$$\dot{n}_e = \beta_1 n_1 - \alpha_1 n_e n_{i1} - \gamma_e n_a j_{ecf} \tag{1}$$

$$\dot{n}_{i1} = \beta_1 n_1 - \alpha_1 n_e n_{i1} - n_a j_{i1cf} - J_{ad} + J_{desp} \tag{2}$$

$$\dot{n}_1 = \alpha_2 n_e n_{i1} - \beta_1 n_1 + n_a \left(1-\gamma_{i1}\right) j_{i1} - n_a \gamma_1 j_1 \tag{3}$$

where subscript 1 denotes the hydrogen, β_1 is the coefficient of ionization of the constituent neutral atoms due to external field, n_2 is the number density of neutral atom, n_e is the electron number density, n_{i1} is the ion number density, $\alpha_j(T_e)=\alpha_{j0}\left(300/T_e\right)^k cm^3/\text{sec}$ is the coefficient of recombination of electrons and positively charged ions, $k=-1.2$ is a constant, η_a is the number of catalyst particle per unit area γ_{i1} is the ion sticking coefficient, γ_2 is the sticking coefficient of neutral atoms J_{ad} is the adsorption flux onto the catalyst substrate surface, J_{desp} is the desorption flux from the catalyst-substrate surface, j_{ecf} , $j_{i\alpha cf}$ and $j_{\alpha f}$ are the collection currents on the film surface due to electrons, ions and neutrals, respectively [4].

Equations (1), (2), and (3) indicate the balance of electrons, ions, and neutral atoms number density in the reactive plasma. The first of equation (1)denotes the gain in electron density per unit time due to ionization of neutral atoms; second and third terms refers to the decay rate due to electron-ion recombination and electron collection current at the film surface, respectively.

The first term of equation (2) describes the gain in ion number density due to ionization of neutral atoms. The second and the third terms represent the decay rate of positively charged ion density due to electron-ion recombination and ion collection current at the surface of the film, respectively. The fourth and fifth terms indicate adsorption and desorption of ions to/from the film surface, respectively

The First and third terms of equation (3) show the growth rate of neutral density due to electron-ion recombination and neutralization of ions collected at the film surface, respectively. The second and fourth terms represents the decay rate of neutral density due to ionization and accumulation of neutrals on the film surface, respectively.

2.2. The power equalization at the catalyst film surface

$$\dot{H}_{input} = \dot{H}_g + \dot{H}_{output} \tag{4}$$

where $\dot{H}_{in}$ is the input energy per unit time towards the catalyst film, $\dot{H}_g$ is the energy gained per unit time during the plasma treatment and $\dot{H}_{out}$ is the output energy or energy loss per unit time [5]. Various plasma processes contribute to these energies.

$$\dot{H}_{input} = \left[j_{ecf}\xi_e^{cf} + j_{i1cf}\xi_{i1}^{cf} + j_{1cf}\xi_1^{cf} \right] + A_{cf}\sigma\left(\left(\vartheta_{srd}\right)^4 \in_{srd} - \left(\vartheta_S\right)^4 \in_{cf}\right) + \frac{3}{2}k_B\vartheta_S(1-\kappa_{i1})j_{i1cf} \tag{5}$$

$$\dot{H}_g = \frac{d}{dt}\left(m_{cf}S_f\vartheta_S\right) = A_{cf}S_{cf}\rho_{cf}\left[\tau_{cf}\frac{d\vartheta_s}{dt} + \eta_a\vartheta_s\frac{d}{dt}\left(\frac{1}{6}\pi d^3\right)\right] \tag{6}$$

$$\dot{H}_{output} = \Lambda_{ef}(1-\kappa_e)\left(\xi_{ecf}^S - \frac{3}{2}k_B\vartheta_S\right)j_{ecf} + \sigma A_{cf}\left(\vartheta_s^4 \in_{cf} - \vartheta_{srn}^4 \in_{srn}\right) + \left[\upsilon_0\sigma\varphi_1 E_B + \varphi_{i1}\left[y_d(E_i)\right](1-\theta)\right]A_{cf}, \tag{7}$$

where A_{cf} is the catalyst film surface area, τ_{cf} is the catalyst film thickness, d is the catalyst nanoparticle diameter (it is assumed that catalyst nanoparticle of spherical morphology are formed as result of plasma treatment of thin catalyst film), ϑ_s is the substrate temperature (in the present model temperature of substrate and thin catalyst film is assumed to be same), ϑ_{cf} is the mass of catalyst film,

κ_{i1} and κ_1 are the ion sticking coefficient and neutral atom sticking coefficient, respectively, s_f is the specific heat of the film, $\in_{srn}$ is the emissivity of the surrounding, $\in_{cf}$ is the emissivity of the film, ξ_e^{cf}, ξ_{i1}^{cf}, and ξ_1^{cf} are the mean energy of electrons, ions and neutrals, respectively collected by film [6], Λ_{ef} is the energy loss during collision with film, υ_0 is the no. of sites available, σ_{ads} is the cross-section area, φ_a and φ_{ia} are neutral and ion fluxes, respectively, E_B is the binding energy of the material, σ is the Stefan's constant and θ is the total surface coverage.

The first term in equation (5) represents the power given to the film surface by the electrons, ions and neutral atoms, the second term denotes power towards the film surface due to heat radiations from surrounding, and the third term is power collected by film surface due to formation of neutrals at the film surface. The equation (6) describes the power gained during the plasma treatment of thin catalyst film. The input plasma power causes the thin catalyst film into nanoparticles and also causes the increase in substrate temperature. The equation (7) describes the power loss due to sticking and elastic collision of electrons with catalyst film surface, power output from the surface to the surrounding in the form of radiations, power loss in etching and sputtering of thin film in the presence of plasma, respectively.

3. Results and discussion

The model acknowledges the different physical process responsible for the formation of catalyst nanoparticle from thin catalyst film in the presence of plasma and hence the model relates the diameter of the catalyst nanoparticle formed with the plasma parameters such that number density of the electrons, ions, and neutrals, plasma power, and thin film thickness.

The figure1 shows the time evolution of catalyst nanoparticle diameter for various thin film thickness. From figure 1, it can be seen that diameter of the catalyst nanoparticle increases with thickness of the thin film.

The figure 2 shows variation of catalyst nanoparticle diameter with plasma power. As plasma power increases the catalyst nanoparticle diameter decreases. This is attributed to the fact that with increase in plasma power highly energetic plasma species (electrons, ions, andneutrals) are created in the reactive plasma that leads to higher etching and sputtering of the catalyst thin film resulting in the lesser catalyst nanoparticle diameter.

The figure 3 shows the variation of substrate temperature during the plasma treatment of the thin catalyst film. The substrate temperature increases during the plasma treatment and achieves the saturation after sometime. The highly energetic plasma species transfer their energy to the film surface during the bombardment and collisional processes with the film surface and thus increase the film surface/substrate temperature. These theoretical findings are in good agreement of the experimental observations of Poa et al. [7], Chang et al. [3], and Srivastava et al. [8].

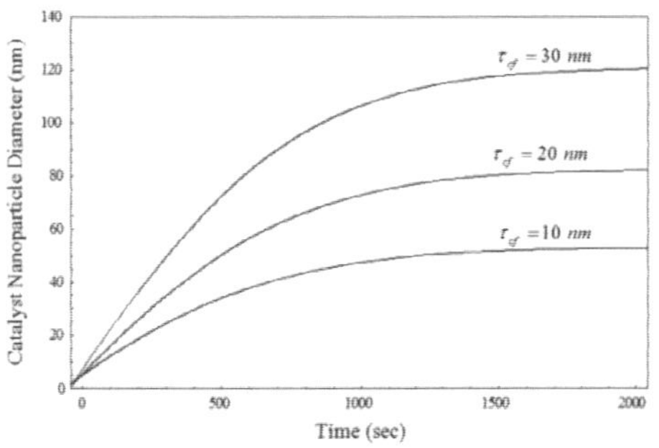

Figure 1. The time evolution of catalyst nano-particle diameter for various catalyst film thicknesses.

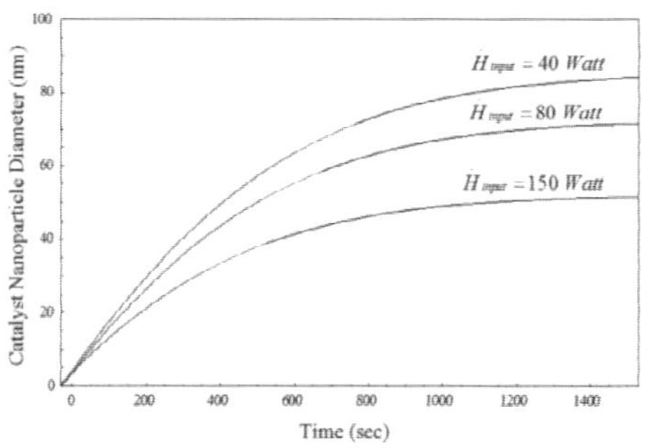

Figure 2. The time evolution of catalyst nanoparticle for different input plasmapower.

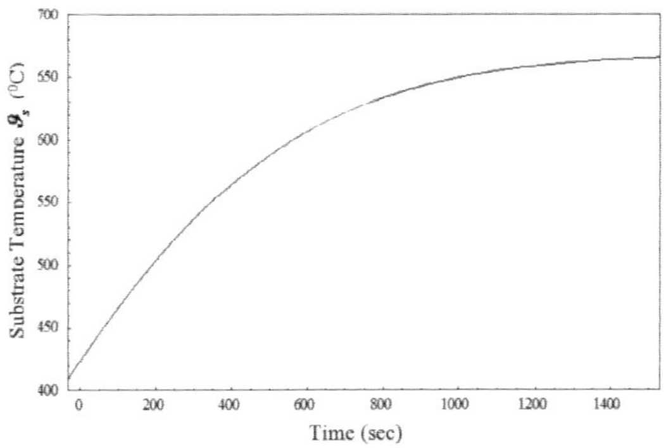

Figure 3. The time variation of substrate temperature during the plasma processing (plasma treatment) of the thin catalyst film in the plasma containing electrons and ions and neutrals of hydrogen.

4. Conclusions

An analytical model to relate the catalyst nanoparticle diameter with plasma parameters (number density of plasma species, their temperatures, and plasma power) and catalyst film thickness during the plasma treatment catalyst film has been developed. It is found that diameter of spherical catalyst nanoparticles increases with catalyst thin film thickness and decreases with plasma power. Moreover, we found that substrate temperature also increases during the plasma treatment of thin film and after some time achieves the saturation. The present work could foresee the catalytic synthesis of the carbon nanostructure in near future.

References

[1] Yang R T and Chen J P 1989 *J. Catal* **115** 52
[2] Yudasaka M, Kikuchi R, Matsui T, Ohki Y, Yoshimura S and Ota E 1995 *Appl. Phys. Lett.* **67** 2477
[3] Chang S C, Lin T C and Pai C Y 2007 *Microelectron. J.* **38** 657
[4] Shrama S C and Gupta N 2015 *Phys. Plasmas* **22** 123517
[5] Kersten H, Deustch H, Steffen H, Kroesen G M W and Hippler R 2001 *Vacuum* **63** 385
[6] Sodha M S and Guha S 1971 *Physics of Colloidal Plasmas* (New York: John Wiley & Sons) p 219
[7] Poa C H P, Henley S J, Chen G Y, Adikaari A A D T, Giusca C E and Silva S R P 2005 *J. Appl. Phys.* **97** 114308
[8] Srivastava S K, Shukla A K, Vankar V D and Kumar V 2005 *Thin Solid Films* **492** 124

IOP Conf. Series: Journal of Physics: Conf. Series **836** (2017) 012025 doi:10.1088/1742-6596/836/1/012025

Effect of temperature anisotropy on Alfven waves in multi-component magnetospheric plasma

V P Ahirwar[a] and G Ahirwar

School of Studies in Physics, Vikram University, UJJAIN (M.P.)-456010, INDIA

E-mail: [a)]ahirwarvishnu02@gmail.com

Abstract. The main objective of the present investigation is to examine the effect of temperature anisotropy on Alfven waves in multi-component plasma by using the method of kinetic approach and using different plasma parameters in auroral acceleration region. In this paper, we are considering the multi- component (Ions and electrons) plasma for the calculations of dispersion relation, growth/damping rate and growth length on Alfven waves. Dispersion relation, growth/ damping rate and growth length of Alfven waves are measured in multi-component plasma consisting of mixture of Hydrogen (H^+), Helium (He^+) and Oxygen (O^+) ions. An Alfven waves is an important electromagnetic wave that transports electromagnetic energy in many space and astrophysical regions. These waves parallel propagate along the magnetic field with parallel wave vector in the x-z plane. Curves show that the damping rate increases of waves with increases temperature anisotropy.

1. Introduction

An Alfven waves is low frequency electromagnetic waves with wave frequency less than ion or electron cyclotron resonance frequency in magnetospheric plasma. In the Alfven wave the background magnetic fieldtension provides the restoring force, whereas the ion mass provides the inertia. Theexistence of such waves was theoretically predicted by [1]. This theoretical prediction was of great importancebecause it opened new possibilities to transport energy in a medium. Theexperimental confirmation of this extremely fruitful idea appeared several yearslater, when experiments in the Alfven laboratory in Stockholm proved theexistence of these waves in magnetosphere plasma. An Alfven waves play an important role in energy transport in particle acceleration, and heating in the magnetosphere [2]. Alfven waves have been invoked as the most promising wave mechanism to explain the heating of Sun's outer atmosphere, or corona, to millions of degrees and the acceleration of the solar wind to hundreds of kilometres per second [3]. Spacecraft measurements, measurement from the FAST satellite [4] and the POLAR spacecraft [5] have been speculated the existence of small-scale Alfven waves that carry a large net pointing flux along magnetic field lines towards the earth.The phenomena associated with the Alfven waves in auroral acceleration region can also be examined by in Situ measurements [6]. Plasma kinetic theory also provides a method of investigation to find the influence of collision among the plasma particle and the calculation of the transport properties of fully ionized plasma. The most fundamental kinetic description of a collision –less plasma system is to employ the Vlasov equation to obtain particle distribution functions for all particle species and compute the electric field and magnetic fields from the particle density and plasma current by the Maxwell's equation.

1. Dispersion relation

We consider plasmas with external magnetic field (B_0), since the Alfven waves propagate parallel to the magnetic field in z-direction with parallel wave vector in the x-z plane. Here, linear dispersion relation of Alfven waves in multi-component plasma using auroral acceleration region are evaluated.

$$\omega^2 = k_{II}^2 v_A^2 k_\perp^2 [1 + (A_{H^+} + A_{He^+} + A_{O^+})] \times \left(\frac{v_{T\perp H^+}^2}{\Omega_{H^+}^2} + \frac{v_{T\perp He^+}^2}{\Omega_{He^+}^2} + \frac{v_{T\perp O^+}^2}{\Omega_{O^+}^2}\right) \quad (1)$$

3. Growth/damping rate

Assuming $\omega \to \omega_r + i\gamma$, with $\gamma << \omega$ we obtain an expression for the collision less damping rate of the Alfven wave from [7-8]and further we extend the effect of temperature anisotropy on Alfven waves in multi- component plasma are evaluated.

$$\gamma = k_{II} v_A^2 k_\perp^2 \frac{1}{v_{TIIe}^3} \sqrt{\frac{\pi}{8}} (A_{H^+} + A_{He^+} + A_{O^+}) \times \left(\frac{v_{T\perp H^+}^2}{\Omega_{H^+}^2} + \frac{v_{T\perp He^+}^2}{\Omega_{He^+}^2} + \frac{v_{T\perp O^+}^2}{\Omega_{O^+}^2}\right) \quad (2)$$

4. Growth length

The growth length γ_L of the Alfven waves (AW's) is derived from [8-9] as

$$\gamma_L = \sqrt{\frac{8}{\pi}} \times \frac{v_{TIIe}^3}{K_\perp K_{II} v_A} \times \frac{[1 + (A_{H^+} + A_{He^+} + A_{O^+})]^{\frac{1}{2}}}{(A_{H^+} + A_{He^+} + A_{O^+}) \times \left(\frac{v_{T\perp H^+}^2}{\Omega_{H^+}^2} + \frac{v_{T\perp He^+}^2}{\Omega_{He^+}^2} + \frac{v_{T\perp O^+}^2}{\Omega_{O^+}^2}\right)^{\frac{3}{2}}} \quad (3)$$

where,

$$A_{H^+} = \frac{T_e}{T_{H^+}}, \; A_{He^+} = \frac{T_e}{T_{He^+}}, \; A_{O^+} = \frac{T_e}{T_{O^+}}$$

5. Results and discussions

In the present analysis, the expression for numerical evaluation of the linearly dispersion relation, growth/damping rate and growth length with electron and (hydrogen, helium, oxygen) ion temperature ratio for homogeneous plasma, the following plasma parameter are used which are relevant for auroral acceleration region [9-10].

$$B_0 = 4300 \times 10^9 T, K_{II} = 10^{-10} cm^{-1}, K_\perp = 10^{-6} cm^{-1}, v_{A,multi} = 4002 \times 10^4, \Omega_{He^+} = 103 \sec^{-1},$$
$$\Omega_{He^+} = 103 \sec^{-1}, \Omega_{H^+} = 412 \sec^{-1}, \Omega_{O^+} = 26 \sec^{-1}, v_{TIIe} = 1.876 \times 10^4 cms^{-1},$$
$$v_{T\perp i} = 4.378 \times 10^4 cms^{-1},$$

Figure 1 shows the variation of wave frequency ω (s^{-1}) with K_{II} (cm^{-1}) for different values of hydrogen, helium, oxygen temperature ratio (A_H^+, A_{He}^+, A_O^+). Here, we notice that the wave frequency increases with K_{II} and increases with the increases of temperature anisotropy. Figure 2 shows the variation of growth/damping rate (γ) versus the parallel wave vector K_{II} for different values of hydrogen, helium, oxygen temperature ratio (A_H^+, A_{He}^+, A_O^+). It is seen that the damping rate increases with the parallel wave vector. Thus, these curves show that the damping rate increases of waves with increases temperature anisotropy.

IOP Conf. Series: Journal of Physics: Conf. Series **836** (2017) 012025 doi:10.1088/1742-6596/836/1/012025

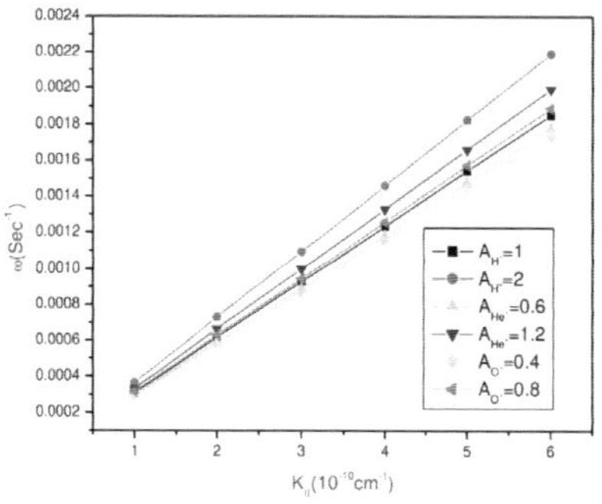

Figure 1. Wave frequency ω versus parallel waves vector K_{II} for different temperature anisotropy.

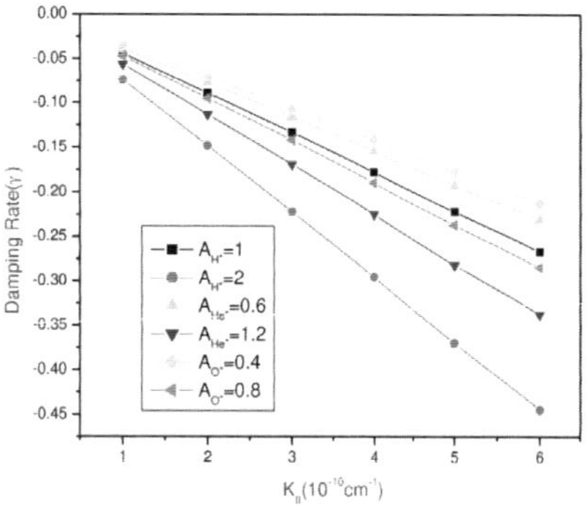

Figure 2. Damping rate (γ) versus the parallel wave vector (K_{II}) for different temperature anisotropy.

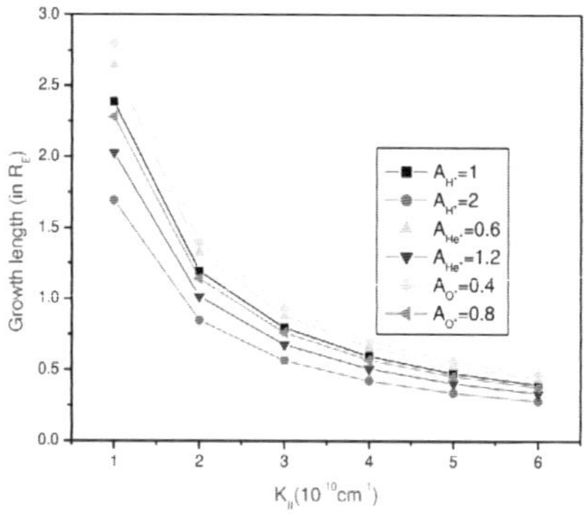

Figure 3. Growth length (γ_L) versus the parallel wave vector (K_{II}) for different temperature anisotropy.

Figure 3 shows the variation of growth length (γ_l) versus the parallel wave vector K_{II} for different values of hydrogen, helium, oxygen temperature ratio (A_{H^+}, A_{He^+}, A_{O^+}). It is found that the growth length exponential decreases with increases the parallel wave vector but decreases with the increases of temperature anisotropy. Here, it is notice that the growth length becomes greater than $1R_E$ on the lower value of K_{II} (cm^{-1}).

6. Conclusions

In this paper, the Alfven waves have been studied in auroral acceleration region using plasma kinetic theory. The wave dispersion relation, growth/damping rate and growth length are evaluated for Alfven waves in multi-component magnetospheric plasma. It is observed that the effect of temperature anisotropy on Alfven wave. The wave frequency increases with K_{II} and increases with the increases of the temperature anisotropy. It is found that the higher wave frequency at lower value of K_{II}. Finally, the wave frequency is much lower than the ion cyclotron frequency on the value of K_{II} (cm^{-1}). The damping rate increases with K_{II} and increases with the increases of the temperature anisotropy. The growth length of wave is exponential decreases with increasesK_{II} and decreases with the increases of temperature anisotropy. It is found that the growth length is a greater than $1R_E$ on the lower value of K_{II}. The findings may be applicable to explain the Alfven waves in space, astrophysical plasma and magnetospheric plasma.

Acknowledgment

Financial assistance by the Rajiv Gandhi National Fellowship, University Grants Commission, Delhi, India is thankfully acknowledged.

IOP Conf. Series: Journal of Physics: Conf. Series **836** (2017) 012025 doi:10.1088/1742-6596/836/1/012025

References

[1] Alfven H 1942 *Nature* **150** 405
[2] Pontieu B De, McIntosh S W and Carlsson M *2001 Science* **318** 1574
[3] Nekrasov A K 1971 *Nucl. Fusion* **10** 4
[4] Wygant J R, Keiling, A and Cattell C 2000 J. *Geophys. Res.* **05** 0148
[5] Rahbarnia K, Ullrich S, Stark A and Grulke O 2009 *J. Plasma Fusion Res.* **8** 0015
[6] Woods L C 1962 *J. Fluid Mech.* **13** 570
[7] Agarwal P, Varma P and Tiwari M S 2011 *Indian J. Pure App. Phys.* **49** 91
[8] Shukla N, Mishra R, Varma P andTiwari M S 2006 *Indian J. Pure App. Phys.* **44** 834
[9] Shrivastav J, Varma P and Baroniya A, 2001 *Indian J.Phys.***75** 117
[10] Agarwal P, Varma P and Tiwari M S 2011 *Indian J. Pure App. Phys.* **49** 91

IOP Conf. Series: Journal of Physics: Conf. Series **836** (2017) 012026 doi:10.1088/1742-6596/836/1/012026

The instability of electrostatic ion cyclotron waves in a multi-component plasma

Vibhooti khaira[a] and G Ahirwar[b]

School of Studies in Physics, Vikram University, Ujjain (M.P.) 456010, India

E-mail: [a)]khaira.vibhuti29@gmail.com, [b)]ganpat.physics@gmail.com

Abstract. The instability of electrostatic ion cyclotron wave in a plasma consisting of isotropic hydrogen ions (H^+), oxygen ions (both positively and negatively charged and denoted by O+ and O^-) and electron. ESIC waves with multi component plasma have been studied by kinetic approach at different plasma densities. The dispersion relation and growth rate of the electrostatic ion-cyclotron waves with multi-ion plasma has been investigated. The effect of different plasma densities on ESIC waves in multi-ions is to enhance the growth rate of ESIC waves. The results are interpreted for the space plasma parameters appropriate to the auroral acceleration region of earth's magneto-plasma.

1. Introduction

The electrostatic ion cyclotron (ESIC) instability is a low-frequency field-aligned current-driven instability that has one of the lowest threshold drift velocities among current-driven instabilities Drummond and Rosenbluth [1]. In a plasma containing electrons and positive ions, the fundamental ESIC wave has frequency of the order of the ion cyclotron frequency Ωi, propagates nearly perpendicular to the magnetic field B, and has a small but finite wave number along B so that it can be destabilized by electrons drifting along B. Electrostatic ion cyclotron waves have been extensively investigated in connection with both space and laboratory plasma. The excitation mechanisms for these waves are the instabilities caused by electron currents [2], ion beams [3] and loss cone distributions of energetic ions [4]. The EIC waves are characterized by frequencies close to the harmonics of ion cyclotron frequency, perpendicular wave number$k_\perp \leq 1/\rho_{li}$, where *l* is the harmonic and $\rho_{li}$is the ion Larmor radius. Also these waves are characterized by parallel wave number $k_\parallel \ll k_\perp$. Thus the wave electric fields are almost perpendicular to the magnetic field. Another important characteristic of EIC waves is that these are typically very narrow banded and consequently turn out to be highly coherent over several oscillation periods [5].

In the present work kinetic theory extended to ESIC instability in a plasma consisting of isotropic hydrogen ions (H^+), oxygen ions (both positively and negatively charged and denoted by O^+ and O^-) and electrons propagating obliquely to static magnetic field. Using the kinetic approach in the presence of ESIC wave in multi ions, the dispersion relation, and growth rate are derived and studied for different distribution indices and different values of plasma densities in multi-ions like H^+, O^+ and O^-, and electron of auroral acceleration region. The advantage of present approach is to its suitability for dealing with auroral electrodynamics involving the heating, acceleration and energy exchange by wave - particle resonant interaction. To determine the basic trajectories we assume left - handed circularly polarized ESIC wave having angular frequency ω[5, 6].

IOP Conf. Series: Journal of Physics: Conf. Series **836** (2017) 012026 doi:10.1088/1742-6596/836/1/012026

2. Basic Equation

The cold plasma theory Khaira and Ahirwar [8] is very valuable for mode identification. But it cannot account for finite temperature effects. Therefore we now consider, in this section the fully magnetized, kinetic dispersion relation for electrostatic waves. The plasma is made up of hydrogen (H), oxygen ions (both positively and negatively charged and denoted by O+ and O-) and electrons. Thus generalizing the well-known dispersion relation for electrostatic waves, we have

$$1+\sum_\alpha \frac{1}{k^2\lambda_\alpha^2}[1+\frac{\omega}{\sqrt{2}k_\parallel v_{T\alpha}} e^{-Y_\alpha^2} \sum_{n=-\infty}^{\infty} Z(\frac{\omega+n\Omega\alpha}{\sqrt{2}k_\parallel v_{T\alpha}})\, I_n(Y_\alpha^2)]+ \frac{1}{k^2\lambda_H^2}[1+\frac{\omega-Uk_\parallel}{\sqrt{2}k_\parallel v_{TH}} e^{-Y_H^2} \sum_{n=-\infty}^{\infty} Z(\frac{\omega+n\Omega H-Uk_\parallel}{\sqrt{2}k_\parallel v_{TH}})\, I_n\,(Y_H^2)] = 0 \quad (1)$$

In this equation, the species summation α is over electrons, O+ and O- , the hydrogen contribution has been written out separately. K is wave vector, which has components $k_\parallel$ *and* $k_\perp$respectively parallel and perpendicular to the magnetic field. Z is the plasma dispersion function which arises from $dv_\parallel$integration and is defined as (Fried and Conte, 1961),

$$Z\,(\xi) = \frac{1}{\sqrt{\pi}}\int_{-\omega}^{\omega} \exp(-x^2)/x^{-\xi}\mathrm{dx} \quad (2)$$

While $I_n\,(Y_H^2)$is the modified Bessel function which arises from the $dv_\perp$integration with an argument

$$Y_\alpha^2 = \frac{k_\perp^2 T_\alpha}{\Omega_\alpha^2 M_\alpha} \qquad \alpha = H, O^+, O^-, or\ e. \quad (3)$$

In temperature T_α are related to the thermal velocity $v_{T\alpha}$by the relation

$$v_{T\alpha}^2 = \frac{T_\alpha}{M_\alpha} \quad \alpha = H, O^+, O^-, or\ e. \quad (4)$$

All the four constituents of our plasma have thus been modelled by isotropic Maxwellian distribution (the hydrogen ions alone beings modelled as an isotropic, drifting component). We assume that ω is complex and can be written as

$$\omega = \omega_r + i\gamma \quad (5)$$

We can separate out real and imaginary contributions of each species to the dispersion relation. Writing out each contribution in full and carrying out the involved algebra, we finally arrive at the expression for the growth / damping rate which, normalized against the O^+ ion gyro frequency, can be written as

$$\frac{\gamma}{\Omega_\alpha^+} = \frac{\sqrt{\pi}\{\gamma_e + \gamma_H + \gamma_{o^+} + \gamma_{o^-}\}}{\Gamma_e + \Gamma_H + \Gamma_{o^+} + \Gamma_{o^-}} \quad (6)$$

where the species indices indicate the contribution from each constituent of the plasma.

3. Results and discussion

We now consider the computation of the expression of the growth / damping rate. The density of the hydrogen ions is $n_H = 3.0$ with a temperature equivalent of $T_H \approx 3.0$ keV and a drift velocity of U = 10 V_{TO}+. The oxygen ions n_o^+ and n_o^-were assigned a value of 1.0 each, and a temperature equivalent of 0.03 eV. In addition the electrons were assigned a temperature equivalent of Te = 1.OkeV; the number densities being calculated from the charge neutrality condition.

$$n_e + n_{o^-} = n_o + + n_H.$$

The magnetic field used was the observed value of 60 X 10^{-5} Gauss. Figure 1 is a plot of the growth rate versus frequency as a function of θ(= k‖/k⊥= 0.6, 0.7 and 0.8) for n_H = 3.0 (T_H = 1 .0 keV, U= l0.0V_{TO}^+), $n_o^+ = n_o^- = 1.0$ ($T_O^+ = T_O^- \approx 0.03$ev) and $T_e \approx 1.0$keV. We find that the instability of the wave shifts towards higher frequencies with increasing θ, or for increasingly parallel propagation. Also the growth rate, in general, increases with increasing θ.

The stability of the wave was also studied as a function of the hydrogen density, n_H. Figure 2 thus depicts the variation of the growth rate versus frequency as a function of n_H (= 2.5, 3.0 and 3.5), with θ= 0.6; the other parameters being the same as in figure 1. We find, from the figure that at the lower frequency and hydrogen has a marked effect on the stability of the wave, with the growth rate increasing with increasing hydrogen densities. As the frequencies increase, the lighter ion hydrogen does not have any effect on the growth rate of the wave. This is in contrast to the results obtained for the ion-acoustic wave where a slight decrease in the growth rate was found with increasing hydrogen densities.

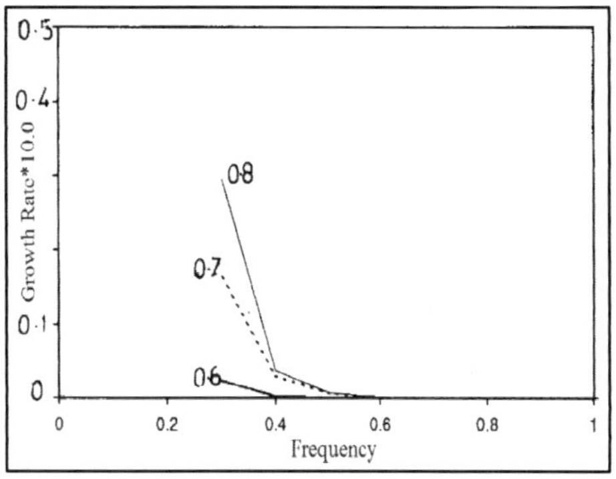

Figure 1

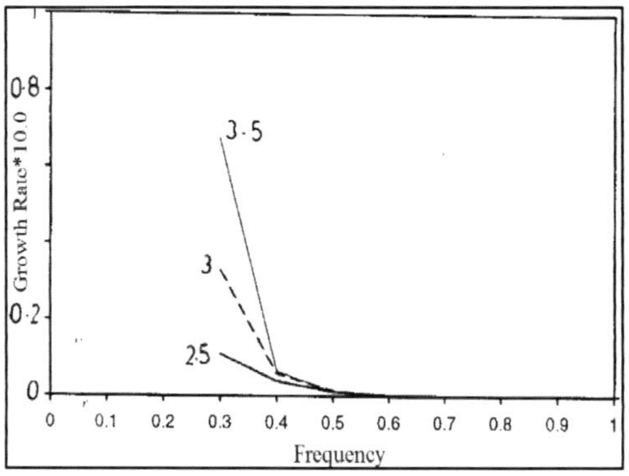

Figure 2

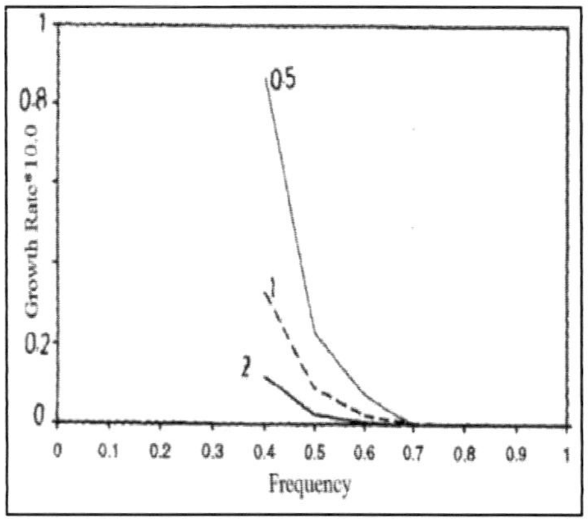

Figure3

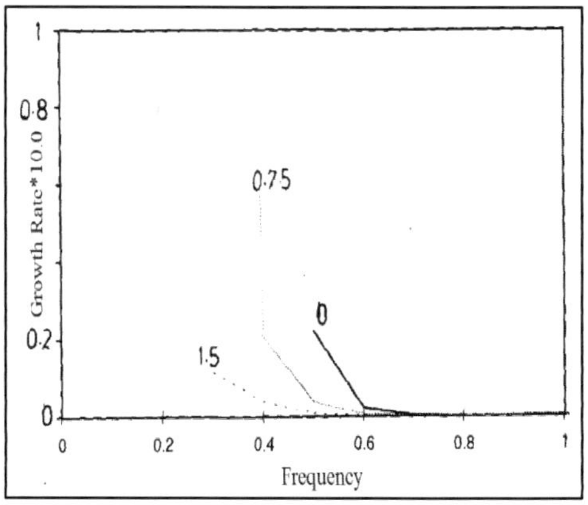

Figure 4

Figure 3 is a plot of the growth rate versus frequency as a function of the ratio of the temperatures T_o^+ to T_O^- (=2.0, 1.0 and 0.5*)* for $n_o^+= n_o^- = 1.0$ and $\theta = 0.6$; the parameters *of* hydrogen and electrons being the same as in figure a. The instability of the wave starts at lower frequencies when positively ionized oxygen is hotter ($T_O^+/ T_O^- = 2.0$); Instability sets in at the same frequency when $T_O^+/ T_O^- = 1.0$ and 0.5; however, and inspection of the three curves shows that the growth rate is, in general, larger when $T_o^+ \neq T_o^-$.Finally, we plot, in Figure 4, the growth rate versus frequency as a function of n_o^-, the negatively ionized oxygen density. The other parameters for the figure are $n_o^+ = 1.0$ ($T_O^+ = T_O^- =.03$eV), $n_H = 3.0$ ($T_H \approx 1.0$keV, $U = 10V_{TO}^+$) and $T_e = 1.0$keV. We find that when $n_o^- = 0$, the growth rate is low with the instability setting in at the higher end of the frequency spectrum. However, when $n_o^- \neq 0$ ($n_o^- = 0.75$ and 1.5) the instability sets in at very low frequencies thus clearly bringing out the influence of negative ions on the stability of the lower hybrid waves.

4. Conclusions

The stability of electrostatic waves in a multi-ion plasma made up of electrons, positively and negatively charged oxygen ions and Hydrogen ions which drift with a velocity U with respect to the 0' ions. Whose frequencies were derived using the cold plasma theory. We find that the stability of the mode increases with increasingly parallel propagation; it increases with increasing hydrogen densities, it is larger when $T_0^+ \neq T_0^-$ than compared to the case where $T_0^+ \neq T_0^-$; it increases with increasing Yo+ where Y_0^+ is the perpendicular wave vector normalized with the 0^+ ion Larmour radius and is driven unstable at lower frequencies in the presence of n_O^-. No variation with respect to U, the drift velocity of the hydrogen ions was detected probably because no resonance conditions were satisfied for the parameters considered.

References

[1] Drummond W E and Rosenbluth M N 1962 *Phys. Fluids* **5** 1507
[2] Kindel J M and Kennel C F 1971 *J. Geophys. Res* **76** 3055
[3] Yamada M S, Sciler S, Hendel H W and Ikezi H 1977 *Phys. Fluids* **20** 450
[4] Ashour Abdulla M and Thorne R M 1978 *J. Geophys. Res* **83** 4755
[5] Singh N, Conrad J R and Schunk R W 1985 *J.Geophys. Res.* **90** 5159
[6] Ahirwar G, Varma P and Tiwari M S 2006a *Indian J. Pure App. Phys.* **80** 1179
[7] Ahirwar G, Varma P and Tiwari M S 2006b *Ann Geophysicae* **24** 1919
[8] Khaira V and Ahirwar G 2015 *AIP Conf. Proc.* **16** 1670

Jeans instability of rotating anisotropic plasma with tensor viscosity

Archana Patidar[1,a], Prerana Sharma[1]

[1] Physics Department, Ujjain Engineering College, Ujjain, M.P. -456010, INDIA

Email: [a)]archanapatidar2@gmail.com

Abstract. In the present work, the role of rotation with viscous tensor in anisotropic self-gravitating plasma has been investigated using magnetohydrodynamic (MHD) model and Chew-Goldberger-Low (CGL) fluid theory. The general dispersion relation is obtained by normal mode analysis theory with the help of linearized perturbation equations and further discussed with some limiting cases. The modified condition for Jeans instability has been obtained. The influence of considered parameters on the growth rate of instability is shown analytically and numerically. The result of present study may be useful in the region of spiral arms of galaxy.

1. Introduction

The research of self-gravitational instability in astrophysical and space plasma has acquired considerable attention for years because of its prominent role in structure formation. Jeans [1] was the first to investigate the self-gravitational instability in plasma system. Since then a lot of investigation has been done to study the self-gravitational instability [2-4]. Recently, Sharma and Rimza [5] have studied self-gravitational instability in two fluid spin quantum plasma. In various astrophysical and space situations the plasmas are collisionless and strongly magnetized. For such a case the pressure becomes anisotropic and separates into parallel and perpendicular components with respect to the direction of magnetic field. A large number of researchers have studied the effect of anisotropic pressure on wave propagation and instabilities in plasma. Gliddon [6] has investigated that self-gravitating plasma with anisotropic pressure distribution also exhibits firehose instability along with gravitational instability. Bhatiya [7] has considered the anisotropic pressure and finite Larmor radius to investigate the Jeans instability of plasma. Along with this, the other transport properties like viscosity of the medium also become anisotropic. In this direction the influence of viscosity on magnetohydrodynamic waves has been examined by Mangesha and Tessema[8]. The self-gravitational instability in anisotropic plasma with tensor viscosity and heat flux has been studied by Mangesha and Tessema [9].

In addition to this, rotation also plays important role in the instabilities of many astrophysical plasma like spiral arms of galaxy. Chandrashekhar [10] has examined the role of rotation and magnetic field on Jeans instability of plasma system. The effect of rotation on self-gravitating instability in anisotropic plasma with finite Larmor radius has been studied by Bhatiya [11].Therefore in the present paper, we have studied the role of rotation and tensor viscosity on the Jeans instability of anisotropic plasma.

The paper is arranged in following manner; using basic set of equations, a general dispersion relation has been derived in section 2. In section 3the dispersion properties and conclusion have been given.

2. Basic equations and dispersion relation

The governing basic set of equations for the self-gravitating viscous anisotropic plasma with angular velocity of rotation $\boldsymbol{\Omega}\left(0,0,\Omega\right)$ can be given as

The continuity equation is given by

$$d_t\rho + \rho\nabla\cdot\boldsymbol{u} = 0 \qquad (1)$$

where ρ is fluid density, $\boldsymbol{u}$ is fluid velocity and $d_t = \left(\partial_t + \boldsymbol{u}\cdot\nabla\right)$ is the convective derivative.

The momentum transfer equation is

$$\rho d_t\boldsymbol{u} = -\nabla.\boldsymbol{P} + \boldsymbol{J}\times\boldsymbol{B} - \rho\nabla\psi + 2\rho\left(\boldsymbol{u}\times\boldsymbol{\Omega}\right) - \nabla\cdot\Pi \qquad (2)$$

In above expression first term of right hand side represents the anisotropic pressure and can be defined as $\boldsymbol{P} = P_\perp\hat{\boldsymbol{I}} + \left(P_\parallel - P_\perp\right)\hat{\boldsymbol{n}}\hat{\boldsymbol{n}}$, where $P_\perp$ is anisotropic pressure perpendicular to the magnetic field and $P_\parallel$ is the anisotropic pressure parallel to the direction ($\boldsymbol{n}$) of magnetic field. The symbol $\boldsymbol{I}$ is identity matrix. The anisotropic pressure $P_\perp$ and $P_\parallel$ can be given as

$$d_t\left(P_\parallel B^2\rho^{-3}\right) = 0 \qquad (3)$$

$$d_t\left(P_\perp B^{-1}\rho^{-1}\right) = 0 \qquad (4)$$

Now, in the second term $\boldsymbol{B}$ is the magnetic field along the z-direction and $\boldsymbol{J}\left(=\nabla\times\boldsymbol{B}\mu_0^{-1}\right)$ is current density, where μ_0 is the magnetic permeability of free space. Hence the fundamental relation of Lorentz force can be written as

$$\boldsymbol{J}\times\boldsymbol{B} = -\frac{1}{\mu_0}\left[\frac{\nabla B^2}{2} - \left(B\cdot\nabla\right)B\right] \qquad (5)$$

The Poisson's equation for gravitational potential is

$$\nabla^2\psi = 4\pi G\rho \qquad (6)$$

and the equation for magnetic force is given by

$$d_t\boldsymbol{B} = \nabla\times\left(\boldsymbol{u}\times\boldsymbol{B}\right) \qquad (7)$$

In equation (7) ψ is the gravitational potential and G is gravitational constant. In expression (1) the last term describes the viscous tensor and the components of viscous stress tensor are given as

$$\Pi_{xx} = -\frac{\eta_0}{2}\left(t_{xx} + t_{yy}\right),\ \Pi_{yy} = -\frac{\eta_0}{2}\left(t_{xx} + t_{yy}\right),\ \Pi_{zz} = -\eta_0 t_{zz} \qquad (8)$$

where η_0 is compressive viscosity coefficient and $t_{ab} = \partial u_a/\partial x_b + \partial u_b/\partial x_a - \left(2/3\right)\delta_{a,b}\nabla.u$. Let us assume that the perturbation in all the quantities is of the form $e^{\left(ik\hat{z} - i\omega t\right)}$, where k is the wave numbers in z-direction and ω is the wave frequency. We linearize equations (1) - (8) using $\boldsymbol{u} = \boldsymbol{u}_0 + \boldsymbol{u}_1$, $\boldsymbol{B} = \boldsymbol{B}_0 + \boldsymbol{B}_1$, $n = n_0 + n_1$ and $\psi = \psi_o + \psi_1$, where $\boldsymbol{u}_1$, $\boldsymbol{B}_1$, n_1 and ψ_1 are the first order linearized quantity and equilibrium quantity $\boldsymbol{u}_0$ and ψ_o are taken equals to zero. We get the following components of momentum transfer equation by using linearized form of equations (1)-(8) and applying considered perturbation, as

$$\left\{\rho\omega^2 + \left(p_\parallel - p_\perp - \frac{B^2}{\mu_0}\right)k_z^2\right\}u_x - 2\rho i\omega\Omega_z u_y = 0 \qquad (9)$$

$$2\rho i\omega\Omega_z u_x + \left\{\rho\omega^2 + \left(p_{\parallel} - p_{\perp} - \frac{B^2}{\mu_0}\right)k_z^2\right\}u_y = 0 \tag{10}$$

$$\left\{\rho\omega^2 + \left(-3P_{\parallel} + \frac{4\pi G\rho^2}{k^2} + \frac{i4\eta_0\omega}{3}\right)k_z^2\right\}u_z = 0 \tag{11}$$

After the usual algebraic manipulation, the general dispersion relation can be obtained as

$$\left[\left\{\omega^2 + \left(\frac{P_{\parallel} - P_{\perp}}{\rho} - \frac{B_0^2}{\mu_0\rho}\right)k^2\right\}^2 - 4\Omega^2\omega^2\right]\left\{\omega^2 + \left(\frac{-3P_{\parallel}}{\rho} + \frac{\omega_J^2}{k^2} + \frac{i4\eta_0\omega}{3\rho}\right)k^2\right\} = 0 \tag{12}$$

Equation (12) represents the general dispersion relation for self-gravitating rotating anisotropic plasma with tensor viscosity. If we ignore rotation and viscosity terms in equation (12) then the obtained form becomes identical to the result given by Gllidon [6]. Thus, the presence of rotation and viscosity modified the general dispersion of self-gravitating anisotropic plasma system.

3. Discussions and conclusions

The general dispersion relation (12) has two factors in which first factor represents the Alfven mode

$$\omega^4 + \left\{2\left(\frac{P_{\parallel} - P_{\perp}}{\rho} - \frac{B_0^2}{\mu_0\rho}\right)k^2 - 4\Omega^2\right\}\omega^2 + \left(\frac{P_{\parallel} - P_{\perp}}{\rho} - \frac{B_0^2}{\mu_0\rho}\right)^2 k^4 = 0 \tag{13}$$

which is modified in the presence of anisotropic pressure and rotation while it remains unaffected due to tensor viscosity. The presence of anisotropic pressure in Alfven mode gives the firehose instability in the rotating self-gravitating plasma system if the parallel anisotropic pressure becomes greater than the sum of magnetic pressure and perpendicular anisotropic pressure i.e., $P_{\perp} + \left(B_0^2/\mu_0\right) < P_{\parallel}$.

The second factor of general dispersion relation (12) represents the self-gravitating mode

$$\omega^2 + \frac{i4\eta_0 k^2}{3\rho}\omega - \frac{3P_{\parallel}}{\rho}k^2 + \omega_J^2 = 0 \tag{14}$$

which is modified due to the presence of tensor viscosity. The presence of rotation and perpendicular anisotropic pressure has no influence on the self-gravitating mode. The condition of Jeans instability obtained from equation (14) is

$$\frac{3P_{\parallel}}{\rho}k^2 < \omega_J^2 \tag{15}$$

Equation (15) shows the condition of Jeans instability for rotating anisotropic viscous plasma. It is clear from above inequality that the presence of rotation and tensor viscosity plays no role in condition of self-gravitational instability but these terms affect the growth rate Jeans instability in rotating anisotropic viscous plasma.

Now, to see the influence of considered parameter on growth rate of self-gravitational instability, we normalize the self-gravitating mode (14) using $-i\omega = \sigma$ as

$$\sigma^{*2} + \frac{4}{3}\Pi^* k^{*2}\sigma^* + 3k^{*2} - 1 = 0 \tag{16}$$

where $\sigma^* = \sigma/\omega_J$, $k^* = \left(k/\omega_J\right)\left(P_{\parallel}/\rho\right)^{1/2}$ and $\Pi^* = \eta_0\omega_J/P_{\parallel}$.

The influence of tensor viscosity (Π^*) on the growth rate of self-gravitational instability against the wave vector (k^*) has been shown in figure 1. The solid, dashed and dotted curves are for Π^* = 0.0, 0.5 and 1.0 respectively. The nature of curves shows that the growth rate of instability decreases with increasing in the tensor viscosity. Hence the viscosity has stabilizing effect on growth rate of self-gravitational instability in anisotropic rotating plasma system.

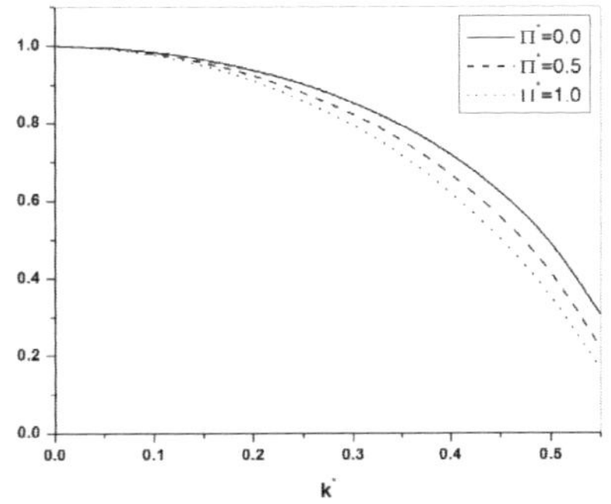

Figure 1. Normalized growth rate of Jeans instability against the normalized wave number for different values of viscosity.

In conclusion the Alfven mode is affected due to the presence of anisotropic pressure and rotation while the gravitating mode is influenced by the presence of compressional viscosity. The condition of Jeans instability is affected byparallel anisotropic pressure.

Acknowledgement
The authors gratefully acknowledge UGC, New Delhi (F. No. 42-820/2013(SR)), MPSCT, Bhopal (R&D File No. A/RD/RP-2/2015-16/243) and SERB DST, New Delhi (SB/FTP/PS-075/2014) for their financial support.

References

[1] Jeans J H 1929 *Astronomy and Cosmology* (Cambridge: Cambridge University Press)
[2] Kumar S S 1960 *Smithsonian Astrophysical Observatory* **12** 4
[3] Kato S and Kumar S S1960 *Pub. Astro. Soc. Japan* **12** 255
[4] Sharma P and Chhajlani R K 2013 *Phys. Plasmas* **20** 9
[5] Sharma P and Rimza T 2017 *Astrphys. Space Sci.* **362** 54
[6] Gliddon J E C 1966 *Astrophys. J.* **145** 583
[7] Bhatiya P K 1968 *Z. Astrophysik* **68** 204
[8] Mengesha A and Tessema S B 2013 *J. Plasma Phys.* **79** 535
[9] Mengesha C A and Tessema S B 2013 *J. Plasma Phys.* **79** 805
[10] Chandrasekhar S1961 *Hydrodynamic and Hydromagnetic Stability* (London: Oxford University).
[11] Bhatiya P K 1968 *Z. Astrophysik* **69** 363

IOP Conf. Series: Journal of Physics: Conf. Series **836** (2017) 012028 doi:10.1088/1742-6596/836/1/012028

Effect of exchange correlation potential on dispersion properties of lower hybrid wave in degenerate plasma

Tripti Rimza and Prerana Sharma[a]

Physics Department, Ujjain Engineering College, Ujjain, M.P. -456010, India

Email: [a)]preranaiitd@rediffmail.com

Abstract. The dispersion properties of lower hybrid wave are studied in electron-ion degenerate plasma with exchange effect in non-relativistic regime. It is found that the combined effect of Bohm potential and exchange correlation potential significantly modifies the dispersion properties of lower hybrid wave. The graphical results explicitly show the influence of degeneracy pressure, Bohm force and exchange correlation potential on the frequency of the lower hybrid mode. Present work should be of relevance for the dense astrophysical environments like white dwarfs and for laboratory experiments.

1. Introduction

Recently, there has been a great deal of interest in investigating the properties of electrostatic waves [1-6] in plasmas incorporating quantum effects. The study of such plasma waves provides important insights about instabilities in various laboratory fusion and astrophysical plasma environments [7-11]. Several authors have studied the propagation characteristics of electrostatic electron and ion waves in quantum plasmas [4-7]. When the de-Broglie wavelength (λ_B) of the particles becomes comparable to the average inter-particle distance (a), these quantum effects cannot be ignored. It is also evident that; when density is high the exchange correlation potential can even dominate the role of particle dispersion effects [11-13]. The general idea that produces this exchange correlation potential is essentially the density fluctuation theory (DFT) [11]. Several authors have studied the influence of exchange potential and Bohm force on the propagation characteristics of different waves [11-13] in plasmas and showed the relevance of exchange correlation potential in context with solid-state plasmas, inertial confinement fusion plasmas and in an astrophysical region such as white dwarf stars. Therefore, it is important to study the combine influence of both the dispersion effects and exchange potential on the dynamics of plasma waves.

The fundamental process responsible for heating and accelerating plasma around white dwarfs, neutron stars and in tokamaks [14] is reconnection. The investigation regarding the electron acceleration and the magnetic reconnection phenomena (via. lower hybrid plasma waves) in laboratory, space, and astrophysical plasmas is widely studied by many researchers [14, 15]. But none of the authors have investigated the instability of the lower hybrid waves in electron-ion quantum plasma taking into account the particle dispersion effects and exchange-correlation potential; as the lower hybrid waves provide the necessary electron acceleration that leads to plasma heating mechanism. These waves are simultaneously in resonance with both the magnetized electrons and un-magnetized ions. The lower-hybrid resonance frequency is given as $\omega_L = \sqrt{\omega_c \Omega_c}$ (where Ω_c and ω_c are the ion and electron gyro frequencies, respectively). Lower hybrid waves have also their importance in space [14] and fusion plasmas [14, 16].

So in the present letter we have studied the effect of exchange correlation potential, Fermi velocity and Bohm force on the propagation characteristics and phase velocity of lower hybrid waves.
The present investigation is done using quantum hydrodynamic (QHD) model. This paper is organized as follows. In Section 2 we have presented QHD set of equations for dense quantum plasmas. In Section 3 we describe contribution of Bohm force and exchange potential on the dispersion of transverse waves in magnetized plasmas. In Section 4 a brief summary of obtained results is presented.

2. Theoretical model

We consider two-fluid quantum plasma system consisting of electron and ion species. We have assumed that the electrons are degenerate, as the Fermi temperature of electrons is much greater than the plasma electron temperature while the ions are assumed to be non-degenerate and are unaffected by quantum corrections. The dynamics of electrons and ions are governed by following set of QHD equations

$$\frac{\partial}{\partial t}n_e + n_e \nabla . \vec{v}_e = 0 \tag{1}$$

$$m_e n_e \frac{\partial}{\partial t}\vec{v}_e = q_e n_e\left(\vec{E} + \frac{1}{c}\left[\vec{v}_e, \vec{B}\right]\right) - \nabla P_{fe} - \frac{\hbar^2}{4m_e}\nabla\left(\nabla^2 n_e\right) - V_{exc}\nabla n_e \tag{2}$$

$$V_{e,xc} = -0.985\frac{n_e^{1/3}e^2}{\varepsilon}\left[1 + \frac{0.034}{a_{Be}^* n_e^{1/3}}\ln\left(1 + 18.37 a_{Be}^* n_e^{1/3}\right)\right] \tag{3}$$

$$\frac{\partial}{\partial t}n_i + n_i \nabla . \vec{v}_i = 0 \tag{4}$$

$$m_i n_i \frac{\partial}{\partial t}\vec{v}_i = q_i n_i\left(\vec{E} + \frac{1}{c}\left[\vec{v}_i, \vec{B}\right]\right) - \nabla P_{ti} \tag{5}$$

where $n_{e,i}$, $\vec{v}_{e,i}$, $m_{e,i}$ and $q_{e,i}$ represents the number density, fluid velocity, mass and charge of species (electrons and ions) respectively. Equation (1) and (2) are the continuity and momentum equations for electrons. Equation (3) shows the expression for exchange correlation potential, where ε is the dielectric constant of material and $a_{Be}^* = \varepsilon\hbar^2/mq_e^2$ is the effective Bohr atomic radius of the species.Since the electrons are fermions they follow Fermi-Dirac statistics. Therefore, the temperature of electrons considered here is Fermi temperature [Sharma and Chhajalani 10] which is given as $T_{fe}^3 = \hbar^6 \times \left(3\pi^2 n\right)^2/\left(2mk_B\right)^3$, Fermi velocity of fermions is $v_{fe}^3 = \hbar^3 \times \left(3\pi^2 n\right)/m^3$ and the corresponding Fermi pressure of degenerate electrons is given as $P_{fe} = \hbar^2 n_e^{5/3} \times \left(3\pi^2\right)^{2/3}/\left(5m_e\right)$. Further the dynamics of non-degenerate ions are governed by following set of continuity and momentum equations (4) and (5) respectively. The ion thermal pressure is given via. equation of state i.e., $P_{ti} = k_B T_i n_i$ in which T_i is temperature of ions and k_B is Boltzmann constant.

3. Dispersion relation and graphical discussion

Let the solution of the system of equation be of the form $\exp\left[ik_x \hat{x} - i\omega t\right]$, where ω is the frequency of perturbation and k_x is the x-component of perturbed wave vector. The perturbation in physical quantities can be ruled according to [10],

$$n = n_0 + n_1,\ \vec{\mathrm{v}} = \vec{\mathrm{v}}_0 + \vec{\mathrm{v}}_1,\ \vec{\mathrm{B}} = \vec{\mathrm{B}}_0 + \vec{\mathrm{B}}_1,\ \vec{\mathrm{E}} = \vec{\mathrm{E}}_0 + \vec{\mathrm{E}}_1 \tag{6}$$

where the subscript "1" denotes the perturbed part and subscript "0" denotes an unperturbed part. At equilibrium, $\vec{\mathrm{v}}_0$, $\vec{\mathrm{E}}_0 = 0$. We restrict ourselves to $\vec{E} = E_x\hat{x}$ and $\vec{B}_0 = B_0\hat{z}$. After linearizing the equations and applying quasi-neutrality condition the dispersion relation for lower hybrid wave in dense astrophysical plasmas can be obtained as

$$\omega^2 = \omega_c\Omega_c + k^2\left(1+\frac{m_i}{m_e}\right)^{-1}\left[v_{fe}^2 - \frac{h^2k^2}{4m_e^2} - \frac{V_{j,exc}}{m_e}\right] + \left(1+\frac{m_e}{m_i}\right)^{-1} k^2 v_{ti}^2 \tag{7}$$

where $\omega_c = eB_0 / m_e c$ and $\Omega_c = eB_0 / m_i c$ are electron and ion cyclotron frequencies respectively. Equation (7) represents the dispersion relation for modified lower hybrid wave. The modification is due to exchange correlation potential, degeneracy pressure and Bohm force of degenerate electrons and thermal pressure of non-degenerate ions.

In the degenerate astrophysical environment like white dwarf stars, neutron stars, pulsar magnetospheres where magnetic field is of the order of $B \approx 10^5 - 10^{10} T$, the number density $n_0 \approx 10^{29} - 10^{34} m^{-3}$ and temperatures ranges $T \approx 10^5 - 10^7 K$ [4, 10-12] the degeneracy effects cannot be ignored. Now in order to analyze the effect of exchange potential and Bohm force in affecting the considered plasma system we write the dispersion relation (7) in normalized form as

$$\omega^{*2} - \omega_c^* - M^*\left[k^{*2}\left(v_{fe}^{*2} - V_{j,exc}^{*2}\right) - H^{*2}k^{*4}\right] - k^{*2} = 0 \tag{8}$$

where we have used $\left[1+\left(m_i/m_e\right)\right]^{-1} \approx \left(m_e/m_i\right)$ and $\left[1+\left(m_e/m_i\right)\right]^{-1} \approx 1$. In deriving equation (8) we have used the following dimensionless parameters

$$\omega^* = \frac{\omega}{\Omega_c},\ k^* = \frac{kv_{ti}}{\Omega_c},\ H = \frac{\hbar^2\Omega_c^2}{4m_e^2v_{ti}^2},\ V_{exc}^* = \frac{V_{j,xc}}{m_ev_{ti}^2},\ v_{fe}^* = \frac{v_{fe}}{v_{ti}},\ M^* = \frac{m_e}{m_i} \tag{9}$$

The normalized parameters can be chosen in a range as $H = 0.0 - 0.4$, $V_{exc}^* = 0.0 - 0.6$, $v_{fe}^* = 0.0 - 0.6$. Now to analyze the dispersion relation numericallyFigure 1 and 2 are plotted to show the influence of exchange correlation potential and Fermi velocity of electrons in affecting the dynamics of lower hybrid waves.

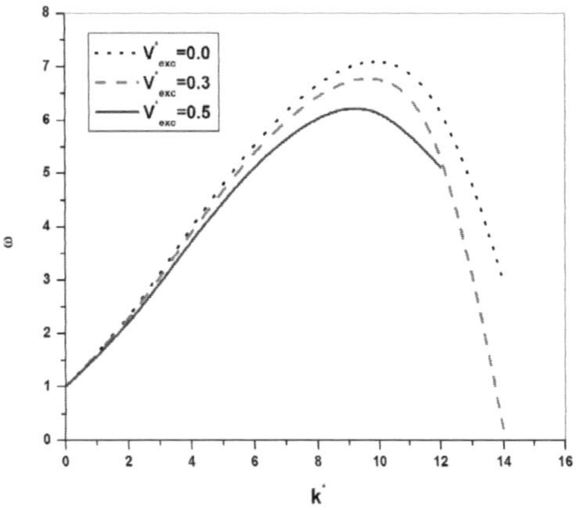

Figure 1. Growth rate of instability ω^* is shown as a function of wave number k^* for arbitrary values of exchange potential parameter ($V_{exc}^* = 0.0, 0.3, 0.5$), keeping the other parameter to be fixed.

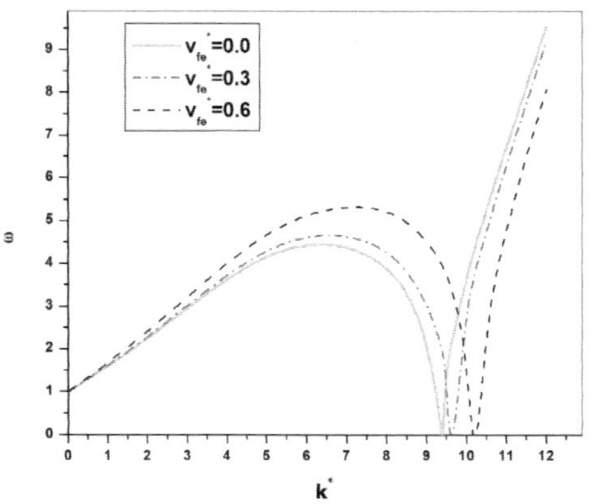

Figure 2. Growth rate of instability ω^* is shown as a function of wave number k^* for arbitrary values of Fermi velocity ($V_{fe}^* = 0.0, 0.3, 0.6$), keeping the other parameter to be fixed.

Figure 1 shows the influence of exchange correlation potential of electrons on the dispersion characteristics of lower hybrid waves. In figure1 the growth rate of lower hybrid wave ω^* is plotted against the normalized wave number k^* in the presence and absence of exchange effects (using equation 8). From figure it is clear that as the value of exchange parameter increases, it decreases the phase velocity of lower hybrid wave. Further, it is quite interesting to study the role of Fermi velocity of electrons in affecting the characteristics of lower hybrid wave. For this purpose we have plotted figure 2 that represents a comparison for higher and lower Fermi velocity of electrons. From figure 2 it is found that the phase velocity of lower-hybrid wave enhances abruptly as Fermi velocity increases. The Fermi velocity of electrons is significantly responsible for the frequent propagation of lower-hybrid waves in degenerate plasmas.

4. Conclusions

In the present work we have studied the role of exchange correlation potential, degeneracy pressure and quantum Bohm force of electrons on the propagation of lower hybrid wave in two-species quantum plasma system. In deriving the dispersion relation we have considered the special case, where the electron-ion plasma is quasi-neutral. The derived dispersion relation is found to be modified due to quantum effect of electrons and thermal effect of ions. It is concluded that the Fermi velocity of electrons and thermal velocity of ions increases the growth rate of lower hybrid waves. The numerical discussion of the results shows that the presence of exchange potential and Bohm force decreases the phase speed of lower hybrid waves while the Fermi velocity of electrons increases it. The present results are applicable to study the propagation characteristics of lower hybrid waves in astrophysical as well as laboratory fusion plasmas.

Acknowledgement

Authors are gratefully acknowledged to SERB, New Delhi, India research project No. SB/FTP/PS-075/2014, UGC, New Delhi (F. No. 42-820/2013(SR)) and MPSCT, Bhopal (R&D File No. A/RD/RP-2/2015-16/243) for their financial support.

References

[1] Haas F 2003 *Phys. Plasmas* **10** 10
[2] Marklund M, Brodin G, Stenflo L and Liu C S 2008 *Euro. Phys. Lett.* **84** 1
[3] Masood W and Eliasson B 2011 *Phys. Plasmas* **18** 034503
[4] Jamil M, Rasheed A, Rozina Ch, Moslem W M and Salimullah M 2014 *Phys. Plasmas* **21** 02070
[5] Sharma Pand Rimza T 2017 *Astrophys. Space Sci.* **362** 54
[6] Kim K S, Kim E H, Lee D H and Kim K 2005 *Phys. Plasmas* **12** 052903
[7] Mukta K N, Zobaer M S, Roy N and Mamun A A 2015 *J. Plasma Phys.* **81** 905810326
[8] Ren H, Wu Z and Chu P K 2007 *Phys. Plasmas* **14** 062102
[9] Yamada M, Ren Y, Ji H, Breslau J, Gerhardt S, Kulsrud R and Kuritsyn A 2006 *Phys. Plasmas* **13** 052119
[10] Sharma P and ChhajlaniR K 2014 *Phys. Plasmas* **21** 072104
[11] Maroof R, Mushtaq A and Qamar A 2016 *Phys. Plasmas* **23** 013704
[12] Jamil M, Rasheed A, Rozina Ch, Jung Y D and Salimullah M 2015 *Phys. Plasmas* **22** 082113
[13] Shahmansouri M and Misra A P 2016 *Phys. Plasmas* **23** 072105
[14] Cairns I H and Mcmillan B F 2005 *Phys. Plasmas* **12** 102110
[15] Bale S D, Mozer F S and Phan T 2002 *Geophys. Res. Lett.* **29** 2180
[16] Qin X G H, Liu J and Fisch N 2013 *J. Phys. Plasmas* **20** 022502

IOP Conf. Series: Journal of Physics: Conf. Series **836** (2017) 012029 doi:10.1088/1742-6596/836/1/012029

Jeans instability of inhomogeneous dusty plasma with polarization force, ionization and recombination

Shweta Jain[1,a], Prerana Sharma[2] and R K Chhajlani[3]

[1]Lokmanya Tilak Science and Commerce College, Ujjain, M. P. 456010, India
[2]Department of Physics, Ujjain Engineering College, Ujjain, M.P.465010, India
[3]School of Studies in Physics, Vikram University Ujjain, M. P. 456010, India

E-mail: [a)]jshweta09@gmail.com

Abstract. The self-gravitational Jeans instability has been studied in dusty plasma containing significant background of neutral pressure and recombination of ions and electrons on the dust surface. The full dynamics of charged dust grains, ions and neutral species are employed considering the electrons as Maxwellian. We have derived the general dispersion relation for collisional dusty plasma with ionization, recombination and polarization force. The general dispersion relation describes the effects of considered parameters which are solved in different dusty plasma situations. Further, the dispersion relation is solved numerically. The present work is applicable to understand the structure formation of interstellar molecular clouds in astrophysical plasma.

1. Introduction

Over the past decades, the field of dusty plasma has become a great deal of interest. The dusty plasmas are ubiquitous in astrophysical plasmas and in low temperature laboratory plasmas. In dusty plasmas, the low frequency dust acoustic wave mode has been studied by many researchers [1,2] in which the thermal pressure effects are provided by the Maxwellian distributed electrons and ions while inertia is provided by dust particles. However, in the molecular clouds, the presence of massive dust and neutral particles give rise to gravitational force which involves the collapse of molecular cloud. The instability mechanisms and dispersion characteristics of dust acoustic wave mode incorporating the effects of free energy sources such as equilibrium ion streaming, collisions with the ions and neutrals, recombination of electrons and ions on the dust grain surface have also been investigated extensively [3,4]. Ostrikov et al. [5] have studied the dust acoustic wave instabilities with ionization and recombination in collisional dusty plasma. Dust acoustic modes have been discussed by Kaw and Singh [6] in presence of recombination and ionization in collisional dusty plasma. The effect of recombination and ionization on dust acoustic instability has been investigated by Bal and Bose [7] in partially ionized dusty plasma with strong coupling limits.

In the study of dust acoustic waves and instabilities, the polarization of plasma ions produces the dust polarization force around the dust grain which significantly modifies the dust dynamics and therefore is an important phenomenon of research. The effect of dust polarization force has been widely discussed by Hamaguchi and Farouki [8,9] and Khrapak et al. [10]. Recently, Jain and Sharma [11] have examined the influence of polarization force on Jeans instability in degenerate and non-degenerate collisional dusty plasma. Therefore, looking to the importance of polarization force, ionization and recombination we have studied the Jeans instability of collisional inhomogeneous dusty plasma incorporating the ionization, recombination and polarization force in present work.

2. Linearized model equations

The model includes a four component unmagnetized self-gravitating partially ionized dusty plasma which consists Maxwellian distributed electrons and ions, negatively charged dust grains and neutral particles to study dust acoustic wave propagation. In the present work, the sufficient background of neutral particles is taken; therefore, for the description of neutral particles we have taken the full dynamics of neutrals. The propagation of dust acoustic waves in collisional self-gravitating dusty plasma is described by the following set of equations.

The linearized electron perturbed density is

$$n_{e1} = \frac{en_{e0}}{k_B T_{e0}} \phi_1 \qquad (1)$$

where n_{e1} and n_{e0} is perturbed and equilibrium electron number density, T_{e0} is electron temperature and ϕ_1 stands for perturbed electrostatic potential.

The continuity and momentum equations for dust and neutrals are

$$D_t n_{j1} + n_{j0} \nabla . \boldsymbol{u}_{j1} = 0 \qquad (j = d, n) \qquad (2)$$

$$D_t \boldsymbol{u}_{d1} = -\frac{\nabla p_{d1}}{m_d n_{d0}} - \frac{q_d \nabla \phi_1 (1-\Gamma)}{m_d} - \nabla \psi_{G1} - \upsilon_{dn} (\boldsymbol{u}_{d1} - \boldsymbol{u}_{n1}) \qquad (3)$$

$$D_t \boldsymbol{u}_{n1} = -\frac{\nabla p_{n1}}{m_n n_{n0}} - \nabla \psi_{G1} - \upsilon_{nd} (\boldsymbol{u}_{n1} - \boldsymbol{u}_{d1}) \qquad (4)$$

The symbol $\boldsymbol{u}_{j1}$, p_{j1} m_j, n_{j0} and n_{j1} represent perturbed velocity, pressure, mass, and equilibrium and perturbed number density of j particle respectively. The collision frequency of dust with neutral and neutral with dust is given by υ_{dn} and υ_{nd} respectively. The polarization force and charge of dust is given by Γ and q_d respectively.

The linearized first order self-gravitational potential is

$$\nabla^2 \psi_{G1} = 4\pi G (m_d n_{d1} + m_n n_{n1}) \qquad (5)$$

where ψ_{G1} expresses the self-gravitation of dust and neutral.

The perturbed ion momentum and continuity equations are

$$D_t (\boldsymbol{u}_{i0} + \boldsymbol{u}_{i1}) = -\frac{\nabla p_{i1}}{m_i n_{i0}} - \frac{e \nabla \phi_1}{m_i} - \upsilon_{in} \boldsymbol{u}_{i1} \qquad (6)$$

$$D_t n_{i1} + n_{i0} \nabla . \boldsymbol{u}_{i1} + \boldsymbol{u}_{i0} \nabla . n_{i1} = -\upsilon_{\mathrm{R}} \left(\frac{n_{dl}}{n_{d0}} \right) n_{i0} \qquad (7)$$

The first order charge neutrality condition is

$$en_{i1} = en_{e1} - q_d n_{d1}. \qquad (8)$$

The equilibrium ion velocity is symbolized by $\boldsymbol{u}_{i0}$, the collision frequencies of ion with neutral is by υ_{in} and the frequency of recombination of ions and electrons on the dust grains surface is by υ_{R}.

3. Dispersion relation and discussions

We now consider that all the perturbed parameters are proportional to $\exp i(\boldsymbol{k}.\boldsymbol{r} + \omega t)$where $\boldsymbol{k}$ is the wavenumber andω is the perturbation frequency. Using this perturbation in equations (1) – (7)we obtained the perturbed densitiesn_{e1}, n_{i1} and n_{d1} for electron, ion and dust respectively which are further combined with first order quasineutrality condition (8) and using $i\omega = \sigma$ we obtained the general dispersion relation for collisional dusty plasma as

$$\sigma^4+\frac{\upsilon_{dn}}{\omega_{Jn}^2}\left(\omega_{Jd}^2+\omega_{Jn}^2\right)\sigma^3+\left(k^2v_{tn}^2-\omega_{Jn}^2+k^2v_{td}^2-\omega_{Jd}^2+k^2\left(1-\Gamma\right)c_D^2\left(1+\beta\upsilon_R\upsilon_{in}/k^2v_{ti}^2\right)\left(1+\alpha\right)^{-1}\right)\sigma^2$$
$$+\frac{\upsilon_{dn}}{\omega_{Jn}^2}\left\{k^2\left(v_{tn}^2\omega_{Jn}^2+v_{td}^2\omega_{Jd}^2+\left(1-\Gamma\right)c_D^2\omega_{Jd}^2\left(1+\beta\upsilon_R\upsilon_{in}/k^2v_{ti}^2\right)\left(1+\alpha\right)^{-1}\right)-\left(\omega_{Jd}^2+\omega_{Jn}^2\right)^2\right\}\sigma \tag{9}$$
$$+\left(k^2v_{tn}^2-\omega_{Jn}^2\right)\left(k^2v_{td}^2+k^2\left(1-\Gamma\right)c_D^2\left(1+\beta\upsilon_R\upsilon_{in}/k^2v_{ti}^2\right)\left(1+\alpha\right)^{-1}\right)-k^2v_{tn}^2\omega_{Jd}^2=0.$$

where $\omega_{Jn}=\left(4\pi Gm_nn_{no}\right)^{1/2}$, $\omega_{Jd}=\left(4\pi Gm_dn_{do}\right)^{1/2}$, $v_{tn}=\left(T_n/m_n\right)^{1/2}$, $v_{td}=\left(T_d/m_d\right)^{1/2}$, $\beta=n_{i0}/Z_dn_{do}$, $v_{ti}=\left(T_i/m_i\right)^{1/2}$, $c_D^2=z_d^2\left(n_{d0}T_{i0}/m_dn_{i0}\right)$ and $\alpha=\left(n_{e0}T_{i0}/n_{i0}T_{e0}\right)$.

We have assumed that the ions are initially at rest (i.e. $u_{i0}=0$) and the ion neutral collision frequency is much less than the ion thermal speed.

Equation (9) represents the modified radiative condensation mode in partially ionized dusty plasma. Equation (9) is modified due to self-gravitation, polarization force and collision frequencies of different species. In the case when $\omega_{Jn}=\omega_{Jd}=0, v_{tn}=0, \Gamma=0$ equation (9) reduces to the dispersion relation obtained by Bal and Bose [7] in weakly coupled state.The Jeans criterion of instability is obtained from the constant term of equation (9) which is given as

$$\left(k^2-\frac{\omega_{Jn}^2}{v_{tn}^2}\right)\left(v_{td}^2+\left(1-\Gamma\right)c_D^2\frac{\left(1+\beta\upsilon_R\upsilon_{in}/k^2v_{ti}^2\right)}{\left(1+\alpha\right)}\right)<\omega_{Jd}^2 \tag{10}$$

The modified Jeans instability criterion for radiative dusty plasma is shown by above equation (10). The modification is due to the dust polarization force, ionization recombination effect; ion and neutral thermal velocity and dust thermal velocity. If we ignore the effects of neutral particles, polarization force and ionization recombination effect the condition (10) recovers with the traditional condition of gravitational instability as described by Chandrasekhar [12].

4. Numerical estimation

We now examine the effects of polarization force and dust neutral collision frequency on growth rate of Jeans instability numerically. For this purpose, we normalize the dispersion relation (9) by neutral Jeans frequency. Therefore, the normalized form of dispersion relation (9) can be written as

$$\sigma^{*4}+\upsilon_{dn}^*\left(1+\omega_{Jd}^{*2}\right)\sigma^{*3}+\left(k^{*2}v_{tn}^{*2}+k^{*2}v_{td}^{*2}-\omega_{Jd}^{*2}-1+k^{*2}\left(1-\Gamma\right)\left(1+\alpha\right)^{-1}\left(1+\beta\upsilon_R^*\upsilon_{in}^*/k^{*2}v_{ti}^{*2}\right)\right)\sigma^{*2}$$
$$+\upsilon_{dn}^*\left(k^{*2}v_{tn}^{*2}+k^{*2}v_{td}^{*2}\omega_{Jd}^{*2}+k^{*2}\omega_{Jd}^{*2}\left(1-\Gamma\right)\left(1+\beta\upsilon_R^*\upsilon_{in}^*/k^{*2}v_{ti}^{*2}\right)\left(1+\alpha\right)^{-1}-\left(\omega_{Jd}^{*2}+1\right)^2\right)\sigma^*$$
$$+\left(k^{*2}v_{tn}^{*2}-1\right)\left(k^{*2}v_{td}^{*2}+k^{*2}\left(1-\Gamma\right)\left(1+\beta\upsilon_R^*\upsilon_{in}^*/k^{*2}v_{ti}^{*2}\right)\left(1+\alpha\right)^{-1}\right)-k^{*2}v_{tn}^{*2}\omega_{Jd}^{*2}=0(34) \tag{11}$$

In above equation we have used the following non-dimensional quantities

$$\sigma^*=\frac{\sigma}{\omega_{Jn}},\quad k^*=\frac{kc_D}{\omega_{Jn}},\upsilon_{dn}^*=\frac{\upsilon_{dn}}{\omega_{Jn}},\ v_{tn}^*=\frac{v_{tn}}{c_D},\ \upsilon_R^*=\frac{\upsilon_R}{\omega_{Jn}},\omega_{Jd}^*=\frac{\omega_{Jd}}{\omega_{Jn}},\ v_{td}^*=\frac{v_{td}}{c_D},\ v_{ti}^*=\frac{v_{ti}}{c_D},\ \upsilon_{in}^*=\frac{\upsilon_{in}}{\omega_{Jn}}.$$

In order to, for numerical estimation of our results we have plotted the normalized growth rate of Jeans instability against the normalized wave number for different values of dust neutral collision frequency and dust polarization force (using equation (11)). The numerical results are shown in figure 1 and 2. In figure 1 the variation of growth rate has been presented for $\upsilon_{dn}^*=0.0, 0.4$ and 0.8. From figure we find that the peak value of growth rate of Jeans instability decreases by increasing the value of collision frequency and therefore the presence of dust neutral collisions moves the system towards stabilization. Further, in figure 2 the influence of polarization force has been shown on growth rate of instability. It is clear from figure that as the value of the dust polarization force increases the growth rate also increases and thus it tries to destabilize the system.

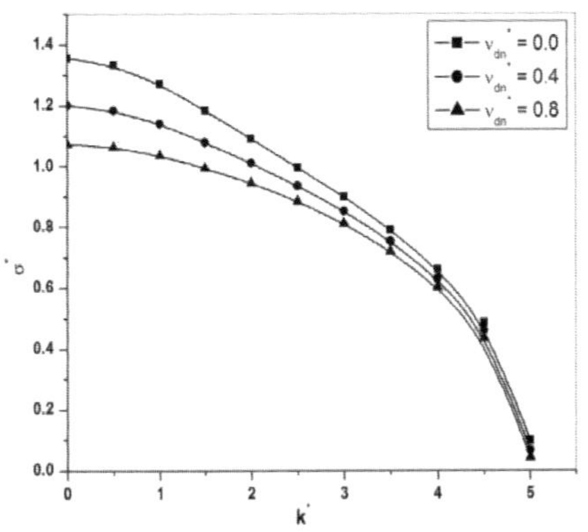

Figure 1. The growth rate of gravitational instability with normalized wave number for different values of dust neutral collision frequency. The values of constant parameters are $v_{td}^*=0.2$, $v_{tn}^*=0.2$, $\upsilon_R^*=0.3$, $\Gamma = 0.4$, $\alpha = 0.1$, $\upsilon_{in}^*=0.1$, $\beta = 0.1$, $v_{ti}^* = 0.2$ and $\omega_{Jd}^* = 0.2$.

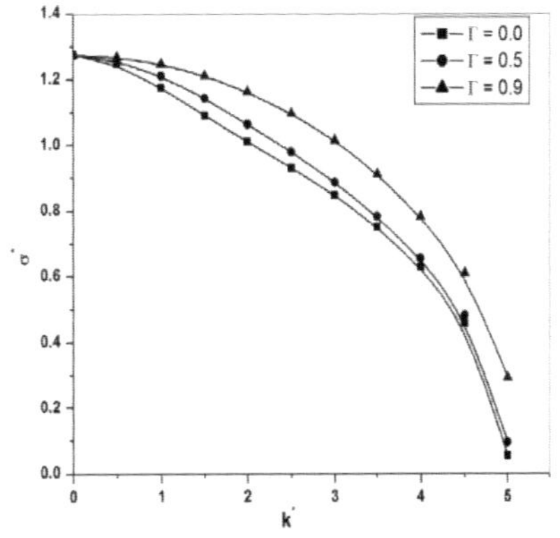

Figure 2. The growth rate of gravitational instability with normalized wave number for different values of polarization force. The values of constant parameters are $v_{td}^*=0.2$, $v_{tn}^*=0.2$, $\upsilon_R^*=0.3$, $\upsilon_{dn}^* = 0.2$, $\alpha = 0.1$, $\upsilon_{in}^*=0.1$, $\beta = 0.1$, $v_{ti}^* = 0.2$ and $\omega_{Jd}^* = 0.2$.

5. Conclusions

In this work, we have examined the Jeans instability of collisional dusty plasma with polarization force, recombination and ionization. The general dispersion relation is obtained using the normal mode analysis method which is further discussed analytically and numerically. It is found that dust neutral collision frequency stabilizes the system whereas dust polarization force has destabilizing effect on dusty plasma system.

Acknowledgement

The authors gratefully acknowledge Madhya Pradesh Council of Science & Technology, Bhopal (R&D File No. A/RD/RP-2/2015-16/243), SERB DST, New Delhi (SB/FTP/PS-075/2014) and UGC, New Delhi (F. No. 42-820/2013(SR)) for their financial support.

References

[1] Rao N N, Shukla P K and Yu M Y 1990 *Planet. Space Sci.* **38** 543
[2] Shukla P K and Silin V P1992 *Phys. Scr.* **45** 508
[3] Pieper J B and Goree J 1996 *Phys. Rev. Lett.* **77** 3137
[4] Rosenberg M 1996 *J. Vac. Sci. Technol. A* **14** 631
[5] OstrikovK N, VladimirovS V, Yu M Y and Morfill G E 2000 *Phys. Rev. E* **61** 4315
[6] Kaw P K and Singh R 1997 *Phys. Rev. Lett.***9** 423
[7] Bal S and Bose M 2013 *Pramana J. Phys.* **80** 643
[8] Hamaguchi S and Farouki R T1994 *Phys. Rev. E* **49** 4430
[9] Hamaguchi S and FaroukiR T 1994 *Phys. Plasmas* **1** 2110
[10] Khrapak S A, Ivlev A V, YaroshenkoV V and Morfill G E 2009 *Phys. Rev. Lett.* **102** 245004
[11] Jain S and Sharma P 2016 *Phys. Plasmas* **23** 093701
[12] Chandrasekhar S 1961 *Hydrodynamic and Hydromagnetic Stability* (London: Oxford University)

Effect of ripple on self-focusing of a cosh-Gaussian laser beam in collisionless plasma

Karishma Jena[1,a], Bharat Vyas[2] and Prerana Sharma[1]

[1]Physics Department, Ujjain Engineering College, Ujjain, M.P. -456010, India
[2]School of Excellence, Ujjain, M.P.- 456010, India

E-mail: [a)]karishmajena189@gmail.com

Abstract. In this paper, the effect of presence of ripple on the cosh-Gaussian beam in collisionless plasma considering relativistic nonlinearity is investigated. Using the Wentzel Kramers Brillouin (WKB) paraxial-ray approximation method and eikonal method, the basic sets of equations has been derived. It is observed that when the initial power of the ripple is greater than the critical power for self-focusing of cosh-Gaussian beam the ripple gets focused over the beam.

1. Introduction

The interaction of electromagnetic (laser) beam with plasmas has generated the outstanding interest in the field of research due to discrete applications such as particle acceleration [1,2], inertial confinement fusion [3,4], X-ray generation [5], ionospheric modificationand optical harmonic generation [6].In these applications, it is found that the laser beams ought to propagate over several Rayleigh lengths in the plasmas.Self-focusing of laser beams in plasma has gained a vulnerable interest to explore these applications theoretically as well as experimentally. The focusing/defocusing phenomena can be disturbed by the small perturbations in the beam. These perturbations or ripples (spikes) are of small amplitude which consistently grows over the surface of beam with propagation of beam in plasma. Nonetheless, many researchers have been studied the effect of presence of ripple on self-focusing/defocusing, variouselectrostatic modes of Gaussian beams [7,8].

In recent years, theinvestigators has paid attention to the dynamics and propagation of decentred Gaussian beam, normally known as cosh-Gaussian beam has become noticeable due its wide range of applications in turbulent atmosphere and optical systems.Patil and Takale[9]has studied the self-focusing of cosh-Gaussian laser beam by considering relativistic-ponderomotive regime. The relativistic self-focusing of cosh-Gaussian beam have been investigated by Patil[10]. Singh and Gupta [11] have examined the second harmonic generation produced by interaction of cosh- Gaussian beam with underdense plasma. In our present work, the self-focusing of cosh-Gaussian beamby considering relativistic nonlinearity has been analysed in the absence and presence of ripple respectively. The governing equations are solved using WKB paraxial approximation andeikonal method.

2. Self-focusing of cosh-Gaussian beam

We consider the propagation of cosh-Gaussian laser beams along the z direction with angular frequency ω in homogeneous plasma. The field distribution of cylindrically symmetric cosh-Gaussian beam is

$$E(r,0)=E_0\exp\left(-\frac{r^2}{r_0^2}\right)\cosh(\Omega_0 r) \quad (1)$$

where r is the radial coordinate of the cylindrical coordinate system, r_0 is the initial beam width of the Gaussian amplitude distribution, E represents the amplitude of the electric field at the center position of $r=z=0$, E_0 is the complex amplitude of the main beam and Ω_0 is the parameter associated with the hyperbolic cosine function also called the cosh factor. On the other hand equation (1) can be expressed as follows:

$$E(r,0)=\frac{E_0}{2}\exp\left(\frac{b^2}{4}\right)\left\{\exp\left[-\left(\frac{r}{r_0}+\frac{b}{2}\right)^2\right]+\exp\left[-\left(\frac{r}{r_0}-\frac{b}{2}\right)^2\right]\right\} \quad (2)$$

where $b=r_0\Omega_0$ is the decentered parameter also called the normalized modal parameter of cosh-Gaussian beam. The total electric vector E of the beam can be written as $E=E_0+E_1$ that satisfies the wave equation

$$\nabla^2E-\nabla(\nabla.E)+\frac{\omega^2}{c^2}\varepsilon E=0 \quad (3)$$

where c is the speed of light in vacuum and ε is the effective dielectric constant of the plasma defined as $\varepsilon=\varepsilon_0+\phi(E.E^*)$, where $\varepsilon_0=1-\omega_p^2/\omega^2$ and $\phi(E.E^*)=\omega_p^2/\omega^2\left[1-\left(1+\alpha E.E^*\right)^{-1/2}\right]$ are linear and nonlinear dielectric constant respectively. In order to solve equation (3) and following Patil[10] one can obtain the equation for dimensionless beam width parameter f as

$$\frac{d^2f}{d\xi^2}=\frac{12-12b^2-b^4}{3f^3}-\frac{\omega_p^2\alpha E_{00}^2r^2}{c^2f^3}\exp\left(\frac{b^2}{2}\right)\left(1+\frac{\alpha E_{00}^2}{f^2}\right)^{-3/2} \quad (4)$$

where $\xi=z/R_d$ and $R_d=k_0r_0^2$. The above differential equation can be solved numerically by using initial boundary conditions, as $f=1$, $df/dz=0$ at $z=0$ for an initial plane wavefront.

3. Self-focusing of rippled cosh-Gaussian beam

The field distribution of a rippled cosh-Gaussian beam is given as

$$E_1(r,0)=E_{10}\exp\left(-\frac{r^2}{r_{10}^2}\right)\left(\frac{r}{r_{10}}\right)^{2m} \quad (5)$$

where r is the radial coordinate of the cylindrical coordinate system, r_{10} is the width of the ripple, E_1 denotes the amplitude of the ripple and E_{10} is the complex amplitude of the ripple. The positive integer m describes the order of the ripple and characterizes the position of ripple. The increment in value of m shifts the maximum of the ripple ($r_{10\max}=r_{10}m^{1/2}$) away from the $z-$axis. The wave equation for ripple can be written as

$$\nabla^2E_1+\frac{\omega_0^2}{c^2}\phi(E.E^*)E_1+\frac{\omega_0^2}{c^2}\left[\phi(E.E^*)-\phi(E_0.E_0^*)\right]E_0=0 \quad (6)$$

The solution of the above equation can be written as $E_1(r,z)=E_{10}(r,z)e^{-i(k_0z+s(r,z))}$, here $E_{10}(r,z)$ is complex amplitude of electric field E_1. Substituting the value of E_1 in equation (6), one can obtain the equation as

$$2\frac{\partial s}{\partial z}+\left(\frac{\partial s}{\partial r}\right)^2=\frac{1}{k_0^2E_{10}}\left(\frac{\partial^2 E_{10}}{\partial r^2}+\frac{1}{r}\frac{\partial E_{10}}{\partial r}\right)+\frac{\phi_{eff}}{\varepsilon_0} \tag{7a}$$

$$\frac{\partial E_{10}^2}{\partial z}+E_{10}^2\left(\frac{\partial^2 s}{\partial r^2}+\frac{1}{r}\frac{\partial s}{\partial r}\right)+\frac{\partial E_{10}^2}{\partial r}\frac{\partial s}{\partial r}+\left[\frac{\omega_0^2}{c^2}\frac{1}{k_0^2}\frac{\omega_p^2}{2\omega_0^2}\alpha E_0^2\left(1+\alpha E_0^2\right)^{-3/2}\sin 2\phi_p\right]E_{10}^2=0 \tag{7b}$$

where $\phi_{eff}=\phi\left(E.E^*\right)+\omega_p^2/2\omega_0^2\,\alpha E_0^2\left(1+\alpha E_0^2\right)^{-3/2}2\cos^2\phi_p$ and ϕ_p is the phase angle between the electric field vectors of the main beam (E_0) and the ripple (E_1). Now, $\phi\left(E.E^*\right)$ can be written as

$$\phi\left(E.E^*\right)=\phi\left(E_0.E_0^*\right)+\left.\frac{d\phi}{dE.E^*}\right|_{E.E^*-E_0.E_0^*}\left(E.E^*-E_0.E_0^*\right)$$

Now, following Akhmanov [12] and Sodha[13], the solution of equations (7) can be written as

$$E_{10}^2=\frac{E_{100}^2}{f_1^2}\left(\frac{r}{r_{10}f_1}\right)^{2m}\exp\left(-\frac{r^2}{r_{10}^2f_1^2}-2\int_0^z k_i(z)\,dz\right) \tag{8a}$$

$$s(r,z)=\frac{r^2}{2f_1}\frac{df_1}{dz}+\phi_1(z) \tag{8b}$$

where f_1 is the dimensionless width of ripple and $\phi_1(z)$ is a constant. On substituting the value of E_{10}^2 and s in equation (7a) and equating the r^2 on both sides one can obtain the following equation as

$$\frac{d^2 f_1}{dz^2}=\frac{c^2}{\omega^2\varepsilon_0 r_{10}^4 f_1^3}-\left(\frac{\omega_p}{\omega_0}\right)^2\frac{\alpha E_{00}^2 f_1}{r_0^2\varepsilon_0}e^{\frac{b^2}{2}}\Bigg\{AD+\frac{E_{100}}{E_{00}}e^{-k_iz}\cos\phi_p m^{m/2}f_1^m$$
$$\exp\left[-\left(\frac{m}{2}+\left(\frac{r_{10}f_1m^{1/2}}{fr_0}+\frac{b}{2}\right)^2+\left(\frac{r_{10}f_1m^{1/2}}{fr_0}-\frac{b}{2}\right)^2\right)\right]\left[3\frac{\alpha E_{00}^2}{4}DB-2A\right]-\frac{2\cos^2\phi_p}{f^4}D\left[A-3CB\right]\Bigg\} \tag{9}$$

here, $$A=\left[1+\frac{\alpha E_{00}^2}{4}e^{\frac{b^2}{2}}\left[\exp\left(-\left(\frac{r_{10}f_1m^{1/2}}{fr_0}+\frac{b}{2}\right)^2\right)+\exp\left(-\left(\frac{r_{10}f_1m^{1/2}}{fr_0}-\frac{b}{2}\right)^2\right)\right]^2\right]^{-3/2}$$

$$B=\left[1+\frac{\alpha E_{00}^2}{4}e^{\frac{b^2}{2}}\left[\exp\left(-\left(\frac{r_{10}f_1m^{1/2}}{fr_0}+\frac{b}{2}\right)^2\right)+\exp\left(-\left(\frac{r_{10}f_1m^{1/2}}{fr_0}-\frac{b}{2}\right)^2\right)\right]^2\right]^{-5/2}$$

$$C=\frac{\alpha E_{00}^2}{4}e^{\frac{b^2}{2}}\left[\exp\left(-\left(\frac{r_{10}f_1m^{1/2}}{fr_0}+\frac{b}{2}\right)^2\right)+\exp\left(-\left(\frac{r_{10}f_1m^{1/2}}{fr_0}-\frac{b}{2}\right)^2\right)\right]^2$$

$$D=\left[\exp\left(-\left(\frac{r_{10}f_1m^{1/2}}{fr_0}+\frac{b}{2}\right)^2\right)+\exp\left(-\left(\frac{r_{10}f_1m^{1/2}}{fr_0}-\frac{b}{2}\right)^2\right)\right]^2$$

The above equation can be solved numerically by applying the initial boundary conditions: $f_1=1$ and $d^2f_1/dz^2=0$.

4. Discussions and conclusions

Figure1. illustrates that the beam-width parameter f decreases for higher values of decentered parameter (b). Thus, as the value of b increases, the nonlinearity of the system increases and the self-focusing will obtain at shorter value of ξ .

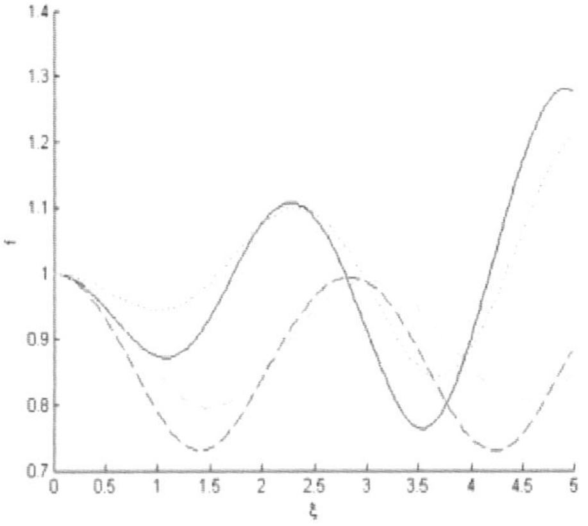

Figure 1. Variation of beam-width parameter f with dimensionless distance of propagation ξ for various values of decentred parameter b. The parameters are: $\omega = 1.778\times10^{20}\, rad/s$, $r_0 = 20\mu m$, $\omega_p/\omega = 2\times10^{-6}$, $\alpha E_{00}^2 = 0.1$. Solid curve $(b=0)$, doted curve $(b=1)$, dashed curve $(b=2)$, dot dashed curve $(b=3)$.

In conclusion, the effect of ripple on self-focusing of cosh-Gaussian laser beam has been studied considering relativistic nonlinearity.The propagation of rippled cosh-Gaussian beam in plasma affects the focusing of beam and also leads to the variation in growth/decay rate of ripple (k_i). The growth or decay rate of ripple depends upon the effective intensity of the main beam in the plasma, electron density in the plasma, frequency of the laser beam and the phase angle ϕ_p .

Acknowledgement

The authors gratefully acknowledge Madhya Pradesh Council of Science & Technology, Bhopal (R&D File No. A/RD/RP-2/2015-16/243), SERB DST, New Delhi (SB/FTP/PS-075/2014) and UGC, New Delhi (F. No. 42-820/2013(SR)) for their financial support.

References

[1] Hora H, Miley G H Ghoranneviss M and Salar Elahi A 2013 *Laser Part. Beams* **31** 249
[2] Bingham R 2006 *Phil. Trans. R. Soc. A* **364** 559
[3] Deutsch C, Furukaw H, Mima K, Murakami M and Nishihara K 1996 *Phys. Rev. Lett.* **77** 2883
[4] Hora H 2007 *Laser Part. Beams* **25** 37
[5] Davis J, Petrov G M and Velikovich A L 2005 *Phys. Plasmas* **12** 123102
[6] Sprangle P and Esarey E 1991*Phys. Rev. Lett.* **67** 2021
[7] Sodha M S, Singh D P and Sharma R P 1979 *Appl. Phys.* **18** 97
[8] Sodha M S, Singh T, Singh D P and Sharma R P 1981 *Phys. Fluids* **24** 914
[9] Patil S D and Takale M V 2013 *Laser Phys. Lett.* **10** 1
[10] Patil S D, Takale M V, Navare S T, Fulari V J, Dongare M B 2012 *Opt. Laser Tech.* **44** 314
[11] Singh A and Gupta N 2016 *Laser Part. Beams* **34** 1
[12] Akhmanov S A, Sukhorov A P, Khokhlov R V 1968 *Sov. Phys. Usp.* **10** 609
[13] Sodha M S, Ghatak A K, Tripathi V K 1976 *Prog. Opt.***13** 169

IOP Conf. Series: Journal of Physics: Conf. Series **836** (2017) 012031 doi:10.1088/1742-6596/836/1/012031

Semi-Classical theory of Nonlinear interaction of circularly polarized optical vortex beam with plasma channel

B.S.Sharma[1,a], **R.C. Dhabhai**[1], **A.Sharma** [2], **and N.K.Jaiman**[3]

[1] Department of Physics, Govt. College Kota, Kota-324001, India.

[2]Department of Electrical Engineering, Indian Institute of Technology Kanpur, Kanpur-208016,India.

[3]Department of Pure and Applied Physics, University of Kota, Kota-324010, India.

E-mail: [a)]`bs_phy@yahoo.com`

Abstract. A semiclassical approach of nonlinear interaction of intense circularly polarized optical vortex Laguerre-Gaussian (LG) beam modes with a plasma channel is analyzed theoretically and numerically. We study an exchange of angular momentum between the vortex beam and plasma channel. The transfer of angular momentum and the generated magnetic field are calculated. We have observed that both the generated magnetic field and angular momentum transfer depend on beam mode, intensity, and the polarization state of beam mode.

1. Introduction

It is now well understood that light beam with helical phase front carries orbital angular momentum (OAM) along their direction of propagation in addition to spin angular momentum that describe their polarization. Photons in a light beam have spin $\sigma_z\hbar$, where $\sigma_z = \pm 1$ refers to the left and right circular polarization states respectively. For linear polarized beam,$\sigma_z = 0$ in the direction of propagation. In polarized light, spin of photons align in the direction of propagation and contribute to a net spin angular momentum. The helical wavefront exists when the wave vectors spiral around the beam axis and constitute to the OAM [1].

The current understanding of transfer of spin and orbital angular momentum suggest that any beam with an amplitude distribution $u(r, \phi, z) = u_0(r, \phi, z) \exp \iota\ell\phi$, carries angular momentum about the beam optical axis. These laser beams carry total angular momentum much greater than that associated with the circularly polarized Gaussian laser beam. This problem was addressed recently in plasma physics where its effect on various phenomenon of laser-plasma interaction were taken into account. Cormier-Michel *et al* [6] and Stupakov *et al* [7] has recently studied the excitation of quasi-magnetic field due to the interaction of LG higher modes with plasma. Ali *et al* [4] pointed out that a linearly polarized beam in OAM state could also generate magnetic field in plasma. The propagation of higher laser modes in plasma channel was studied by York *et al* and has explained direct acceleration of electrons in a corrugated plasma channel [8]. Nesterov *et al* [9] has analyzed the importance of transfer of OAM by CPVBs to inhomogeneous plasma. Andersen *et al* [10] has demonstrated the coherent transfer of the orbital angular momentum of a photon to an atom in quantized units of $\hbar$ using a 2-photon stimulated Raman process with Laguerre-Gaussian beams to generate an atomic vortex state in the Bose-Einstein condensation of sodium atoms. Kyosuke Sakai *et al* [11] has recently studied the excitation of multipole plasmons by an optical vortex beam and explained the transfer of angular momentum between photons and plasmons.

In the present work, we have proposed a model that allow transfer of OAM of LG_p^ℓ modes and magnetic field generation in plasma channel at relativistic limit. We have theoretically examined that a normal incident CPVBs with a specific azimuthal mode can excite the magnetic field and couple to the plasmons of spiraling plasma electrons which results into the transfer of angular momentum of CPVBs to plasma electrons. We have demonstrated that the transfer of OAM and magnetic field generation depend on the mode and the polarization states of the vortex beam.

The paper is organized as follows: In section 2, the governing Equations of the angular momentum transferred to plasma electrons is derived using semiclassical approach. The conclusions are presented in section 3.

2. The Model

The OAM of LG_p^ℓ modes impart helicoidal motion to the plasma electrons and form vortices like structure in a plane perpendicular to the direction of beam propagation. Evolution of these structures result into the excitation of asymmetric quasi-static magnetic fields which depend on the order of beam mode and an axial phase velocity $v_{ph,z} = c(1 - 1/k(\partial\theta_{lp}/\partial z))$. Hence, the average torque received by electrons per unit volume also be depend on the beam mode.
An Equation for the average rate of change of angular momentum of the electrons per unit volume is given as

$$m_e n \frac{d(rv_\phi)}{dt} = -enr(E_\phi + v_z B_r - v_r B_z) - \frac{dM_z}{dt}, \tag{1}$$

where E_ϕ is the azimuthal electric field, B_r and B_z are respectively radial and axial magnetic fields. And M_z is the axial angular momentum density of photons per unit wavelength. The term $m_e v_\phi r$ is the angular momentum of electrons in the z direction and dM_z/dt refers to the pressure like term of quasi-static magnetic field. The relation for E_ϕ is given by

$$E_\phi \simeq \frac{-1}{e\omega L}\frac{\partial I}{\partial r} \tag{2}$$

We observed that the total angular momentum is non-zero for the plane polarized light ($\sigma_z = 0$). Following [?] and Faraday's law, the time derivative of the generated axial magnetic field is given as

$$\frac{\partial B_z}{\partial t} = \frac{1}{enr_{ch}^3 L}\frac{d}{dt}\left(\frac{1}{2\omega c}\frac{\partial}{\partial r}\left(r\frac{\partial I(r,z,\phi)}{\partial r}\right)\times \right. \left. (\ell_{pz} + p \pm \sigma_z)\hbar\cos\theta_{lpz}\right). \tag{3}$$

Integrating Eq.(6) within the time limit t=0 and t and assuming that the damping rate of the laser energy over this period is almost negligible, the generated magnetic field in plasma channel turns out to be of the following form

$$B_z = \frac{\eta(r)}{enr_{ch}^3 \omega L c}\left(\frac{\partial}{\partial r}\left(r\frac{\partial I}{\partial r}\right)\times \right. \left. (\ell_{pz} + p \pm \sigma_z)\hbar\cos\theta_{lpz}\right), \tag{4}$$

It is important to note here that for linearly polarized light B_z is nonzero.

To estimate B_z in Mega Gauss (MG), we assume $\partial/\partial r = 1/r_{ch}$. Thus Eq.(7) can be written as

$$B_z = \eta(r)\left(\frac{\lambda^2}{r_{ch}^2}\right)\left(\frac{n_c}{n}\right)[(\ell_{pz} + p \pm \sigma_z)\hbar\cos\theta_{lpz} \;]\left(\frac{I(r,\phi)\lambda^2}{7.3\times 10^{22} Wm^{-2}(\mu m)^2}\right), \tag{5}$$

IOP Conf. Series: Journal of Physics: Conf. Series **836** (2017) 012031 doi:10.1088/1742-6596/836/1/012031

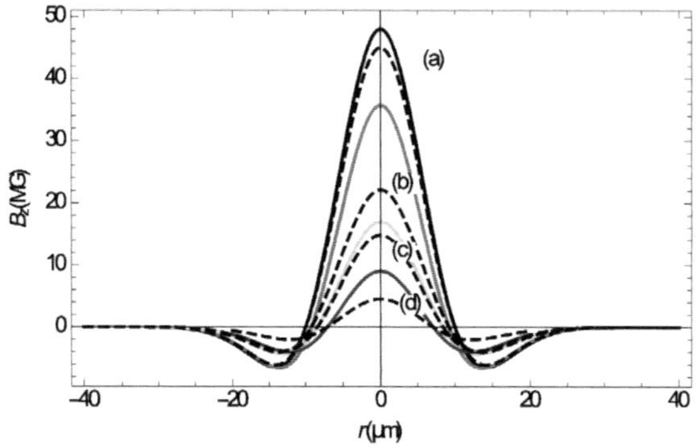

Figure 1. The generated axial magnetic field B_z of mode LG_0^1 at different values of the phase shifts $(a)\theta_{lpz} = 0(b)\theta_{lpz} = \pi/3$ (c) $\theta_{lpz} = \pi/4$ and (d)$\theta_{lpz} = \pi/6$. The continuous and dashed lines respectively refers to the variation right circular and left circular polarized beam.

where λ is wavelength of the laser beam, n_c is critical density (which is approximately $1.1 \times 10^{15} m^{-3}$).

The intensity profile of LG beam modes in the focal plane (z=0) is given as

$$I(r,\phi) = I_0 \frac{(-1)^{2\ell} p!}{(\ell + p)!} \exp(-\rho^2) \times (\sqrt{2}\rho)^{2\ell} \times L_p^\ell(\rho^2) \cos^2(\theta_{lpz}), \tag{6}$$

where $I_0 = a_0^2(4\omega^2/k^2\pi^2)(1/(w_0^2(1+\delta_{0\ell}^2))$ is the maximum intensity of LG beam. It is clear from Eqs.(8-9) that the generated axial magnetic field depends on the beam mode, plasma density, channel width, the phase shift θ_{lpz}, and polarization state of the beam. The generated axial magnetic field for the mode $LG_0^0(\ell = 0, p = 0)$ can be obtained via Eq.(8) and Eq.(9) and read as

$$B_z \approx \frac{I_0 \eta(r) \cos\theta_{pl}}{enr_{ch}^3 \omega c} \left[\sigma_z \hbar \cos\theta_{lpz} \exp(-\rho^2) \right], \tag{7}$$

for $LG_0^1(\ell_{pz} \approx 1$ and $p = 0)$, we have

$$B_z \approx \frac{I_0 \eta(r)}{enr_{ch}^3 \omega c} \left[(1 \pm \sigma_z)(1-\rho^2)\hbar \cos\theta_{lpz} \right] \exp(-\rho^2), \tag{8}$$

Figure 1 shows the variation of the generated magnetic field as a function of r for different values of beam phase transfer and polarization states for right and circularly polarized beams of LG_0^1 mode. The continuous and broken curves show the variation for right circular and left circular polarized beams respectively. We have observed that the magnitude of the generated magnetic field for right circular polarized beam mode is higher than the left circularly polarized beam.

Figure 2 shows the variation of an angular momentum density M_z of the plasma electrons as a function of r for the different beam modes LG_0^0, LG_0^1, and LG_0^2. It is observed that the angular momentum density transfer increases with increasing mode order and also depends on the polarization state of the beam modes.

3. Conclusion

In the present work, we have examined theoretically and numerically the effect of polarization states on the transfer of angular momentum and the generated magnetic field of CPVBs in the

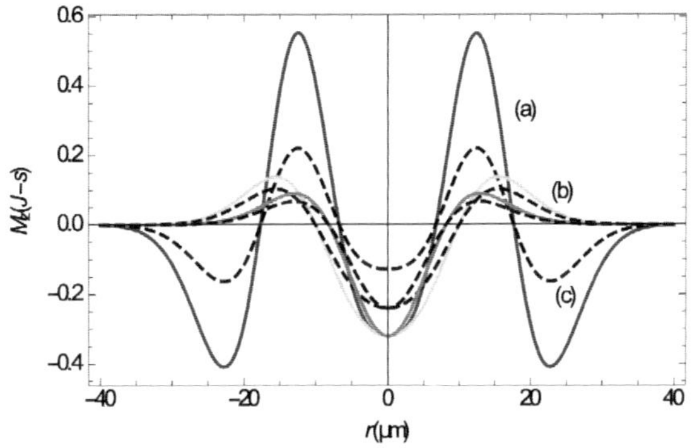

Figure 2. The variation of orbital angular momentum density M_z as a function of r for different modes (a)LG_0^2(b)LG_0^1, and (c) LG_0^0..The continuous and broken curve refers to the variation of right circular and left circular polarized beam respectively.

plasma channel using a semi-classical approach. We note that both the generated magnetic field and the angular momentum transfer depends on the LG_p^ℓ mode order, polarization state of beam and the intensity of beam. The present study may be useful in various context of laser plasma interaction such as in angular dependent wake fields, various laser acceleration schemes, laser fusion research, generation of electron vortex beams and in photonic science.

References

[1] Allen L, Beijersbergen M W, Spreeuw R J C and Woerdman J P 1992 Orbital angular momentum of light and the transformation of Laguerre-Gaussian laser modes *Phys. Rev. A* **45** 8185

[2] Marrucci L, Manzo C and Paparo D 2006 Optical spin-to-orbital angular momentum conversion in inhomogeneous anisotropic media *Phys. Rev. Lett.* **96** 1639305

[3] Leach J, Padgett M J, Barnett S M, Franke-Arnold S and Courtial J 2002 Measuring the orbital angular momentum of a single photon *Phys. Rev. Lett.* **88** 257901

[4] Ali S, Davies J R, and Mendonca J T 2010 Inverse Faraday Effect with Linearly Polarized Laser Pulses *Phys. Rev. Lett.* **105** 035001

[5] Mendonca J T, Thide B, Bergman J E S, Mohamadi S M, Eliasson B, Baan W and Then H 2008 Photon orbital angular momentum in a plasma vortex *arXiv:0804.3221v3.* Najmudin Z, Tatarakis M, Pukhov A, Clark E L, Clarke R J, Dangor A E, Faure J, Malka V, Neely D, Santala M I K and Krushelnick K 2001 Measurements of the Inverse Faraday Effect from Relativistic Laser Interactions with an Underdense Plasma *Phys. Rev. Lett.* **87** 215004

[6] Cormier-Michel E, Esarey E, Geddes C G R,Schroeder C B, Paul K, Mullowney P J, Cary J R and Leemans W P 2011 Control of focusing fields in laser-plasma accelerators using higher-order modes *Phys. Rev. ST Accel. Beams* **14** 031303.

[7] Stupakov G V and Zolotorev M S 2001 Ponderomotive Laser Acceleration and Focusing in Vacuum for Generation of Attosecond Electron Bunches *Phys. Rev. Lett.* **86** 5274.

[8] York A G, Milchberg H M, Plastro J P, and Antonsen T M 2008 Direct Acceleration of Electrons in a Corrugated Plasma Waveguide *Phys. Rev. Lett.* **100** 195001

[9] Nesterov A N and Niziev V.G 2000 Laser beams with axially symmetric polarization *Journal of Phys. D:Applied Phys.* **33** 1817.

[10] Andersen M F, Ryu C, Clade Pierre, Natarajan, Vazir A, Helmerson K, and Phillips W D 2006 Quantized Rotation of Atoms from Photons with Orbital Angular Momentum *Phys. Rev. Lett.* **97** 170406.

[11] Sakai Kyosuke, Nomura Kensuke, Yamamoto Takeaki, and Sasaki Keiji 2015 Excitation of multipole plasmons by optical vortex beams *Scientific Reports* **5** Article 8431 .

IOP Conf. Series: Journal of Physics: Conf. Series **836** (2017) 012032 doi:10.1088/1742-6596/836/1/012032

Study of synthesis and optical properties of Cu nanoparticles

Jaiveer Singh[1,a], Pavitra Devi Lodhi[2], K. K. Choudhary[3] and Netram Kaurav[2,b]

[1] Department of Physics, ISLE, IPS Academy, Rajendra Nagar, Indore 452012, India
[2] Department of Physics, Government Holkar Science College, A. B. Road, Indore 452017, MP, India
[3] Department of Physics, Indian Military Academy, Dehradun (Uttarakhand) 248007, India

E-mail: [a)]jaiveer24singh@gmail.com, [b)]netramkaurav@yahoo.co.uk

Abstract. Nanoparticles of Copper (Cu) have attracted great interest in recent years because of their unique physical and optical properties that are of industrial importance. To understand their basic properties, Cu nanoparticles were synthesized by Polyol method. The synthesized powder was characterized by X-ray diffraction (XRD) and Fourier Transform Infrared Spectroscopy (FTIR). The average particle size and lattice parameter estimated by XRD were found to be ~42.5 nm and 3.617 Å respectively. The results suggest suitability of these nanoparticles as dopants in other materials such as polymer materials and oxides.

1. Introduction

Noble metal nanoparticles have been the subject of focused research due to their unique optical, electronic, mechanical and chemical properties that are significantly different from those of bulk materials [1-2]. For these reasons, metallic nanoparticles have found uses in many applications in different fields, such as catalysis, photonics, and electronics. Preparation of copper nanoparticles has attracted considerable attention due to their diverse properties and uses, like catalysis, antimicrobial and antibacterial activities, and surface-enhanced Raman scattering (SERS) [1-3]. For the last few years, many efforts have been made on the synthesis of metallic copper nanoparticles in condensed phases with shape, size and growth control.

2. Experimental

All the chemical reagents used for the synthesis of Cu nanoparticles were of analytical grade and we have used them as received without further purification. In the polyol method, 2 g of Copper (II) acetate monohydrate (99.9% Merk make) was mixed with 50 ml of ethylene glycol in a round bottom flask. The mixture was then heated at 180°C for 1 hour 30 minutes. This gave rise to bricks colour precipitate due to the formation of cobalt nanoparticles after the completion the reaction. The solution was cooled to 27°C, and washed by adding 20 ml ethanol (99.9% Jiangsu Huaxi) to extract and wash the nanoparticles many times. The particles were then dried at 60°C and used directly for characterizations.

3. Results and discussions

The XRD patterns of copper nanoparticles obtained by polyol method are shown in figure 1. In comparison to a copper standard (JCPDS 04-0836),the characteristic diffraction peaks of copper located at 43.17°,50.47°, 74.15°,89.9° and 95.27° were observed. They correspond to the (111), (200), (220), (311) and (222) planes of the fcc structure, respectively. No other impurity peak was detected in the sample. This result confirms that with this process it is possible to obtain pure copper NPs. The XRD average size was found 42.5 nm. The FTIR spectrum of copper nanoparticles obtained by polyol method process is shown in figure 2.This analysis was used to determine the functional organic groups in the surface of the nanoparticles generated by oleic acid. Two bands at 2929 cm−1 and 2351cm−1 can be seen, and they are attributed to symmetric and asymmetric stretching of CH2 group and terminal groups −CH3 and =CH which correspond to oleic acid. Other absorption bands of carboxylate coordination and metals located at 1623 cm^{-1} and 1450 cm^{-1} for asymmetric stretching COO− and bands located at 1123 cm^{-1} and 847 cm^{-1} corresponding to symmetric stretching of COO− group can be seen[4-6].

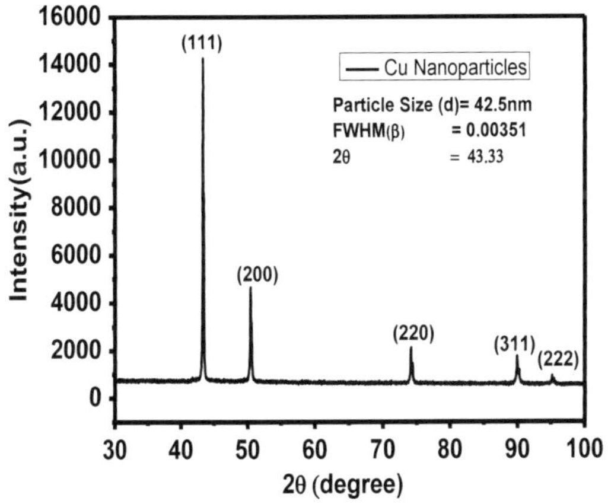

Figure 1. XRD of Cu nanoparticles

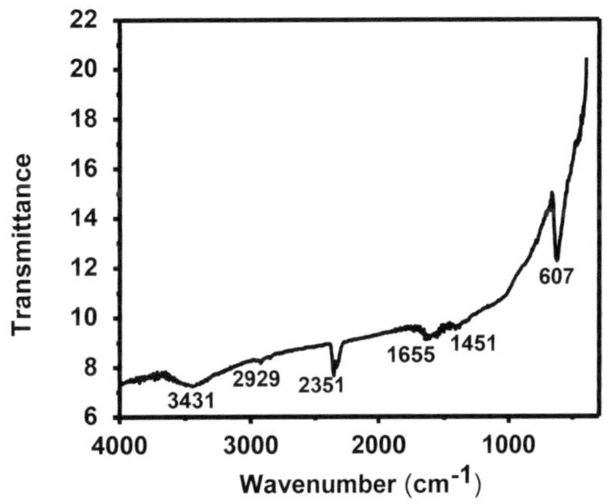

Figure 2. FT-IR of Cu nanoparticles

4. Conclusions

The Cu nanoparticles were prepared by polyol method. The average particle size and lattice parameter estimated by XRD were found to be ~42.5 nm and 3.617 Å respectively. The FT-IR results suggested the good optical properties of the Cu nanoparticles.

Acknowledgements

Dr. Mukul Gupta and Dr. V. Ganesan, Centre Director, UGC-DAE-CSR, Indore for providing XRD facility and financial support from Madhya Pradesh Council of Science and Technology, Bhopal are gratefully acknowledged.

References

[1] Singh S C, Swarnkar R K and Gopal R 2009 *J. Nanosci. Nanotech.* **1** 9
[2] Balamurugan B, Mehta B R and Shivprasad S M 2001 *Appl. Phys. Lett.* **79** 3176
[3] Khanna P K, Gaikwad S and Marimuthu R 2007 *Mater. Lett.* **61** 4711
[4] Park B K, Jeong S, Kim D, Moon J, Lim S and Kim J S 2007 *J. Coll. Inter. Sci.* **311** 417
[5] Kalenskii A V, Zvekov A A, Nikitin A P and Anan'eva M V 2015 *Russ. Phys. J.* **58** 1098
[6] Tilaki R M, Irajizad A and Mahdavi S M 2007 *App. Phys. A* **88** 415

Structural, electronic and magnetic properties of metal thiophosphate $InPS_4$

Priyanka Rajpoot [a], Vikas Nayak, Meena Kumari, Priya Yadav, Shashank Nautiyal and U P Verma

School of Studies in Physics, Jiwaji University, Gwalior-474011 M.P., India

Email: [a)] priyankarajpoot47@gmail.com

Abstract. The non-centrosymmetric crystal, $InPS_4$,has been investigated by means of density functional theory (DFT). In this paper we have calculated the structural parameters, electronic band structures, density of states plot and magnetic properties using full potential linearized augmented plane wave (FP-LAPW) method. The exchange correlation has been solved employing the generalised gradient approximation due to Perdew-Burke-Ernzerhof. The calculations are performed both without spin as well as spin polarized. The results show that $InPS_4$ is an indirect band gap semiconductor with (N-Γ) energy gap of 2.32eV (without spin) and 1.86eV in spin up and down channels.The obtained lattice parameters and energy gap agree well with the experimental results. Our reported magnetic moment results show that the property of $InPS_4$is nonmagnetic.

1. Introduction

Modern technological developments require more research in the physical properties of material such as mechanical properties, which are closely linked to crystalline and electronic structure of materials. Chalcogenide materials show rich structural and compositional diversity [1]. The ternary semiconducting compounds of chemical formula $M^{III}PS_4$($InPS_4$) crystallize in a non-centrosymmetric structure with a crystallographically ordered array of vacancies in the *I-4* space group. Their crystallographic structure belongs to the so called "*twice defective chalcopyrite*" [2]. Several groups of researchers have made these compounds the object of their experimental studies. And to the author's knowledge, no theoretical calculations of magnetic properties of our compounds have been reported till so far.The crystallographic structure was determined by Carpentier et al. [3]. It is a three dimensional structure made of corner-connected [PS_4] and [InS_4] tetrahedrons. In 1993 Bolcatto et al. used tight-binding method to calculate the electronic structures of the MPS_4 (M=B, Al, Ga, and In) thiophosphate family [4]. This paper is organized as follow; after introduction, gives the computational details, calculated results and through discussion on structural, electronic and magnetic properties of our material and in last conclusions is presented.

2. Computational details

We have carried out structural, electronic and magnetic properties of the $InPS_4$with the framework of density functional theory (DFT) [5] while taking into account the spic effect as well using the full potential linear augmented plane wavemethod as implemented in wien2k code [6]. The generalized gradient approximation (GGA) is used employing the scheme of Perdew-Burke-Ernzerhof (PBE) [7] to find the solution of exchange correlation functional. We have performed non-magnetic calculation and spin polarize calculations. We have taken the atomic positions of In, P and Sand take their convergent muffin tin radius. The muffin-tin radius is chosen in such a way that they do not overlap each other and core charge do not leak out the spheres. We expand the basis function up to $R_{MT}*k_{max}=7$, where R_{MT} is the radius of muffin tin and k_{max} is the maximum value of the reciprocal lattice vectors.The dependency of the total energy on the number of k-points in the irreducible wedge of Brillouin zone was optimized and the size of mesh has been set of 512 K-points. The equilibrium lattice constants for $InPS_4$ compound are determined by minimization of the total energy with respect to volume of the unit cells. The self-consistent calculation stops only when the total energy and the charge of the system meet the convergent limit of the order 10^{-4}Ry and $10^{-3}e^{-}$, respectively.

3. Results and discussions

3.1Structural properties

The structural properties are very important for understanding the solid properties from microscopic view point. $InPS_4$ crystallizes in the body-centered tetragonal space group *I-4* with the following cell parameters: a=5.70Å and c=9.18Å. The unit cell structure of $InPS_4$is shown in figure 1.We have investigated the structural parameters of $InPS_4$ at ambient condition. In order to compute the ground state properties we made symmetric structural optimization. The total energy has been calculated corresponding to different cell volume and fitted to theMurnaghan equation of state [8]. Our obtained structural lattice parameters, bulk modulus and its pressure derivative are listed in table 1. Earlier reported theoretical results have also been given for the comparison purpose.

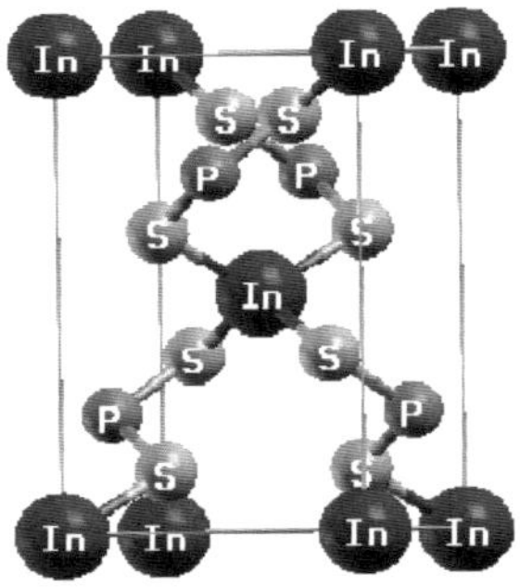

Figure 1. The unit cell structure of $InPS_4$ in space group I-4.

Table 1. Optimized results for equilibrium lattice constants (a, c), c/a ratio, unit cell volumeV_0, bulk modulus B and its first pressure derivative B′ for $InPS_4$in BCT structure.

Spin effect	a(Å)	c(Å)	c/a ratio	V_0 (a.u.^3)	B(GPa)	B'	Results
Without Spin	5.70	9.18	1.61	1007.8771	62.0490	5.000	P.W.
	5.60	9.02	1.61	....	....	....	Other work
With Spin	5.72	9.21	1.61	1019.5823	61.9169	4.5535	P.W.

3.2 Electronic properties

The electronic properties of solids are important to determine the energy distribution of electrons in the valence band and the conduction band. The equilibrium constants are used to calculate the electronic band structures and density of states for $InPS_4$ compound. First of all we have used non-spin polarized

GGA (PBE) approach and after that spin-polarized to calculate the electronic band structure of the $InPS_4$ in their most stable ground state at ambient conditions. The self consistent electronic band structure in non-spin and spin-up and spin-down channels of $InPS_4$for BCT phase are presented in figure 2. In both cases $InPS_4$ compound shows semiconductor nature with (N-Γ) indirect band gap 2.32eV and spin-up and spin-down 1.86eV, 1.86eV, respectively. In order to explicate band structures, total and partial densities of state (DOS) for In, P and S atoms are shown in figure.3. DOS plots show that the valence band near Fermi energy level is formed by In-*p*and *d*, S-*p* and P-*p*, states. While a prominent peak of P-*p* state is found in valence band. Near the Fermi level, the spin-up and spin-downDOS has a band gap between occupied states and unoccupied states. The In atom is relatively dominating in lower valence band and upper conduction band. In the energy range -4 eV to 0 eV In-*p* and In-*d* states are strongly degenerate and hybridize with S-*p* states. The top of the valence band corresponds mainly to S-*p* states.

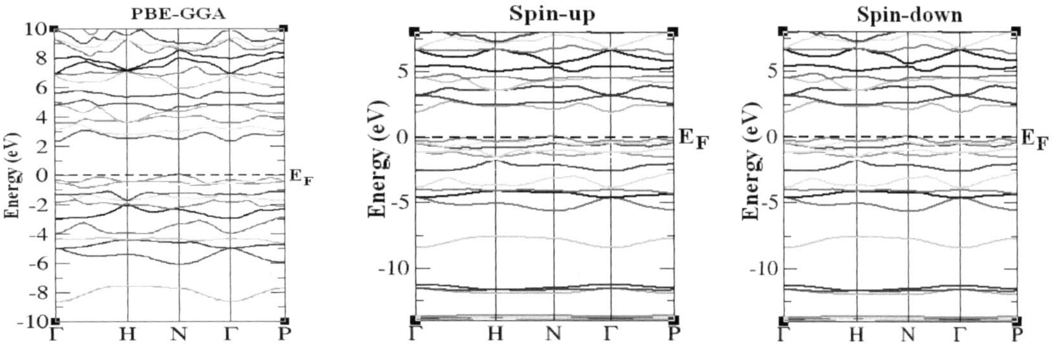

Figure 2. Electronic band structures of $InPS_4$ in BCT phase at ambient pressure and temperature (a) non-spin (b) spin-up and (c) spin-down channel.

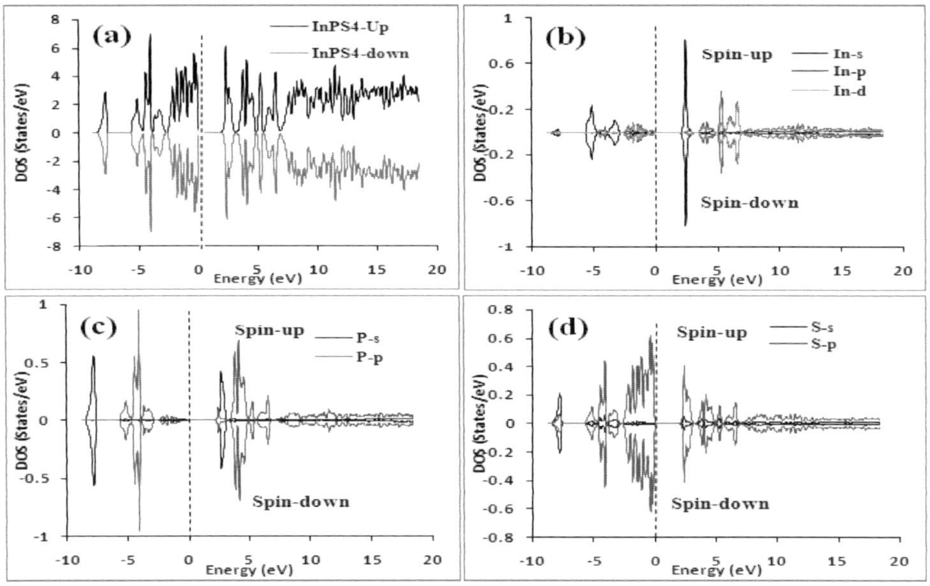

Figure 3. Total and partial DOS (including *s, p* and *d*-state of In, *s* and *p*-state of P and S atom) for $InPS_4$ compound in body-cantered tetragonal structure.

3.3 Magnetic properties

Magnetic moment of any compound shows internal contribution of spin electrons. Magnetic moments of interstitial region, atoms In, P and S, total magnetic moment of $InPS_4$ has been analysed at different cell volumes using PBE-GGA scheme. The calculated values of magnetic moments for $InPS_4$ compound are listed in table 2. The interstitial magnetic moments are negative. The electronic configuration of In is [Kr] $4d^{10}$, $5s^2$ $5p^1$, P is $1s^2$, $2s^2 2p^6$, $3s^2$ $3p^3$ and S is $1s^2$, $2s^2 2p^6$, $3s^2$ $3p^4$. In spite of differences in the number of spin up and spin down valance electrons in constituent atoms of $InPS_4$its total magnetic moment is found to be zero.

Table 2. Calculated values of magnetic moments of interstitial region, atoms In, P and S, total magnetic moment for $InPS_4$.

Lattice Constant		Magnetic moment (μB)				
a(bohr)	c(bohr)	m^{int}	m^{In}	m^{P}	m^{S}	M^{total}
10.39	16.74	-0.00191	0.00061	0.00113	-0.00003	-0.00030
10.58	17.05	-0.00097	-0.00011	-0.00010	0.00036	0.00026
10.77	17.34	-0.00118	-0.00006	-0.00013	0.00039	0.00020
10.94	17.63	-0.00129	-0.00005	-0.00013	0.00039	0.00007
11.11	17.90	-0.00139	-0.00004	-0.00013	0.00036	-0.00011

4. Conclusions

Present paper includes theoretical investigation of structural, electronic and magnetic properties of $InPS_4$ using FP-LAPW method. The PBE-GGA is opted to perform all the calculations related to exchange-correlation functional. The optimized results for equilibrium lattice constants, c/a ratio for stabilisation, unit cell volumes, bulk modulus and its first pressure derivative have been reported. The electronic band structures and density of states (DOS) plots show semiconductor nature in both spin-up and spin-down channels. Our reported results show that its character is non magnetic.To the best of our knowledge, the theoretical predictions on magnetic properties (spin polarized band structure and DOS) of metal thio phosphate ($InPS_4$) have been reported for the first time.

References

[1] Chung I and Kanatzidis G M 2014 *Chem. Mater.* **26** 849
[2] Lavrentyev A A, Gabrelian B V, Nikiforov I Ya, Rehr J J and Ankudinov A L 2003 *J. Phys. Chem. Solids* **64** 1251
[3] Carpentier C D, Diehl R and Nitsche R 1970 *Naturwiss.* **8** 393
[4] Bolcatto P G, Evelina A, Garcia and Sferco S J 1994 *Phys. Rev. B* **49** 15
[5] Hohenberg P and Kohn W 1964 *Phys. Rev. B* **136** 864
[6] Blaha P, Schwarz K, Madsen G K H, Kvasnicka D and Luitz J 1990 *Phys. Commun.* **59** 399
[7] Perdew J P, Burke K and Ernzerhof M 1996 *Phys. Rev. Lett.* **77** 3865
[8] Murnaghan F D 1994 *Proc. Natl. Acad. Sci. U.S.A.* **30** 244

IOP Conf. Series: Journal of Physics: Conf. Series **836** (2017) 012034 doi:10.1088/1742-6596/836/1/012034

Computational study of Cr substituted single-walled GaN nanotubes

Uma Shankar Sharma [a] and Rashmi Shah

Nano Materials Lab, Rustamji Institute of Technology, BSF Tekanpur, Gwalior, India-475005

Email: [a)] umashankar_rjit@yahoo.co.in

Abstract. In present study, we have investigated electronic, optical and transport properties of Cr substituted SW GaNNT. We find that the (6, 0) nanotubes of GaN are stable and semiconducting with the band gap of 2. 4 eV. Doping with Cr appears to reduce the band gap of GaN nanotubes. The V-I characteristics show typical tunnel diode behaviour for the device where the tunnelling current reaches a maximum value at the applied bias of 0.4 V and we observed a negative slope region when forward bias is applied. Because of this unusual behaviour, it can be used in number of special applications in electronic devices.

1. Introduction

In present scenario, interest in GaN based 1D structures are high in commercial and scientific level, because of the potential to apply in electronic devices[1]. Though 2D structure GaN materials were synthesized successfully for several years, research on (1D) GaN materials is limited due to the difficulties associated with their synthesis. But, pure BN nanotubes have been successfully produced by arc discharge[2,3], therefore, it opens the possibility of synthesis of other CN and GaN nanotube.

In this paper, electronic, optical and transport properties of GaN NT were carried out within the framework of density-functional theory (DFT)combined with NEGF, as implemented in the TranSiesta module within SIESTA-3.1 code[4]. We have also studied the issue of Cr droping on their electronic structure, morphological, optical and Transmission Coefficients of (6,0) GaNNTs will be talked about.

2. Computational detail

The transport properties were carried out utilizing the first-principles method implemented in the TranSiesta. The exchange–correlation energy are described by the RPBE[5] and generalized gradient approximation (GGA) implemented in calculations.TranSiesta provide the ability for the model open boundary conditions it is based on the NEGF i.e. is constructed using the DFT Hamiltonian obtained from a given electron density.The device setup for a (6,0) GaN NTwas constituted by three parts: the left (L) and right (R) semi-infinite GaN leads (electrodes) and the scattering region. The size of the scattering regiontakes place enough to prevent the interaction between the left and right electrode and current through the scattering region was calculated using Landauer-Buttiker Formula [6].

IOP Conf. Series: Journal of Physics: Conf. Series **836** (2017) 012034 doi:10.1088/1742-6596/836/1/012034

3. Results and discussions

We begin with electronic structure calculations of pristine and doped (6, 0) GaN nanotube to determine their stability. The binding energy is calculated as follows:

$E_b = -[E_{tot} - (x\ \ E_{Ga} + y\ \ \ E_N + z\ \ E_{Cr})] / (x+y+z)$,where E_{tot} is the total energy of GaN nanotube, E_{Ga}, E_N, E_{Cr} are energies of Ga, N and Cr atoms, respectively and x, y, and z are number of Ga, N and Cr atoms, respectively considered in the supercell. The formation energy E_{form}in terms of energies of constituents is calculated as follows:

$E_{form} = [E_{doped\ \ GaNNT} - (\ E_{PrisGaNNT} + z\ \ E_{Cr})] / N$, table 1 lists the calculated binding energy (E_b) and formation energy (E_{form}) together with band gap of pristine and doped GaN nanotubes.

Table1. The calculated structural and electronic properties of pristine and doped (6,0) GaN nanotubes.

System	Bond length (Å)	E_b (eV)	E_{form} (eV)	Bandgap (eV)
Pristine	1.84	6.51	-	2.41
Doped– 8% Cr	1.84	6.81	11.6	metallic
Doped - 17% Cr	1.84	7.11	22.22	metallic
Doped - 25% Cr	1.84	7.40	32.84	metallic

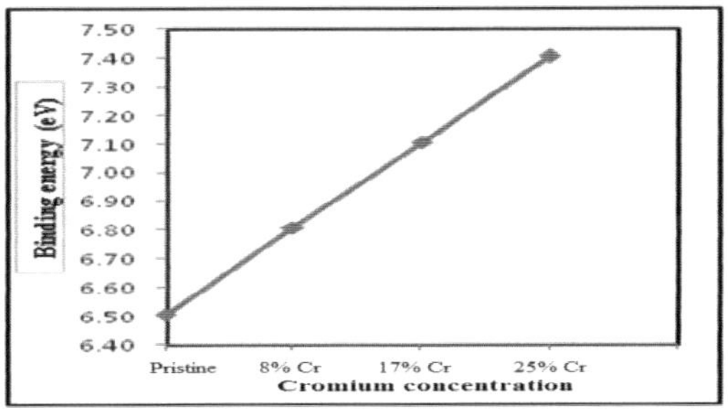

Figure 1. Plot of Binding energy V/s Cr concentration for (6, 0) GaN nanotubes

Figure 1 shows thatincrease in the binding energy with the increase in the Cr concentration for (6, 0) GaN nanotubes, are related to the fact that the electronegativity of Cr (1.66) lies between Ga (1.81) and N (3.04), thereby facilitating formation of covalent bondsin the doped lattice.

The calculated band structure and density of states of the pristine and chromium doped GaN nanotubesare shown in figure 2. The pristine GaN nanotube is found to be direct-gap material which is consistent with previous study [7]. Doping of nanotube with Cr induces an additional state in the band structure. Initially,the valence band maximum (VBM) shifts towards Fermi level as concentration of Cr increase in the lattice. Later, VBM shifts away from Fermi level, and the induced states starts shifting towards Fermi level yielding GaN nanotube to be metallic.

The reason for decrement of energy bandgap simply reflects the ionic bonding character of the doped lattice, i.e. s and p bands localized to Ga and N respectively are well separated. Due to doping of Cr atoms, more p states near the valence band edge and more s states near the conduction band edge appear to hybridize to metallicity of the system.

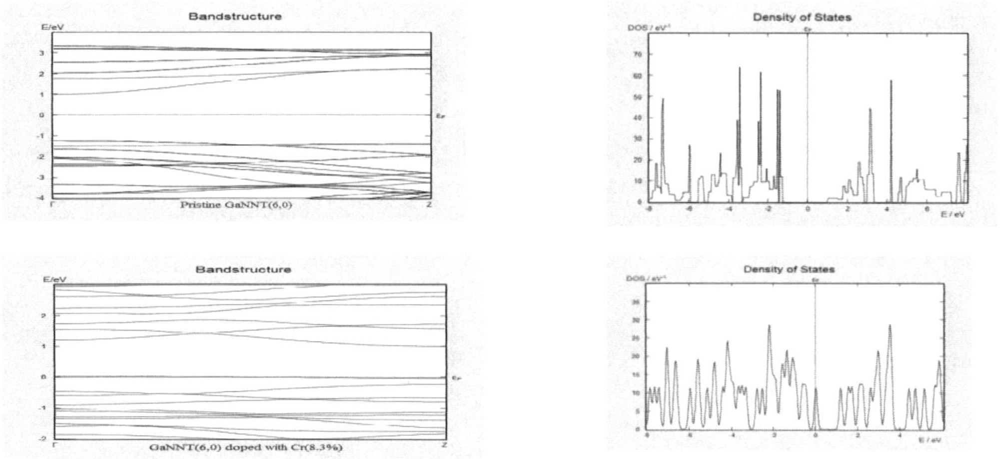

Figure 2. Band structure and density of states of (6,0) GaNNTs pristine and 8% Cr doped.

This is further confirmed by DOS showing that the lower part of the valance band is dominated by the N-4s states, and N-4p states near the Fermi level, while Ga-4s states reside just between them. On the other hand, Ga-4p states dominate in the conduction band of the tubular lattice. In the doped configuration,Crassociated additional peak near Fermi level leads to the metallic behaviour of GaN nanotube.

Next, we were calculated the imaginary parts of dielectric function then, get real part of dielectric functions and other optical properties by using the Kramers-Kroning relations, for incident light polarized parallel to the nanotube. TranSiesta. Figure 3 shows the absorption coefficient that is corresponding with imaginary dielectric curves of the configuration. Major peaks in the absorption coefficient of pristine GaN nanotubes are found at 1.75, 2.75, 6.25and 9.6 eV, and are associated with transitions of electrons from different energy states of the valence band to the conduction band, The location of these peaks depend on the Cr concentration in the doped GaN nanotubes.

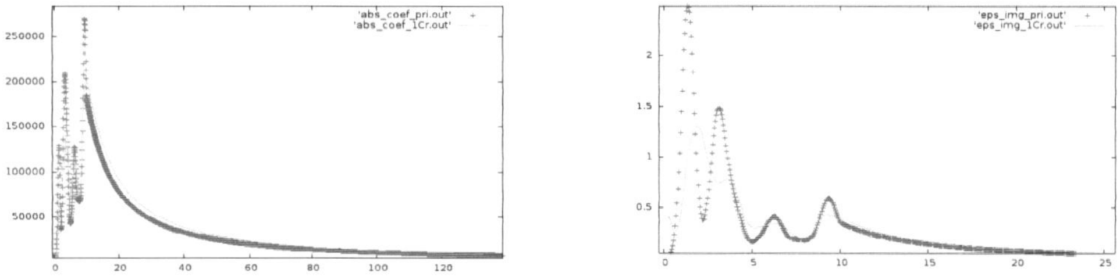

Figure 3. Absorption coefficient and imaginary part of dielectric coefficient for pristine and 8% Cr-doped (6, 0) GaN nanotube.

The electronic transport properties of the pristine and doped (6,0) GaN nanotubesfor different Cr concentration are determined under equilibrium conditions in figure 4. The transmission probability reaches high when the electrons resonate with the covalent π orbital of the atomic [8]. It is also proves thetransmissioncoefficient T (E) is zero at Fermi level confirming the semiconducting nature of the GaNNT. As the Cr concentration increases nature of transmission spectrum change and the transmission spectrum has a series of peaks and valleys. It is clear that the peaks and valleys of transmission spectrum and the density state curves have a similar trend.

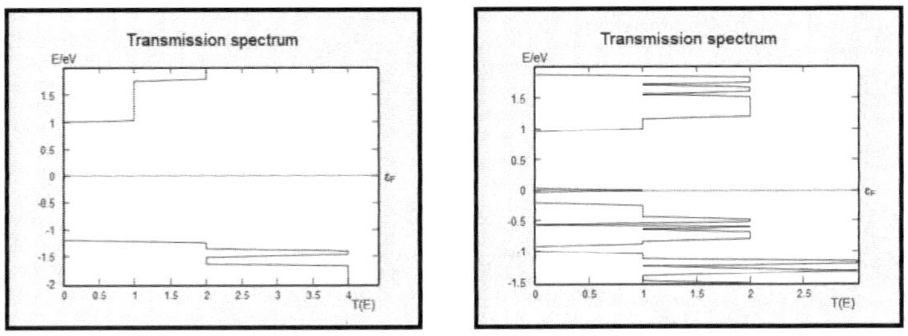

Figure 4. Transmission spectrum (6,0) pristine and 8% Cr doped GaNNT.

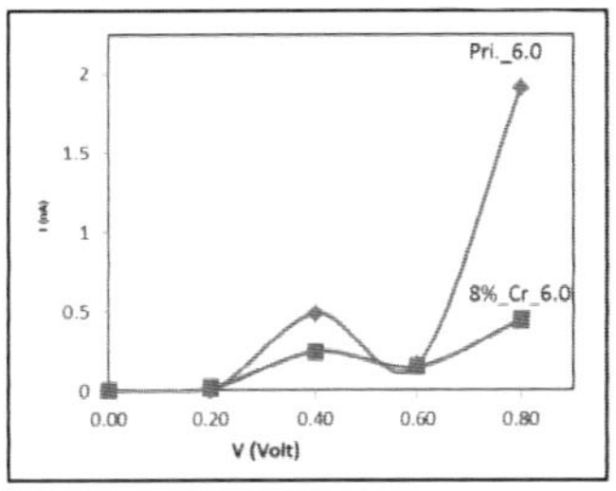

Figure 5. V-I plot for pristine and Cr-doped (6, 0) GaN nanotubes.

It is clear from figure 5, the current is almost zero in between 0.0 – 0.2 V, which means that the transmission probability T (E) in the integral windows is zero. Next,in the bias voltage in the range 0.2 to 0.4 V, the current is increased with voltage. From 0.4 to 0.6V it gives ultimate result, as the voltage increases current decreases and valley occurs at 0.6 V. This region is very important because resistance is negative showing that pristine and doped GaN nanotubes. This is the ultimate behavior of (6,0) SWGaNNT, due to which it can be used in number of special applications in electronic devices, it can act asa tunnel diode in the considered device architecture.

4. Conclusions

We conclude that, the pristine GaNNT are semiconductor having a direct band gap 2.41 eV and doping with Cr makes the tube to be metallic. This behavior is supported by density of states plots and transmission coefficient calculations.V-Icurve exhibits ultimate behavior of GaN nanotubes. For the low biasregion, resistance becomes negative leading to tunnel diode characteristics for pristine and doped GaN nanotubes. The calculated results thus suggest that GaN nanotubes may be used for full color displays in the nanoscale optoelectronic devices.

Acknowledgement

Authorsarethankful to Shri Mukesh Tyagi, Commandant, CSMT/Chief Administrator, RJIT for providing valuable support and Dr. Arvind Jain, Principal RJIT for motivation and inspiration.

References

[1] Sharma U S, Bisht P S and Verma U P 2009 *J. Phys. Condens. Matter* **21** 025501
[2] Loiseau A, Pascard H, Willaime F, Demoncy N and Hug G 1996 *Phys. Rev. Lett.* **76** 4737
[3] Suenaga K, Colliex C, Demoncy N, Loiseau A, Pascard H and Willaime F 1997 *Science* **278** 653
[4] Ordejon P, Sanchez-Portal D, Garcia A, Artacho E, Junquera J and Soler J 2000 *RIKEN Rev.* **29** 42
[5] Dreizler R M and Gross E K U 1990 *Density Functional Theory* (Berlin: Springer)
[6] Dutta S1995 *Electronic Transport in Mesoscopic Systems* ed Ahmed H, Pepper M and Broers A (Cambridge, England: Cambridge University Press)
[7] Srivastava A, Khan M I and Khare P S, 2014 *The Scientific World Journal* **2014** 984591
[8] Moradian R, Azadi S and Farahani S V 2008 *Phys. Lett. A* **372** 935

IOP Conf. Series: Journal of Physics: Conf. Series **836** (2017) 012035 doi:10.1088/1742-6596/836/1/012035

Theoretical approach to predict the phase transition of NaH under pressure

Aayushi Jain[a] and R. C. Dixit[b]

Department of Physics, Government Holkar Science College, A. B. Road, Indore 452017, MP, India

E-mail: [a)]aayushijain2810@gmail.com, [b)]dixitrc@rediffmail.com

Abstract. This paper is dealing with the high pressure behavior of alkali halide (NaH) and a structural phase transition from NaCl to CsCl phase is predicted under high pressure. We have adopted an effective interionic interaction potential, which includes the long range Coulomb, van der Waals (vdW) interaction and the short-range repulsive interaction upto second-neighbor ions within the Hafemeister and Flygare approach. The thermodynamic requirement, for the equality of free energies, at a phase boundary, suggests that the transition pressure of NaH occur at 37 GPa. The transition involves the coordination number increasing from 6 for the NaCl-type structure to 8 for the CsCl-type structure.

1. Introduction

The alkali-metal hydrides (CsH, RbH, KH, NaH, and LiH) represent a series of partly ionic materials whose high-pressure behavior is expected to parallel that of the alkali-metal halides. The alkali metal hydrides are ionic compounds which can be treated as analogues to alkali metal halides. In alkali matal hydrides hydrogen considered as the first member of the halogen group of elements and formally it behaves as an H -ion. Support for this view can be obtained from the observed crystal structures of these compounds which are typical for ionic systems (NaCl and CsCl). But at normal condition alkali metal hydrides have NaCl or *B*1-type crystal structure, while the alkali metal halides can have both NaCl or *B*1-type structure or CsCl or *B*2 type crystal structure [1]. Pressure is most important thermodynamical quantity to bring out the mechanical properties of solids and alloys. The phase change in solids at high pressure usually involves a change in crystal structure. Co ordination number is important in phase transition due to high pressure. This is because thermodynamically the molar volume of high pressure phase must be smaller than low pressure phase. Hochheimer et al. [2] has reported the transition pressures for CsH, RbH, and KH, as well as the equation of state of NaH to 28 GPa. The change in volume can be estimated using known equation of state (EOS) [3]. Bashkin et al. [4] have pointed out, using a hard-sphere ion model, that the *B*1 to *B*2 transition is thermodynamically possible in NaH, KH, RbH, and CsH compounds. The Duclous *et al* determined the specific volumes of each phase by summing the metal and hydride ion radii, which depend on the structure and coordination number, to determine the *B*2 lattice constant [5].

In particular, NaH show a structural phase transition from NaCl to CsCl. For example, Ahuja et. al. [6] calculated transition pressure for NaH to be 37 GPa. In search of B1-B2 transition Duclous et al. [5] submitted sodium hydrides to pressure as high as 54 GPa which is in good agreement with the experiment value of around 51 GPa.

IOP Conf. Series: Journal of Physics: Conf. Series **836** (2017) 012035 doi:10.1088/1742-6596/836/1/012035

To estimate the possibility of the NaCl to CsCl transition in NaH under pressure, we consider the relationship of the Gibb's free energy of both phases. The phase transition pressures and the volume collapses were obtained from the effective interionic potential model (EIOP), which incorporates the long range Coulomb, the short-range vdW attraction and the overlap repulsive interactions operative up to the second neighbor ions within the Hafemeister and Flygare framework. The vander Waals coefficient are calculated by the Slater Kirkwood variation method as a first step. Then, we evaluate volume collapse.

2. Theory and method of computation

The effective interionic potential is expressed as

$$U(r)=\sum_{ij}\frac{Z_m e^2}{r_{ij}}+\sum_{ij} b\beta_{ij}\exp\left(\frac{r_i+r_j-r_{ij}}{\rho}\right)+\sum_{ij} c_{ij}r_{ij}^{-6}+\sum_{ij} d_{ij}r_{ij}^{-8}. \quad (1)$$

Here, long-range Coulomb is represented by first term, second term correspond to Hafemeister and Flygare [7] form of short-range repulsive energies and van der Waals multipole are represented by third and fourth terms respectively. The symbols cij and dij represents the van der Waals coefficient and βij being the pauling coefficients, respectively. Z_m is the modified ionic charge and parametrically includes the effect of Coulomb screening effect, *b* and ρ are short-range parameters. Thus, the effective interionic potential contains only three free parameters (Z_m, *b* and ρ) which can be determined from the crystal properties [8]. Further, an isolated phase is stable only when its free energy is minimum for the specified thermodynamic conditions. As the pressure or any other variable acting on the systems is changed, the free energy of the system changes smoothly and continuously. A phase transition is said to occur when the changes in structural details of the phase are caused by such variations of free energy. The test materials transform from their initial *B*1 to *B*2 structure under pressure. The stability of a particular structure is decided by the minima of Gibbs's free energy, $G = U + PV - TS$. Here *U* is internal energy, which at 0 K corresponds to the cohesive energy, *S* is the vibrational entropy at absolute temperature *T*, pressure *P* and volume *V*.

The Gibbs's free energies $G_{B1}(r) = U_{B1}(r) + 2Pr^3$ for NaCl (*B*1) phase and $G_{B2}(r') = U_{B2}(r') + [8/3\sqrt{3}]Pr'^3$ for CsCl (*B*2) phase become equal at the phase-transition pressure *P* and at zero temperature i. e., ΔG (= $G_{B1} - G_{B2}$). Here, U_{B2} and U_{B1} represent cohesive energies for *B*1 and *B*2 phases, and are

$$U_{B1}(r)=-1.7475\frac{e^2Z_m^2}{r}+6V_{ij}(r)+6V_{ii}(r)+6V_{jj}(r), \quad (2)$$

$$U_{B2}(r')=-1.7627\frac{e^2Z_m^2}{r'}+8V_{ij}(r')+3V_{ii}(r')+3V_{jj}(r'). \quad (3)$$

Here *r* and *r'* are nearest-neighbor (nn) separations corresponding to NaCl and CsCl phases, respectively.

$$V_{ij}(r)=b\beta_{ij}\exp\left(\frac{r_i+r_j-r_{ij}}{\rho}\right)-c_{ij}r_{ij}^{-6}-d_{ij}r_{ij}^{-8}; \quad i,j=1,2. \quad (4)$$

Here, c_{ij} and d_{ij} being the vdW coefficients due to dipole-dipole (*d-d*) and dipole-quadrupole (*d-q*) interactions. We follow SKV method [9] to derive c_{ij} and d_{ij} as

$$c_{ij}=\frac{3}{2}\frac{e\hbar}{\sqrt{m_e}}\alpha_i\alpha_j\left[\left(\frac{\alpha_i}{N_i}\right)^{1/2}+\left(\frac{\alpha_j}{N_j}\right)^{1/2}\right]^{-1}, \quad (5)$$

and

$$d_{ij}=\frac{27}{8}\frac{\hbar^2}{m_e}\alpha_i\alpha_j\left[\left(\frac{\alpha_i}{N_i}\right)^{1/2}+\left(\frac{\alpha_j}{N_j}\right)^{1/2}\right]^2\alpha_i\alpha_j\left[\left(\frac{\alpha_i}{N_i}\right)+\frac{20}{3}\left(\frac{\alpha_i\alpha_j}{N_iN_j}\right)^{1/2}+\left(\frac{\alpha_j}{N_j}\right)\right]^{-1}. \quad (6)$$

Here, m_e being the electron mass. α_i is the electronic polarizability and N_i denotes the effective number of electrons of the ith ion, respectively. The values of the overall vdW coefficients (C and D) are obtained following Tosi approach [10] in terms of appropriate lattice sums [S6(0), S6(r), S8(0) and S8(r)].

Table 1. Van der Waals coefficients c_{ij} $(i, j = 1,2)$ [in units of 10^{-60} erg cm^6], d_{ij} $(i, j = 1,2)$ [in units of 10^{-76} erg cm^8] and overall van der Waals coefficients (C, D) for NaH compound.

	c_{11}	c_{12}	c_{22}	C	d_{11}	d_{12}	d_{22}	D
NaH	9.236	8.482	9.632	72.982	3.142	5.717	8.31	39.742

Table 2. Crystal data and model parameters for NaH compound.

Solid	Material Parameters					Model Parameters		
	r_i (Å)	r_j (Å)	a (Å)	B_T (GPa)		Z_m^2	ρ (10^{-1} Å)	b(10^{12} erg)
NaH	0.9	1.28	4.88	23		0.81	2.41	3.089

Table 3. Calculated (reported) transition pressures and volume collapses in NaH compound.

Compound	Transition pressure P_t (GPa)	Volume collapses (%)
NaH	37(28-32 [4], 37 [6], 30.7 [13], 29.3 [14], 32 [15])	4.3

3. Results and discussions

The effective interionic potential is constructed in a hierarchical and easy generalizable manner. We have undertaken such structural and elastic properties in an ordered way. For such purposes we have then three free parameters, namely, modified ionic charge, range and hardness parameters (Z_m, ρ and b). To estimate the free parameters, we first deduce the vdW coefficients from the Slater-Kirkwood variational method [11] and are listed in table 1. It is perhaps worth to remark that we have deduced the values of free parameters modified ionic charge (Z_m), range (ρ) and hardness (b) from the knowledge of equilibrium distance and the bulk modulus following the equlibrium conditions [12]. The input data along with their relevant references and the model parameters for NaH compounds are given in table 2.

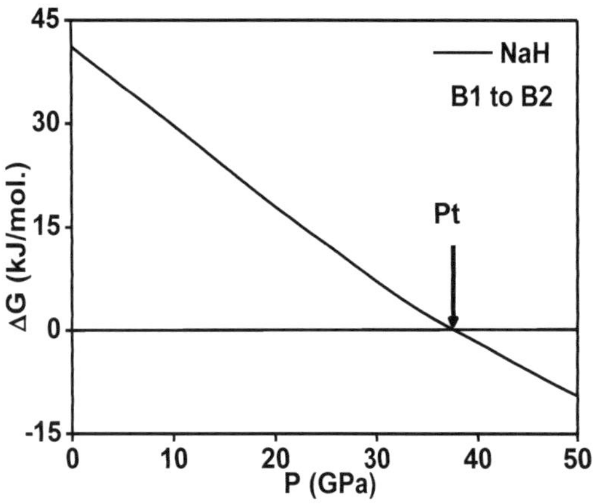

Figure 1. Variation of Gibb's Free energy with pressure

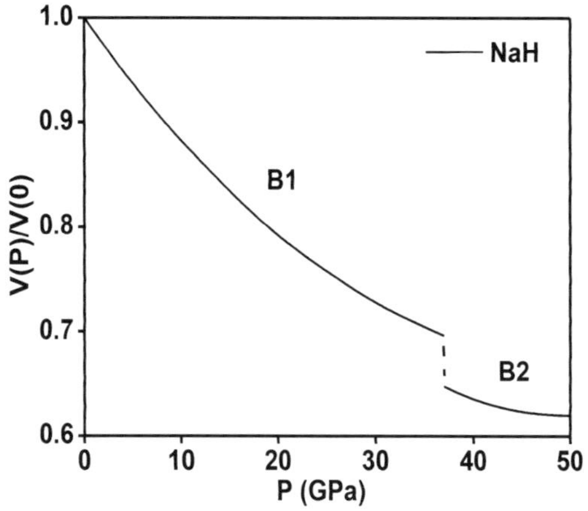

Figure 2. Variation in relative volume with pressure

In an attempt to reveal the structural phase transition of the test materials, we minimise the Gibbs's free energies $G_{B1}(r)$ and $G_{B2}(r')$ for the equilibrium interatomic spacing (r) and (r'). The Gibbs's free energy difference ΔG [= $G_{B2}(r)$ - $G_{B1}(r')$] have been plotted as functions of pressure (P) in figure 1 using the interionic potential discussed above for NaH.

Let us now estimate the values of relative volumes associated with various compressions following Murnaghan equation of state.

$$\frac{V}{V_0} = \left(1 + \frac{B'}{B_0} P\right)^{-\frac{1}{B'}}, \tag{7}$$

V_0 being the cell volume at ambient conditions. The estimated value of pressure dependent radius for both structures, the curve of volume collapse with pressure to describe the phase diagram is exemplified in figure 2 for NaH. It is noticed from the graph that our access has predicted correctly the relative stability of competitive crystal structures, as the values of ΔG are positive. The magnitude of the discontinuity in volume at the transition pressure is obtained from the phase diagram and their values are tabulated in table 3, which is in fair good agreement with those revealed from experiments [4, 6, 13] and other theoretical works [14, 15].

3. Conclusions

A high pressure structural phase transition from NaCl ($B1$) to CsCl ($B2$) in alkali metal hydride (NaH) was predicted following the developed EIOP model. It has been found that the $B1$ type structure of NaH is unstable at after pressure 37 GPa. This was established by calculating the Gibb's free energies in both the structures under consideration. Further, the associated volume collapses at the phase transitions have been found to be 4.3% for sodium hydride.

References

[1] Majewski A and Vogl P 1986 *Phys. Rev. Lett.* **57** 1366
[2] Hochheimer H D, Strossner K, Honle W, Baranowski B and Filipek F 1985 *J. Phys. Chem.* **143** 139
[3] George L and Saxsena S K 2010 *Int. J. Hydr. En.* **35** 5454
[4] Basshkin I O, Dymova T N and Powyatovskii E G 1980 *Phys. Stat. Sol.* **100** 87
[5] Duclos S J, Vohra Y K , Ruoff A L, Fillipek S and Baranowski B 1987 *Phys. Rev. B* **36** 7664
[6] Ahuja R, Eriksson O and Johansson B 1999 *Phys. B* **265** 87
[7] Hafemeister D W and Flygare W H 1965 *J Chem. Phys. Soc.* **43** 795
[8] Varshney D, Kaurav N, Sharma P, Shah S and Singh R K 2004 *Phase Trans.* **77** 1075
[9] Singh R K 1982 *Phys. Rep.* **85** 259
[10] Tosi M P 1964 *Solid State Phys.* **16** 1
[11] Slater J C and Kirkwood J G 1931 *Phys. Rev.* **37** 682
[12] Varshney D, Sharma P, Kaurav N, Shah S, and Singh R K 2005 *J. Phys. Soc. Jpn.* **74** 382
[13] Rodriguez C O and Methfessel M 1992 *Phys. Rev.* **45** 11
[14] Martins J L 1990 *Phys. Rev. B* **41** 11
[15] Sun X W, Chen Q F, Chen X R, Cai L C and Jing F Q 2011 *J. Solid State Chem.* **184** 427

Preparation and characterization of bismuth oxichloride (BiOCl) nanoparticles and nano zerovalent iron (nZVI)

Bhawna Sarwan[1,a], Brijesh Pare[1] and Aman Deep Acharya[2]

[1] Laboratory of Photocatalysis, Department of Chemistry, Govt. Madhav Science P. G. College, Ujjain 456010, MP, India,
[3] Lovely Professional University, Phagwara 144411, Punjab, India

Email: [a)] sarbhawna@gmail.com

Abstract. In this work, we have synthesized nano scale zerovalent iron (nZVI) particles by borohydride reduction method and bismuth oxichloride (BiOCl) by a hydrolysis method. X-ray powder diffraction (XRD) was used for the structural and chemical characterization, while scanning/transmission electron microscopy (SEM/TEM) were employed to determine the physical properties of the nanoparticles. The reactivity of synthesized nanoparticles was compared by decolorization of nile blue (NB) dye under visible irradiation.

1. Introduction

Nano zero-valent iron nano particle (nZVI) having zero valency with primary particle size less than 100 nm is moderately reactive in water and can serve as an excellent electron donor, which makes it a versatile remediation material [1].Zinc oxide, with its unique physical and chemical properties, such as high chemical stability, high electrochemical coupling coefficient, broad range of radiation absorption and high photostability, is a multifunctional material.The physical properties of as prepared materials were characterized by XRD, SEM, TEM. The efficiency of prepared nanoparticles was examined by the decolorization of nile blue (NB) under visible light irradiation.

2. Experimental

BiOCl powder was synthesized by a hydrolysis method. In this synthesis reagent Bi_2O_3 (AR, Aldrich) was dissolved in hydrochloric acid (AR, Merck) to get a $BiCl_3$-HCl aqueous solution. By adjusting the pH value of the solution about 3 with ammonia (AR, Shanghai), the surface of Bi_2O_3 particle converted to BiOCl [2-3]

For synthesis of NZVI, 0.8 M borohydride ($NaBH_4$)(AR, Merck) aqueous solutionwas added drop wise to a 0.6 M ferric chloride ($FeCl_3$.$6H_2O$) (AR, Merck) aqueous solution with stirring at ambienttemperature, black particles immediately formed. Synthesized nano particles were filtered by vacuum filtration and rinsed three times with ethanol to remove all the water. Then particles were stored in a bottle covered with ethanol and placed in a desiccators till further use.

3. Results and discussions

XRD patterns of the NZVI and BiOCl particles are shown in figure 1. In figure 1a, the strong and sharp diffraction peaks confirmed the highly crystalline nature of BiOCl. XRD patterns of the NZVI figure 1b demonstrated that the peak reveals the existence of an iron. Apparent peaks at the 2θ of 45.9° and 32.2° indicate the presence of both zero-valent iron (α-Fe) and iron oxide (FeO) crystalline phases.

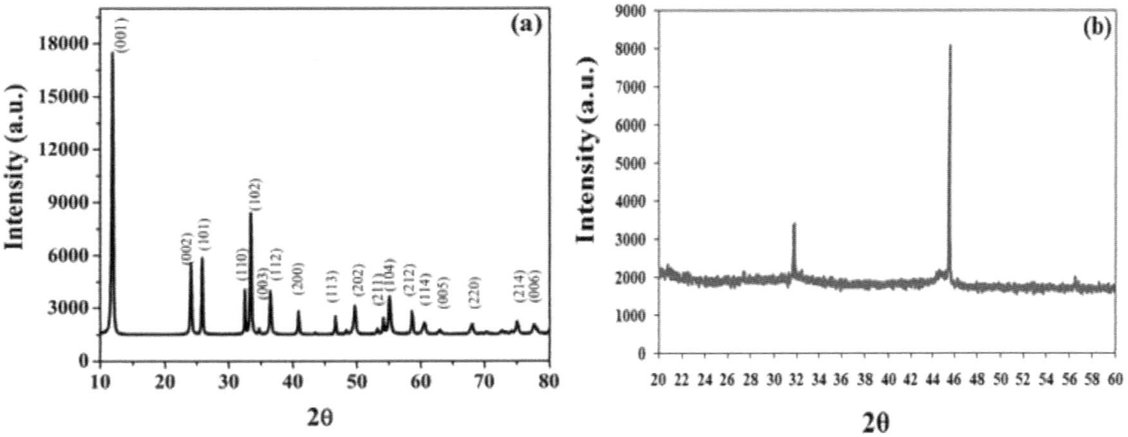

Figure 1. XRD scans of BiOCl and ZVI nanoparticles.

The SEM image of BiOCl figure 2a shows the morphology of BiOCl is dominated with plates. The NZVI figure 2b reveals that iron nanoparticles tend to form a chainlike aggregate due to the magnetic attractive force between particles. The particle size was found to be in the range of 50-100 nm.

Figure2. SEM images of (a) BiOCl and (b) ZVI nanoparticles

The TEM image of the BiOCl in figure 3a indicates that plates have a flat surface and form an octahedral shape with a diameter of about 100 nm. The prepared iron particles figure3bwas found to be largely spherical, a few particles had size as large as 200–250 nm, whereas most particles were less than 100 nm. TEM image (see figure3b) also shows that most of the particles formed chain-like aggregates.

IOP Conf. Series: Journal of Physics: Conf. Series **836** (2017) 012036 doi:10.1088/1742-6596/836/1/012036

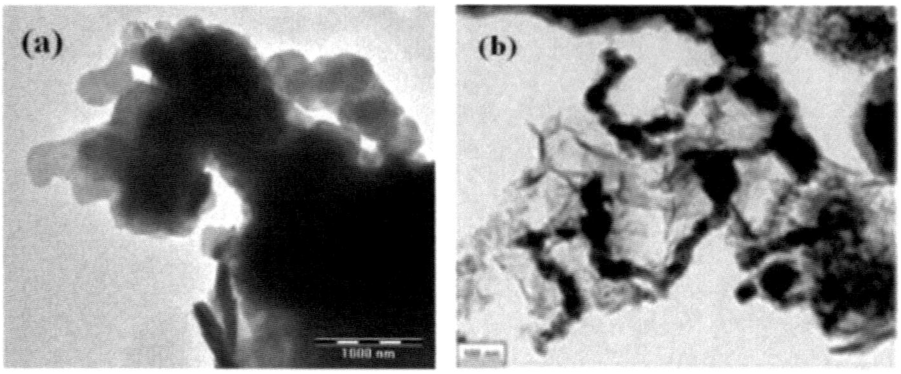

Figure 3. TEM images of (a) BiOCl and (b) ZVInanoparticles

4. Photocatalytic activity

The experiments were carried out by irradiating (with visible light) the aqueous solutions of NB dye (20ppm) containing BiOCl and ZVI nanoparticles (2.0g/L). The rate of decolorization was estimated from residual concentration spectrophotometrically. 76% and 97% of 20ppm NB was degraded with undopedBiOCl and ZVI nanoparticles, respectively, within 120 min (figure4). However, the results revealed that the activity of NZVI was better than BiOCl/visible.

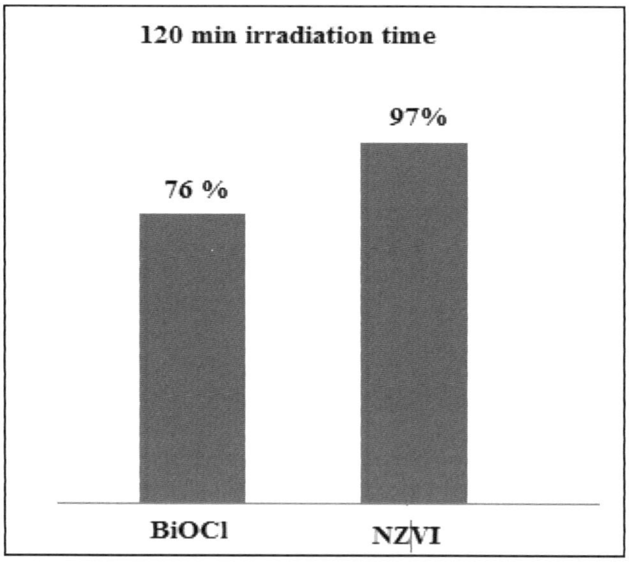

Figure 4. Photocatalytic degradation of NB

5. Conclusions

XRD data demonstrated the crystalline nature of BiOCl and ZVI nanoparticles.SEM and TEM results confirmed chain-like aggregates of NZVI andoctahedral shape of BiOCl synthesized particles. The experimental results indicated that the performance of NZVI was more effective than BiOCl/visible.

IOP Conf. Series: Journal of Physics: Conf. Series **836** (2017) 012036 doi:10.1088/1742-6596/836/1/012036

Acknowledgements
Authors acknowledge support and laboratory facilities provided by Department of Chemistry, Madhav Science College, Ujjain, India. Also the authors would like to thank the UGC-DAE CSR, Indore for their help in XRD experiment and SEM and TEM measurements.

References
[1] Zazuli M A, Safarpour M, Veisi F and Gholilou M A 2016 *Arch. Hyg. Sci.* **5** 153
[2] Pare B, Sarwan B and Jonnalagadda S B 2011 *App. Surf. Sci.* **258** 247
[3] Zainal Abedin N R and Hossain M A 2014 *Am. J. Environ. Sci.* **10** 157

IOP Conf. Series: Journal of Physics: Conf. Series **836** (2017) 012037 doi:10.1088/1742-6596/836/1/012037

Density functional investigation on structural, elastic, thermal and mechanical properties of NiTi intermetallic compound

Gitanjali Pagare

Department of Physics, Sarojini Naidu Govt. Girls College, Bhopal- 462016, India

Email: gita_pagare@yahoo.co.in

Abstract. Theoretical study of structural, elastic, mechanical and thermal properties of B_2-type binary intermetallic NiTi is performed using full-potential linearized augmented plane wave (FP-LAPW) method. In this approach the generalized gradient approximation and local density approximation is used for exchange-correlation (XC) potential. We have calculated the ground state properties such as lattice constant (a_0 = 3.0140 Å and 2.9439 Å), bulk modulus (B =161.58 GPa and 191.92 GPa) and pressure derivative of bulk modulus (B'= 4.21 and 4.15) using PBE-GGA and LDA approximations respectively for NiTi. Our calculated lattice constants are in good agreement with the experimental data available. A special attention has been paid to the determination of the second order elastic constants. The second order elastic constants (C_{11} = 308.58 GPa, C_{12} = 87.97 GPa and C_{44} = 57.90 GPa) have been calculated using PBE-GGA at ambient condition. In addition Poisson's ratio (σ), Young's Modulus (E), Shear modulus (G_H) and the ratio of anisotropy factor (A) are also reported. Ductility/brittleness of this compound is further analyzed by calculating the B/G_H ratio and Cauchy pressure (C_{12}-C_{44}). The studied compound is found to be ductile in nature. Sound wave velocities with Debye Temperature (θ_D) are also investigated.

1. Introduction

The B_2 structured (CsCl-type structure) intermetallic compounds have been extensively studied for several decades due to their high oxidation resistance, high stiffness and high strength for elevated temperature for engineering applications [1]. Intermetallics are the compounds prepared with the combination in the certain proportions of two or more metals may be non-metals and react to produce a solid-state phase which is completely different from its elements of combination. A brief discussion is mentioned concerning the experimental and theoretical work carried out related to these kinds of intermetallic compounds in the literature till now. Due to their above-mentioned properties, these intermetallics are used in aircraft engine. Cheng et. al. [2] presented the first principles study of elastic properties and electronic structure of NiTi, CoTi and FeTi. Baranov [3] predicted structural stability of pure elements and binary intermetallic compounds assuming the density of atoms as stationary and un-deformable in internal and external spherical symmetrical shell. The B_2-type NiTi intermetallic compound motivated our interest because B_2-type is the simplest and most common cubic structure and there is no systematic study of mechanical and thermal behaviour of NiTi is found in the literature. In the present paper, we have performed theoretical FP-LAPW study of structural, elastic, mechanical and thermal properties of NiTi intermetallic compound using GGA and LDA approximation. The paper is organized as follows. The method of calculation is briefly described in Section 2. Section 3 deals with the results with discussion of the present work.

IOP Conf. Series: Journal of Physics: Conf. Series **836** (2017) 012037 doi:10.1088/1742-6596/836/1/012037

2. Method of calculation

For the calculations of total energy and electronic structure, FP-LAPW method based on density functional theory is used. It is implemented in the WIEN2k code [4]. PBE-GGA [5] and LDA [6] was used as exchange correlation energy calculations. All the calculations were performed with the non-spin polarized density functional theory. Generally, the convergence is achieved by expanding the basis function up to $R_{MT}*K_{max} = 7$, where R_{MT} is the smallest atomic radius in the unit cell and K_{max} gives the magnitude of the largest k vector in the plane wave expansion. The maximum value for partial waves inside the atomic sphere is $l_{max} = 10$ while the charge density is Fourier expanded up to $G_{max} = 12\ (Ryd)^{1/2}$. The self-consistent calculations are converged when the total energy of the system is stable within 10^{-4} Ry. A dense mesh of 1000 k points and the tetrahedral method [7] have been employed for the Brillouin zone integration. The total energies are fitted to third order Birch equation of state [8] to obtain the ground state properties.

3. Results and discussions

3.1 Structural properties

In order to calculate the ground state properties of NiTi, the total energy is calculated as a function of reduced volume in B_2-type (CsCl) structure using full potential linearized augmented plane wave (FP-LAPW) method. The plot of total energy with the reduced volume is shown in figure 1. The calculated total energies are fitted to the Birch equation of state [8] to investigate the volume-energy relationship. The ground state properties, such as equilibrium lattice constant (a_0), bulk modulus (B) and pressure derivative of bulk modulus (B') for NiTi have been calculated using GGA and LDA approximations. The calculated values for these properties are presented and compared with available experimental [9] and other theoretical [3] data in table 1. As a comparison of our calculated lattice constants with the available experimental and other previous theoretical calculations, we found PBE-GGA overestimate the lattice constants while LDA underestimate the same [10]. The over binding in LDA underestimate the volume and overestimate the bulk modulus while the enhancement factor $F_x(s)$ for PBE-GGA rises more rapidly, give more accurate atomization energy [11]. Since the lattice constants obtained by PBE-GGA is more realistic and closer to experimental results and because of its advantage over LDA, the further calculations have been carried out using only PBE-GGA. It is also observed that our calculated bulk modulus (B) overestimates the other theoretical data [9] for the same compound which may be due to the different approximations used in the calculation. Due to the unavailability of any data both experimental and theoretical regarding B', it could not be compared.

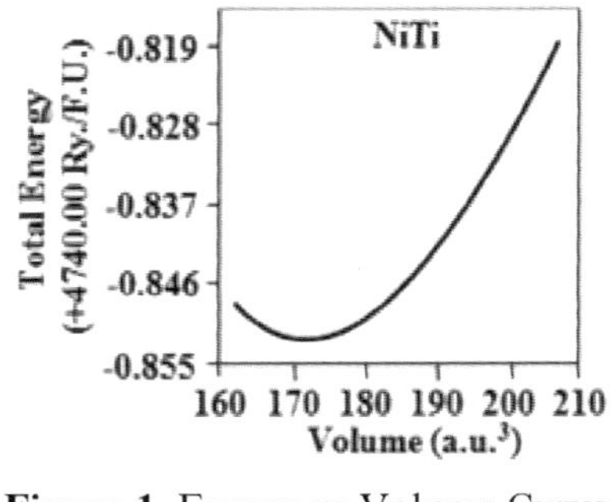

Figure 1. Energy vs Volume Curve

3.2 Electronic properties

In the present work, the self-consistent band structures (BS) along the high symmetry directions, total and partial density of states (TDOS and PDOS) are presented in figure 2(a) and figure 2(b) to have insight into electronic behaviour of NiTi. In figure 2(a), one can see that the Fermi level is fixed at the origin. In the same figure the lowest lying band is observed due to '*s*'like states of Ti atom at Γ-point. From the Figure 2(b), it is very clear that some of states are crossing Fermi level from valence band to conduction bands, means there is no band gap at E_F, fulfil the metallic character of NiTi compound.

IOP Conf. Series: Journal of Physics: Conf. Series **836** (2017) 012037 doi:10.1088/1742-6596/836/1/012037

The peaks are observed at -2 eV and -4 eV in valence band due are to '*d*' like state of Ni. The metallicity of NiTi is due to the hybridization '*d*' like states of Ni and Ti at the Fermi level. The conduction band is mainly dominated by '*d*' like state of Ti. As can be seen from the Table 1 that due to metallic character the value of finite DOS at Fermi level for NiTi is obtained 1.57 states/eV/F.U.

Table 1. The calculated ground state properties and density of states at the Fermi level $N(E_F)$ for NiTi at ambient pressure.

Solid	Work	Approx.	a_o (Å)	B (GPa)	B'	$N(E_F)$ (States/eV)
NiTi	Present	PBE-GGA	3.0140	161.58	4.21	-
		LDA	2.9439	191.92	4.15	1.57
	Exp.	-	3.013[a]	-	-	-
	Oth. The.	-	2.985[b]	139.00[b]	-	-

Exp. - Experimental, Oth.The. - Other Theory, Approx. – Approximations
[a]Ref [9], [b]Ref [3]

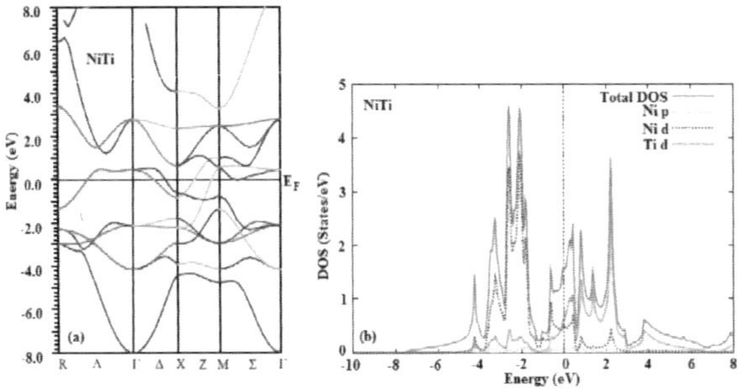

Figure 2. (a) Band Structure and (b) Density of States of NiTi using PBE-GGA

3.3 Elastic properties

The elastic constants provide a link between the mechanical and dynamical behavior of crystal. We have calculated the elastic constants for NiTi using PBE-GGA at ambient pressure using the method developed by Thomas Charpin and integrated it in the WIEN2k package. The calculated values of elastic constants at ambient pressure are given in table 2. We have also analyzed the ductility of investigated compound using Pugh's criteria [12] and Cauchy's pressure. If the value of B/G_H is greater than 1.75, the investigated compound is predicted as ductile in nature. Our calculated result indicates that the studied compound is ductile in nature with B/G_H ratio as 4.90. The ductile nature of NiTi can also be correlated to the positive Cauchy pressure and thereby metallic nature of the studied compound. It can be noted that our calculated elastic constants satisfy the stability criterions: $C_{11}-C_{12} > 0$, $C_{44} > 0$, $C_{11} + 2C_{12} > 0$, $C_{12} < B < C_{11}$, which clearly indicate the stability of NiTi in CsCl structure.

Table 2. Calculated elastic properties of NiTi.

Solids	Approx.	C_{11} (GPa)	C_{12} (GPa)	C_{44} (GPa)	B/G_H	C_{12}-C_{44}
NiTi	PBE-GGA	308.58	87.97	57.90		
	LDA	207.53	186.15	85.61	4.9	100.54

3.4 Mechanical properties

Elastic constants can be used to determine mechanical properties such as Density (ρ), anisotropic factor (A), shear modulus (G_H) Young's modulus (E) and Poisson's ratio (σ) for useful applications.

These are fundamental parameters which are closely related to many physical properties like internal strain, thermo-elastic stress, sound velocity, fracture, toughness etc. We have calculated these properties of NiTi and presented in Table 3. The calculated value of Poisson's ratio reveals the metallic nature of the studied compound. Neither experimental nor theoretical results of these properties are available for NiTi, so we could not compare our results with other measured or theoretical values.

Table 3. The calculated mechanical properties of NiTi at ambient pressure using LDA.

Solids	ρ x 10^3 (kg/m^3)	A	G_H	E	σ
NiTi	6.935	8.00	39.07	109.82	0.40

3.5 Thermal properties

We have calculated the Debye temperature (θ_D) by using the average sound velocity v_m and elastic constants. The values of v_m and (θ_D) for B_2 phase are found to be 3190.50 m/s, 240.65°K respectively. In the absence of any measured data in the literature, they could not be compared. Hence, our results can be considered as a prediction for these properties of intermetallic compounds and it will testify future experimental work.

4. Conclusions

We have systematically performed the FP-LAPW calculations to evaluate the structural and electronic properties of NiTi binary intermetallic compound using density functional theory (DFT). We have applied generalized gradient approximation of PBE-GGA and local density approximation (LDA) for the exchange correlation calculations. The total energy is fitted to the Birch equation of state. The ground state properties like lattice constant, bulk modulus and pressure derivative of bulk modulus are calculated and compared. The obtained lattice constants are in good agreement with the available experimental and theoretical values. Due to the overlapping of valence band and conduction band the metallic nature of NiTi is satisfied. The calculated elastic constants show that NiTi is elastically stable in B_2 phase. Using these elastic constants, the shear modulus (G_H), Poisson's ratio (σ), Young's modulus (E) and anisotropy factor (A) are also reported. In the present study, we found $B/G_H > 1.75$ and C_{12}-C_{44}> 0 which implies that all these compounds are ductile in nature and NiTi have an excellent ductility.

References

[1] Shi Y J, Du Y L, Chen G and Chen G L 2008 *Mater. Trans.* **492** 480
[2] Cheng D, Zhao S, Wang S and Ye H 2001 *Philos. Mag. A* **81** 1625
[3] Baranov M A *EPhTJ* 2006 **1** 49
[4] Blaha P, Schwarz K, Madsen G K H, Kuasnicka, DJ Luitz and Wien2k 2001 *An augmented plane wave+local orbitals program for calculating crystal properties*, (Wien: K. Schwarz Technical Universitat)
[5] Perdew J P, Burke K and Ernzerhop M 1996 *Phys. Rev. Lett.* **77** 3865
[6] Hohenberg P and Kohn W 1964 *Phys. Rev. B* **136** 864
[7] Monkhorst H Jand Pack J D 1976 *Phys. Rev. B* **13** 5188
[8] Birch F 1938 *J. Appl. Phys.* **9** 279
[9] Elliott R P 1968 *Constitution of binary alloys* (New-York: McGraw-Hill Book company)
[10] Kurth S, Perdew J P, Blaha P 1999 *Int. J. Quantum Chem.* **75** 889
[11] Pagare G, Chouhan S S, Soni P, Sanyal S P and Rajagopalan M 2010 *Comput. Mater. Sci.* **50** 538
[12] Pugh S F 1954 *Philos. Mag.* **45** 823

IOP Conf. Series: Journal of Physics: Conf. Series **836** (2017) 012038 doi:10.1088/1742-6596/836/1/012038

Study of structural properties of alloy $Tl_xGa_{1-x}N$

Madhu Sarwan[a] and Sadhna Singh

High Pressure Research Lab, Department of Physics, Barkatullah University, Bhopal 462026, M.P., India

Email: [a)] madhusarwan@gmail.com

Abstract. The structural properties of $Tl_xGa_{1-x}N$ have been examined by the use of interaction potential model (IPM). Interaction potential model (IPM) is made up of Coulomb interaction, three-body interaction, van der Waal interaction and short range overlap repulsive interactions. The structural properties namely phase transition, volume collapse for the present alloy have been computed via vegard's law. We have also investigated the elastic properties i.e. second order elastic constants, bulk modulus, shear modulus, Pugh ratio and anisotropy factor. These elastic properties are the first attempt for the present alloy

1. Introduction

The combination of III group and V group reveal the new III-V alloys. These alloys are more suitable for the purpose of electronic and optoelectronic devices. Various III-V semiconductor alloys have been studied but there is no experimental study available for the synthesis of alloy of thallium containing nitride. Also the theoretical reports are very less on them. The $Tl_xGa_{1-x}N$ is regarded as wide band gap semiconductors ranging from ultraviolet to visible regions of the spectrum. It has a very high thermal conductivity and large bulk modulus and makes them important material [1]. Thallium containing nitride bears the advantage more than II-VI compounds due to which they are good alternates to II-VI compounds. Houat et al. [1] have offered the structural and optical properties of alloy $Tl_xGa_{1-x}N$ through first-principle calculations. To expand the structural and elastic properties we present the systematic and comparative study. In order to depict the present alloy we employ the theoretical approach viz interaction potential model (IPM) in the present study. We applied this IPM to study the phase transition, volume collapse and elastic properties of $Tl_xGa_{1-x}N$. In the present model we have incorporated the Coulomb interaction, three-body interaction (TBI), van der Waal (vdW) interaction and short range overlap repulsive interaction. All the properties have been calculated by using Vegard's law.

2. The potential model and method of computation

The chief part of our theoretical method (IPM) is TBI. To understand TBI we first define the many body interaction (MBI). We know that as the pressure is applied on crystal then its volume decrease owing to this there is exchange of charges between the overlapping electron shells of the adjacent ions. This overlapping provides the transfer of charges and these charges interact with other charges which are present in the columbic field. This mechanism is called many body interactions (MBI), the important part of MBI is three-body interactions TBI [2].

The Gibbs free energy (G=U+PV-TS) has included the TBI effects. Noted that, U is the lattice energy, P is pressure, V is volume and S is the entropy at absolute temperature T. At zero temperature and pressure, the Gibbs free energy for the zinc-blend (B3) and NaCl (B1) structures are

$$G_X(r) = U_X(r) + PV_X \tag{1}$$

$$G_Y(r) = U_Y(r') + PV_Y \tag{2}$$

Here X and Y stands for B3 and B1 phase. The primary terms $U_X(r)$ and $U_Y(r')$ are the lattice energy and secondary are pressure-volume term of both phases. We have added several interactions in lattice energies to develop IPM.

$$U_X(r) = U_X(Col) + U_X(TBI) + U_X(vdW) + U_X(OR) \tag{3}$$

$$U_Y(r') = U_Y(Col) + U_Y(TBI) + U_Y(vdW) + U_Y(OR) \tag{4}$$

Where former is lattice energy of coulomb interaction, next is lattice energy of TBI, after that lattice energy of van der Waal interaction and last is overlap repulsive energy.

3. Results and discussions

For studying the problems of alloy of III-V semiconductors we generally take the Vegard's law [3] which states that the atoms are located at the ideal sites and the lattice constants vary linearly with composition x.

$$a(A_xB_{1-x}C) = xa_{AC} + (1-x)a_{BC} \tag{5}$$

here AC and BC stands for TlN and GaN, x is concentration and "a" is lattice parameter for pure mixed compound.

First we have calculated the model parameters, following method given in our previous work [4]. By the use of Vegard's law and model parameters we have computed the structural properties viz phase transition pressure and volume collapse for present alloy $Tl_xGa_{1-x}N$. Fig.1 (a-b) shows the variation of phase transition pressure and volume collapse with concentrations. It is obvious from this figure that as the concentrations increases the phase transition pressure and volume collapse decreases.

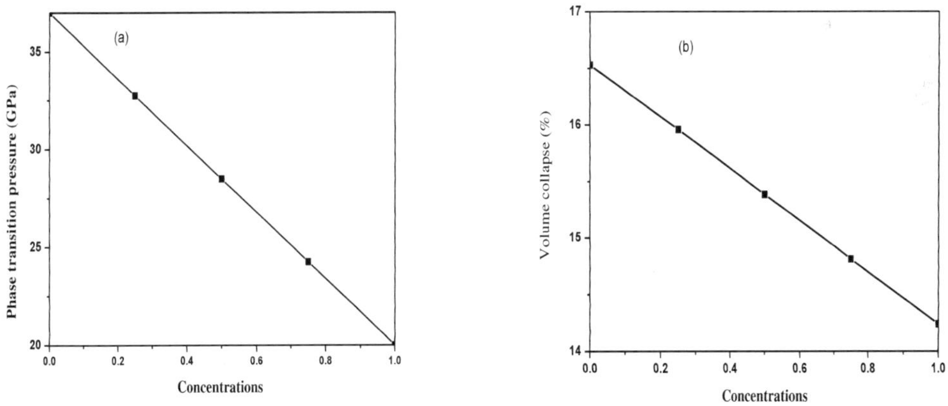

Figure 1(a-b). Variation of phase transition pressure and volume collapse with concentration.

Next we have studied the elastic behavior with concentrations of this alloy to enhance the applicability of our model. To get this we have determined the second order elastic constants (C_{11}, C_{12}, C_{44}) for the pure compounds after this through the Vegard's law we obtained the second order elastic constants for mixed compound i.e. $Tl_xGa_{1-x}N$ at different composition x. The method for calculations of second order elastic constants is same which we have reported in our earlier work [4]. The calculated values of C_{11}, C_{12} and C_{44} are listed in Table 1and they are compared with available theoretical results for pure compounds.

Table 1. Elastic constants - bulk and shear modulus of $Tl_xGa_{1-x}N$.

Compounds	C_{11} (GPa)	C_{12}(GPa)	C_{44}(GPa)	B(GPa)	G (GPa)
GaN	278.96 274.2[5]	177.0 166.1[5]	186.97 199.0 [5]	210.98 202.0 [5]	66.31
$Tl_{0.25}Ga_{0.75}N$	251.63	163.26	160.85	192.05	57.12
$Tl_{0.5}Ga_{0.5}N$	224.30	149.53	134.73	174.45	47.94
$Tl_{0.75}Ga_{0.25}N$	196.97	135.80	108.61	155.19	38.75
TlN	169.65 194.3[6]	122.07 114.5[6]	82.5 103.2[6]	137.93 -	29.56

The variation of second order elastic constants (SOECs) with pressure at different concentrations (x) is shown in figure 2. It is clear from this figure as the pressure increased the SOECs also increased at different concentrations (x=0, 0.25, 0.5, 0.75, 1.0). This influence is more pronounced in C_{11} as compared to C_{12} and C_{44}.

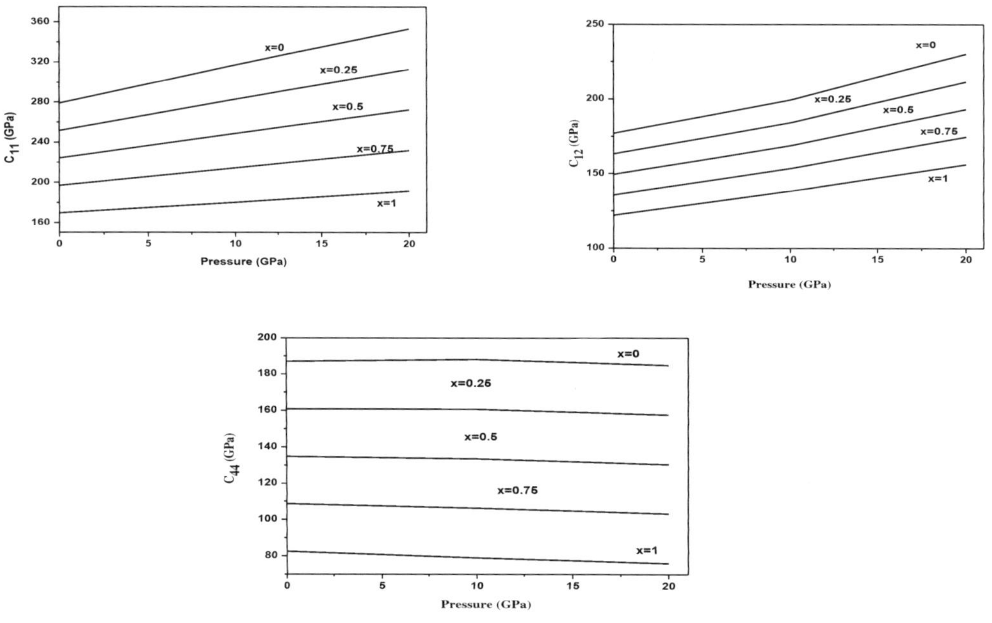

Figure 2. Variation of C_{11}, C_{12}, C_{44} with pressure at different concentrations X.

The variation of bulk modulus and shear modulus at different concentrations are shown in table 1. It is noticed from this Table that as the concentrations of TlN increased the bulk modulus and shear modulus decreased. No compression could be made due to unavailability of experimental and theoretical results.

To identify the ductile and brittle nature of materials there are various way to obtain this. For the sake of simplicity we computed the pugh ratio B/G. It has a decisive value for separating the brittle and ductile nature i.e. 1.75. Ductile material having the value equal or above 1.75 while brittle material possess the value below 1.75. Looking the Fig. 3 we can say that our studies represent the highly ductile nature of mixed compound and this behaviour enhance with concentrations.

Since the anisotropic factor A is a property related to the micro cracks in the materials and one important feature of this is measurement of isotropic and anisotropic nature of materials. It is one for isotropic and greater than one for anisotropic materials. Fig.3 shows the anisotropic behaviour of present alloy with concentrations.

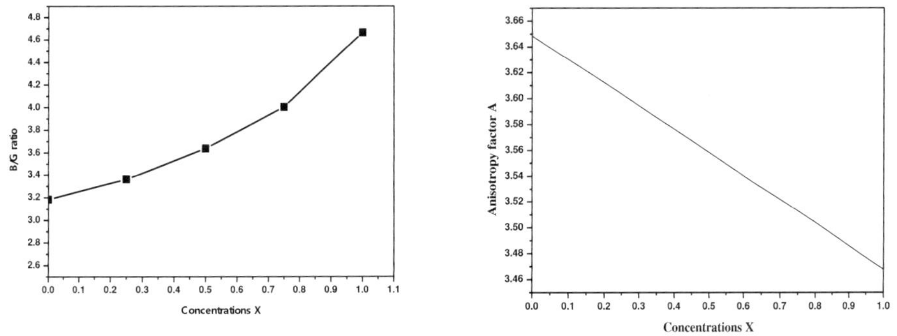

Figure 3. Variation of Pugh ratio and anisotropy factor with concentrations X.

4. Conclusions

The structural properties like phase transition and volume collapse of alloy $Tl_xGa_{1-x}N$ are analysed. The pressure dependence of SOECs will provide valuable information on the valuable quality of the solids. The Bulk (B) modulus, Shear modulus (G), Pugh ratio (B/G) and anisotropic factor for the present mixed compound are analysed for the first time. Ultimately it may be concluded that, the IPM approach is sufficiently suitable for structural and elastic properties of $Tl_xGa_{1-x}As$ alloy. Our results may be tested with different theoretical and experimental workers.

Acknowledgement

The authors are thankful to the U.G.C. New Delhi for financial support.

References

[1] Saidi Houat N, Zaoui A, Belabbes A and Ferhat M 2009 *Mat. Sci. Eng. B* **162** 26
[2] Singh R K 1982 *Phys. Rep.* **85** 259
[3] Vegard L 1921 *Z. Physik* **5** 17
[4] Sarwan M and Singh S 2012 *Comp. Mat. Sci.* **58** 167
[5] Kanoun M B, Merad A E, Merad G, Cibert J and Aourag H 2004 *Solid State Electron* **48** 1601
[6] Wang S Q and Ye H Q 2003 *Phys. Stat. Sol. (b)* **240** 45

IOP Conf. Series: Journal of Physics: Conf. Series **836** (2017) 012039 doi:10.1088/1742-6596/836/1/012039

X-ray diffraction and X-ray K absorption near edge studies of copper (II) complexes with amino acids

P K Sharma[1,a], Ashutosh Mishra[2], Varsha Malviya[2], Rashmi Kame[2,,b] and P K Malviya[3]

[1]Holkar Science College, Indore 452017, India
[2]School of Physics, Devi Ahilya Vishwavidyalaya, Indore 452001, India
[3]Govt. College, Nagda 456335, India

E-mail: [a)]sharma29762@yahoo.co.in, [b)] rashmikame23@gmail.com

Abstract. Synthesis of copper (II) complexes $[CuL_1L_2X].nH_2O$, where n=1, 2,3 ($X=Cl,Br,NO_3$) (L_1 is 2,2'-bipyridine and L_2 is L-tyrosine) by the chemical root method. The XRD data for the samples have been recorded. EXAFS spectra have also been recorded at the K-edge of Cu using the dispersive beam line BL-8 at 2.5 Gev Indus-2 Synchrotron radiation source at RRCAT, Indore, India. XRD and EXAFS data have been analysed using the computer software. X-ray diffraction studies of all complexes indicate their crystalline nature. Lattice parameter, bond length, particle size have been determined from XRD data.

1. Introduction

X-ray structural studies have revealed that serine phosphorylation of the active and less active forms of an allosteric enzyme muscle phosphorylase induces a conformational transition due to hydrogen bonds involving the phosphate moiety and histidine and arginine residues [1, 2]. In its complex with a substrate, glycyl-L-tyrosine, it fixes the substrate by electrostatic interactions between the carboxylate group of the substrate and the arginine guanidinium group of the enzyme [3]. Ternary Cu(II) complexes containing aromatic heterocycles such as phen and aromatic amino acids like phenylalanine(phe) and L-tyrosine(Tyr) have been shown to be stabilized by stacking interaction [4]. This paper describes the synthesis and characterization of three Cu(II) complexes$[CuL_1L_2X].3H_2O$ where L_1 is 2,2'-bipyridine and L_2 is L-tyrosine [5].

2. Experimental

The required chemicals were of analytical reagent grade and were purchased from Merck, all manipulation were performed using materials as received the content of the metallic ion was elemental analyzer. Copper (II) complexes have been prepared by chemical root method. Elemental analyses were performed after drying the complexes at 600C. All the sample were dissolve in 0.2M HCl (25ml) by heating and after cooling the solution of L-Tyr in 1 M NaOH was added the mixture was concentrated and kept at room temperature.The seprated crystal were collected and crystallized from water.
The powder (debye-scherrer) method is the standard method for obtaining crystal lattice spacing. This experiment can give lattice parameters to about four-figure accuracy. The powdermethod assumes that all orientation is present in the sample. In practice, this would be a time consuming operation to reorient the crystal, measure the angle quad determine the d-spacing for all atomic plane. A faster way is to use a method called the powder method.

IOP Conf. Series: Journal of Physics: Conf. Series **836** (2017) 012039 doi:10.1088/1742-6596/836/1/012039

3. Results and discussions

Figure 1 Shows the X-ray diffraction pattern of [Cu(bpy)(L-Tyr)Cl].3H2O, Cu(bpy)(L-Tyr)Br].3H2O, Cu(bpy)(L-Tyr)SO4]3H2O. The diffraction pattern of complexes recorded between 2θ ranging from 10^0 to 80^0. The interplanar distance d (Å) are calculated using Brag's law. The crystalline size for each composition are calculated width of the using Scherrer Formula [6-9]. The phase confirmed by XRD studies. The particle size and lattice parameter are shown in the table 1 . Crystal structure has found cubic. The table 2 presents the results for the K- absorption (E_K) and the energy of the principal absorption maximum (E_A) of copper metal and its complexes. EXAFS data shown in figure 2. The bond length of copper complexes evaluated with the help of Athena software [10].

Table 1: XRD for Copper (II) complexes with amino acid

S. No.	Sample name	2θ	h k l	Particle size (nm)	Lattice parameter (Å)
1	[Cu(bpy)(L-Tyr)Cl].3H_2O	12.50	1 1 1	9.18	12.01
2	[Cu(bpy)(L-Tyr)Br].3H_2O	12.07	1 1 1	10.56	12.40
3	[Cu(bpy)(L-Tyr)SO_4].3H_2O	20.95	2 2 2	10.59	14.82

Table 2: XANES parameters and bond length for the K absorption edge of copper in the complexes

S.No.	Sample name	E_K-edge (eV)	E_A (eV)	Bond length
1	[Cu(bpy)(L-Tyr)Cl].3H_2O	8984.6	8997.8	0.9
2	[Cu(bpy)(L-Tyr)Br].3H_2O	8988.6	8998.4	1.4
3	[Cu(bpy)(L-Tyr)SO_4].3H_2O	8989.0	8997.5	1.3

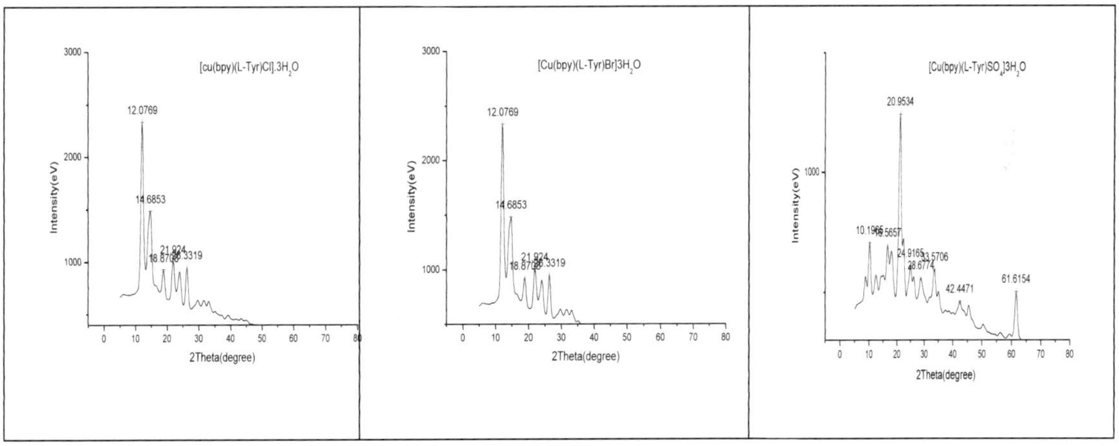

Figure 1. XRD Pattern for Cu complexes with amino acids

IOP Conf. Series: Journal of Physics: Conf. Series **836** (2017) 012039 doi:10.1088/1742-6596/836/1/012039

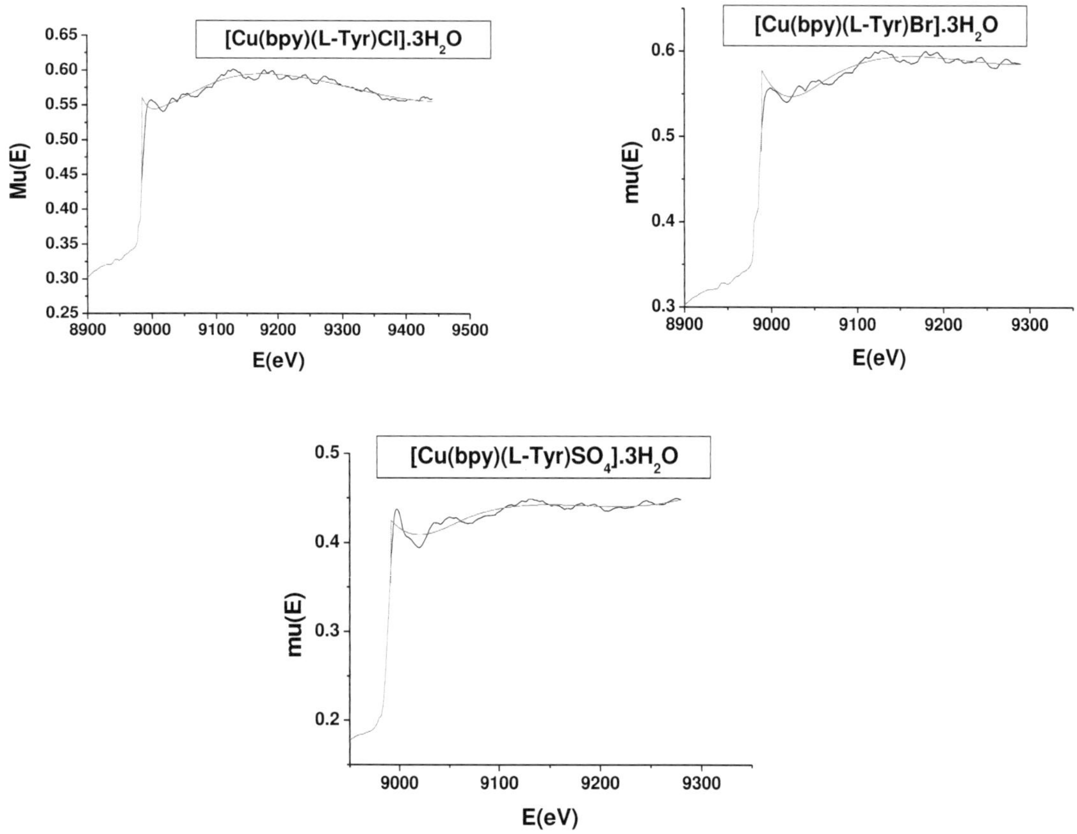

Figure 2. EXAFS spectra [(μ(E) Vs E] for copper complexes

References

[1] Sprang S R, Acharya K R, Goldsmith E J, Stuart D I,Varvill K, Flettrick R J, Madsen N B, Johnson L N 1988 *Nature* **336** 215
[2] Hurley J H, Dean A M, Sohl J L, Koshland D E Jr and Stroud R M 1990 *Science* **249** 1012
[3] Christianson D W and Lipscomb W N 1989 *Acc. Chem. Res.* **22** 62
[4] Yamauchi O and Odani A 1985 *J. Am. Chem. Soc.* **107** 5938
[5] Rosu T, Negoiu M, Carcu V, Gulea A 2006 *Revue Roumaine de Chimie* **51** 391
[6] Mishra A, Dwived J, Shukla K and Malviya P K 2014 *J. Phys.: Conf. Series* **534** 012014
[7] Hinge V K, Joshi S K and Shrivastava B D, Prasad J and Srivastava K 2012 *J. Phys.: Conf. Series* **365** 012033
[8] Mishra A, Parsai N and Dagonkar N 2009 *Ind. J. Pure App. Phys.* **47** 337
[9] Malviya P K, Sharma P K and Mishra A 2013 *Int. J. Sci. Res. App. Phys.* **1** 2348
[10] Sharma P and Mishra A 2005 *Ind. J. Chem. A* **44** 307

IOP Conf. Series: Journal of Physics: Conf. Series **836** (2017) 012040 doi:10.1088/1742-6596/836/1/012040

Synthesis and thermogravimetric analysis of non-stoichiometric nickel oxide compounds

Paras Dubey[a] and Netram Kaurav[b]

Department of Physics, Government Holkar Science College, A. B. Road, Indore 452017, MP, India

E-mail: [a)]paras_dubey74@yahoo.co.in, [b)]netramkaurav@yahoo.co.uk

Abstract. Non-stoichiometric nickel oxide was synthesized by the thermal decomposition method at different sintering temperatures upto 1100 °C. The samples were analyzed by thermogravimetric analysis (TGA). Wherein, the stoichiometry of the sample was analyzed by calculating the excess oxygen. It was found that as the sintering temperature of the precursor increases the excess oxygen present in the sample decreases, and at temperature above 700 °C, the sample becomes stoichiometric. These results were interpreted as the decomposition temperature increases, which heals the defects present in the non-stoichiometric nickel oxide. Thus, highly defective nickel oxide at lower decomposition temperature approaches stoichiometric nickel oxide as the temperature is increases above 700 °C.

1. Introduction

Nickel oxide (NiO) is the most comprehensively investigated transition metal oxide over the decades [1]. It is a anti-ferromagnetic oxide semiconductor which crystallizes in NaCl-type crystal structure [2]. NiO has been considered as a promising functional material for varieties of applications, such as, smart windows [3], spin valves giant magneto resistance (GMR) sensor [4], solar cells [5], gas sensors [6], high permittivity dielectric material and electrochromic material for displays [7], photoelectrolysis [8], etc. In fact, bulk material of NiO has a rhombohedral crystal structure and it shows antiferromagnetic behavior below 250 °C and above the Neel temperature of 250 °C, the structure is transformed into a cubic one with paramagnetic behavior [9]. Pure stoichiometric NiO crystals are perfect insulators [10], several efforts have been made to explain the insulating behavior of NiO, were in, appreciable conductivity can be achieved in NiO by creating Ni vacancies or substitution at nickel side, thereby creating Ni vacancies and the composition ratio of Ni and O changes and NiO becomes non-stoichiometric. Hence, in a non-stoichiometric NiO, various property changes, this can be attributed to presence of excess oxygen. Nickel oxide can been prepared by various techniques that involve: thermal decomposition [11], anodic oxidation [12], chemical deposition [13], and sol–gel [14]. In fact, Gray [11] and Iida [15] reported that the properties of nickel oxide also changes with the preparation method. All these characteristics in nickel oxide provide an ideal opportunity and in present case non-stoichiometric nickel oxide was prepared by the thermal decomposition method at different sintering temperature.

2. Experimental

Following [11, 15] nickel nitrate hexa-hydrate [$Ni(NO_3)_2.6H_2O$] sintered at different temperature, which was decomposed to produced nickel oxide samples of various stoichiometries. Nickel nitrate hexa-hydrate was heated in presence of air for 3 hr. at temperatures range from 400 °C to 1100 °C. By sintering at four different temperature samples NiO(400 °C), NiO(500 °C), NiO(700 °C) and NiO (1100 °C) were then prepared. X-ray diffraction was conducted with the help of the Bruker D8 Advance X-ray diffractometer with Cu Kα radiation (0.154 nm) in the angle range 10°–90°. Thermo gravimetric analyses (TGA) was done to classify the stoichiometry of the samples and to obtain the oxygen content of the samples. In our experiment, TGA with the top-of-the-line METTLER TOLEDO ultra-micro balance with unique built-in calibration weights ensures with an accuracy of 0.1 μg.

3. Results and discussions

3.1. X-ray diffraction analysis

XRD analysis, which is the most useful technique for identification of crystalline structure, was employed to investigate the crystallinity and purity of the solid product obtained by thermal conversion in air of the poly-nuclear coordination compounds. The XRD data of nickel oxide sample prepared at different temperature is shown in Fig. 1, which shows the presence of the characteristic peaks for NiO at 2θ in the range of 10° – 90°, in accordance with JCPDF File 78-0643. The lattice parameters were obtained from the following relationship analytically: $\sin^2\theta = C\,(h^2 + k^2+l^2)$, where $C= \lambda^2/4a^2$. The results of the X-ray structural analysis are given in Table 1 for one of the representative samples sintered at 400 °C. Our analysis indicates good agreement between Bragg's reflection peaks obtained from analytical analysis data and the structural data of with JCPDF File 78-0643. This indicates that the NiO sample obtained by thermal conversion in air of the polynuclear coordination compounds, revealed the NiO with face-centered cubic phase, also known as the bunsenite structure (lattice constant *a* of cubic unit cell, 0.4177 nm). For others samples, there was also no appreciable deviation in structural data within the limit of accuracy of analytical method. No other additional peaks corresponding to other phases were emerged, suggested NiO phase is stable.

Further, the mean crystallite size of the NiO sample can be calculated using Sherrer formula: $L = k\lambda/\beta\cos(\theta)$, where L is the crystallite size, k is the Sherrer constant, usually taken as 0.89, λ is the

Table 1. Crystallographic data using analytical analysis for NiO samples.

Intensity (%)	2θ	$\sin^2\theta$	$h^2 + k^2 + l^2$	C	(hkl)	a(Å)	$\bar{a}$(Å)
61.69	37.279	0.1022	3	0.034	(111)	4.1772	4.1775
100	43.472	0.1365	4	0.034	(200)	4.1727	
44.83	62.901	0.2722	8	0.034	(220)	4.1772	
15.68	75.416	0.3741	11	0.034	(311)	4.1772	

Table 2. Rietveld parameters of different NiO samples.

Compounds	a(Å)	V (Å^3)	Rf factor	Bregg's R-factor	R_p	R_{wp}	R_e	χ^2	L(nm)
NiO (400 °C)	4.176	72.82	2.39	3.94	24	16.7	14.9	1.261	240
NiO (500 °C)	4.174	72.74	4.44	4.33	29.6	19.6	16.7	1.381	290
NiO (700 °C)	4.172	72.64	4.04	4.51	22	15.8	13.3	1.416	350
NiO (1100 °C)	4170	27.58	20.4	31.7	56.3	37.4	22.3	2.673	460

Table 3. Calculated excess oxygen present in different samples.

Samples→	NiO (400 °C)	NiO (500 °C)	NiO (700 °C)	NiO (1100 °C)
Excess oxygen %	7.8	2.8	1.2	0

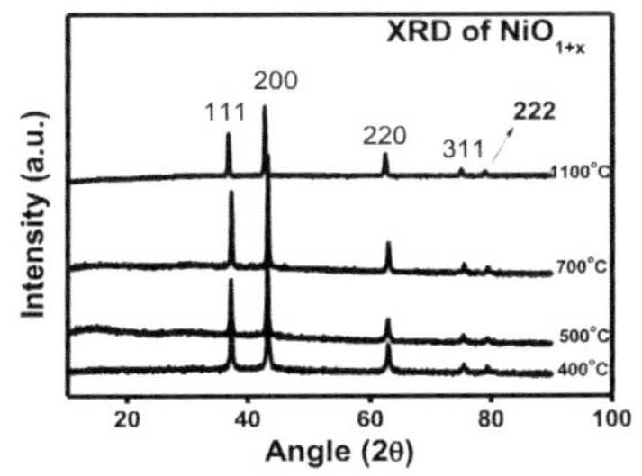

Figure 1. X ray diffraction pattern of non stiochiometric NiO

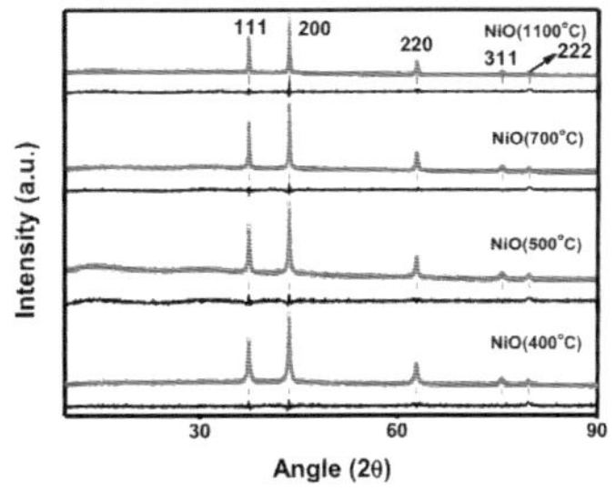

Figure 2. Rietveld refinement of x-ray diffraction of non-stoichiometric NiO.

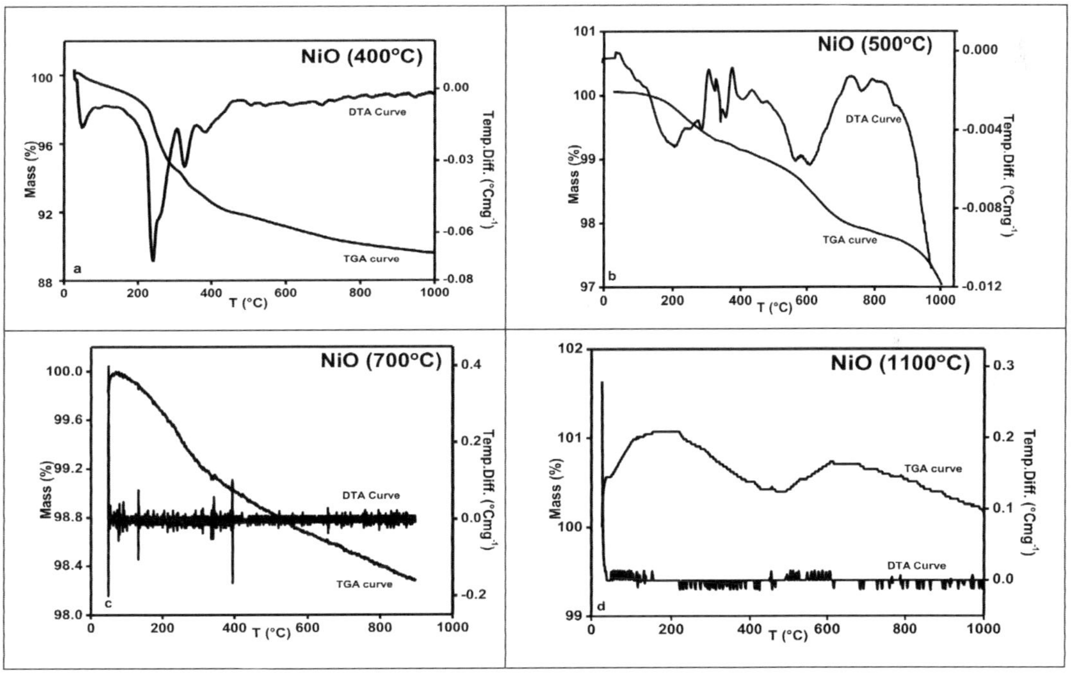

Figure 3. TGA and DTG curves of NiO samples prepared at different temperature

wavelength of the x-ray radiation (0.154056 nm for Cu Kα), and β is the full width at half maximum (FWHM) of diffraction peak measured at 2θ. The deduced values of crystallite size of the analyzed NiO samples are reported in Table 2, which indicates that as the sintering temperature of precursor increases the crystallite size of NiO also increases.

We have also performed Rietveld refinements for all the samples using Fullprof software. Fig.2 shows the refinement profile of all the samples and the various parameters obtained from refinement data are given in Table 2. It can be concluded that the crystal parameters obtained from analytical calculation from XRD graph and fitting the curve by Rietveld refinement corroborates each other.

3.2. Thermogravimetric analysis (TGA)

TGA measures the amount of weight change of a material, either as a function of increasing temperature, or isothermally as a function of time, in an atmosphere of nitrogen, helium, air, other gas, or in vacuum. In order to establish the estimation of excess of oxygen, TGA was performed on the samples prepared at different temperature. In present investigation, the weight of the non-stoichiometric oxygen is obtained directly by obtaining the weight loss. The thermo-gravimetric (TG) curve and its differential curve (DTG) are shown in Fig. 3 for NiO (400 °C). It is clear from Fig. 3 that a very sharp peak is present between 200 and 300 °C on the DTG curve, which is a measure of amount of excess oxygen released in the heating process. The corresponding weight loss reflects desorption of oxygen in non-stoichiometric sample. On the other hand, the peak below 100°C is due to the evaporation of the physically bound water in the sample and does not contribute in the estimation of excess oxygen. Further, it is noted that, for NiO (500 °C) and above samples, the reduction of mass with temperatures decreases and is found to be negligible above NiO (600 °C) sample as shown in Fig. 3 by TG and DTG curves. Consequently, composition ratio of O/Ni is equal to 1 for above the NiO (700 °C) sample and crystallizes with Ni deficiency, establishing stoichiometry in nickel oxide. The weight percentage of excess oxygen is tabulated in Table 3. From above TG curves it is clear that as the temperature of preparation of sample increases the stoichiometry of the sample changes and at above 600 °C the excess oxygen present in the sample completely evolved out from the sample and becomes stoichiometric.

4. Conclusions

Nickel oxide of various stoichiometry were prepared by using a simple and inexpensive thermal decomposition technique at different sintering temperatures. The XRD studies show that all the nickel oxide samples were crystallize in cubic phase. Thermogravimetric analysis of samples prepared at different temperature indicates that as sintering temperature of precursor increased, the stoichiometry of the sample changes, the samples which prepared at low temperature have more excess oxygen as compare to sample prepared at higher temperature.

Acknowledgements
Dr. Mukul Gupta and Dr. V. Ganesan, Centre Director, UGC-DAE-CSR, Indore for providing XRD facility and financial support from SERB, DST, New Delhi are gratefully acknowledged.

References

[1] Rao C N R 1989 *Annu. Rev. Phys. Chem.* **40** 291
[2] Rutchings I and Samuelson S 1972 *Phys. Rev. B* **6** 3447
[3] Browne M P *et al.* 2016 *J. Electrochem. Sci.* **11** 6636
[4] Pinarbasi M *et al.* 2000 *J. Appl. Phys.* **87** 5714
[5] Bandara J and Weerasinghe H 2005*Solar Energy Mate. and Solar Cells* **85** 385
[6] Hotovy I, Huran J, Hascik S and Rehacek V 2007 *Sensors and Actuators B* **57** 147
[7] Lampert C M 1984 *Solar Energy Mater.* **11** 27
[8] Liu H, Zheng W, Yan X, and Feng B 2008 *J. Alloys Compd.* **462** 356
[9] Smart J S and Greenwald S 1951 *Phys. Rev.* **82** 113
[10] Bosman A J and Crevecoeur C 1966 *Phys. Rev.* **144** 763
[11] Gray T G 1956 *J. Phys. Chem.* **60** 201
[12] Hotovy I, Bue D, Hascik S and Nennewitz O 1998 *Vacuum* **50** 41
[13] Pramanik P and Bhattachraya S 1990 *J. Electrochem.Soc.* **137** 3869
[14] Chigane M and Ishikawa M 1998 *Faraday Trans.* **94** 3665
[15] Iida Y and Ozaki S 1959 *J. Am. Ceram.Soc.* **42** 219

IOP Conf. Series: Journal of Physics: Conf. Series **836** (2017) 012041 doi:10.1088/1742-6596/836/1/012041

Structural and electronic properties of rare earth skutterudites $EuRu_4P_{12}$: a first principle study

Priya Yadav[a], Meena Kumari, Priyanka Rajpoot, Shashank Nautiyal and U P Verma[b]

School of Studies in Physics, Jiwaji University, Gwalior-474011, M.P., India

Email: [a)]priyayadavambah@gmail.com, [b)]upv.udai@gmail.com

Abstract. Ternary skutterudites materials exhibit a large variety of electronic properties due to the unpaired 3d and 4f electronic configuration of the transition and rare-earth elements respectively. In this communication we have performed structural optimization and electronic structure calculation of the Rare Earth Skutterudite $EuRu_4P_{12}$, using FP-LAPW method. The optimized lattice parameter is in good agreement with the available experimental value. No band gap between the valence band and the conduction band has been observed which indicates that its character is metallic.

1. Introduction

Recently, the utilization of industrial waste heat into electrical energy from the thermoelectric materials has been a growing research interest [1]. These materials exhibit a large variety of electronic and magnetic properties, which are determined by the unpaired 3d and 4f electronic configuration of the transition and rare-earth elements respectively. The strongly correlated electron phenomena at low temperatures and potentially useful thermoelectric properties at high temperatures of these materials have attracted much attention of researchers over the last few years [2]. The materials with dense energy band and low energy gap at EF are the key features of thermoelectric materials, which can be achieved by either doping of heavy elements or from a thin film by controlling the lattice vibration (decreasing the thermal conductivity) [3-5]. This can also be achieved by the structural optimization of binary skutterudites MX_3 (M = Fe, Ru or Os; X = P, As or Sb) filled with rare earth elements, which gives ternary skutterudites RM_4X_{12} (R = rare earth) belonging to the space group *Im-3* (No. 204) [6,7]. In their structure the M ions are located inside the cages formed by X ions, and the M ions are believed to show random motion (rattling) around the equilibrium positions [8,9]. Bauer et al. [10] have reported the ferromagnetic behavior of some of the Nd and Eu-filled Skutterudites and Krishnamurthy et.al. [11] have reported the ferromagnetic nature of $EuFe4Sb_{12}$. Their FP-LAPW [12] study suggests the ferromagnetic nature of the ground state of $EuFe_4Sb_{12}$. The experimental study suggest the anti-ferromagnetic nature of $GdOs_4P_{12}$ [13], whereas the theoretical study by Kihou et al. [14] indicates the ferromagnetic ground state with magnetic moment 8.54 μB. Because of the technological importance of $EuRu_4P_{12}$ in various devices it is worthwhile to present detailed structural and electronic properties of this material. Thus, in the present work, we have performed the structural and electronic properties of $EuRu4P_{12}$ using the concept of the density functional theory.

IOP Conf. Series: Journal of Physics: Conf. Series **836** (2017) 012041 doi:10.1088/1742-6596/836/1/012041

2. Computational details

All the calculation of present work is carried out within the DFT framework. We use the self-consistent full potential linear augmented plane wave (FP-LAPW) method to solve Kohn-Sham equation as implemented in the Wien-2k code.$EuRu_4P_{12}$ compounds crystallize in a unique body centered cubic (BCC) structure with *Im-3* space group (No.204), where one Eu atom takes the atomic position of 2a (0, 0, 0) and one Ru atom located at 8c (0.25, 0.25, 0.25), while one P atom is located at the position of 24g (0, 0.35, 0.16). The positions of the remaining atoms in the unit cell are determined by symmetry operations associated with the space group. The unit cell crystal structure of $EuRu_4P_{12}$ is presented in Fig. 1. In this paper all the calculations performed within the density functional theory implemented in the WIEN2k code work. The exchange correlation potential is treated within the generalized gradient approximation (GGA) by the density functional of Perdew–Burke–Ernzerhof (PBE) for the calculation of structural and electronic properties. The energy eigenvalues are converged expanding the wave function in the interstitial region and the plane wave cut-off was RMT×Kmax = 7, where R_{MT} is the smallest atomic muffin-tin sphere radius and K_{max} is the maximum value of the wave vector in the plane wave expansion. The R_{MT} (muffin tin radii) were selected to be 2.31, 2.10 and 1.9a.u.for Eu, Ru and P, respectively. The self-consistent potentials were calculated on a 10× 10×10 k-mesh in the Brillouin zone which corresponds to 1000 k points in the irreducible Brillouin zone and the convergence criterion is set to 10^{-4} Ry.

3. Results and discussions

3.1 Structural properties

We have performed the structural optimization of $EuRu_4P_{12}$ using PBE-GGA, we obtained lattice parameter ***(a)***, bulk modulus (B), the equilibrium volume (V_0) corresponding to minimum energy (E_0) and the pressure (P) are calculated by fitting the total energy to the Murnaghan's equation of states [15]. All physical properties are related to the total energy. The lattice constant that minimizes the total energy is the equilibrium lattice constant of a crystal. The detailed values of the optimized lattice parameter and bulk modulus are given in table 1. The optimized equilibrium lattice parameters were slightly different than the experimental lattice parameter (in table 1). Bulk modulus is a measure of resistance to volume change by an applied pressure. A series of calculated total energies as a function of volume can be fitted to an equation of states according to Murnaghan's equation of state .

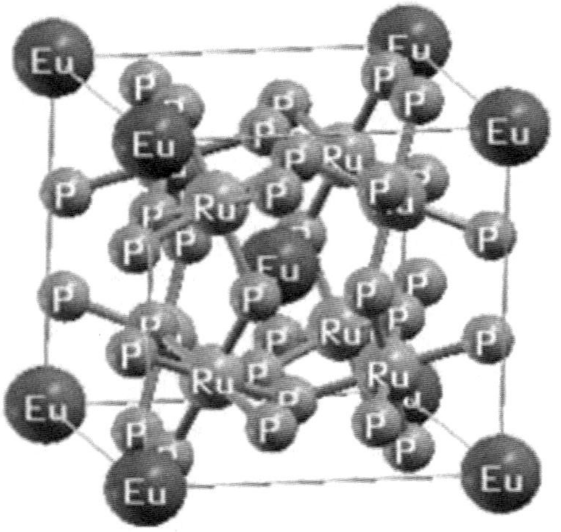

Figure 1. The unit cell structure of $EuRu_4P_{12}$ in space group *Im-3*.

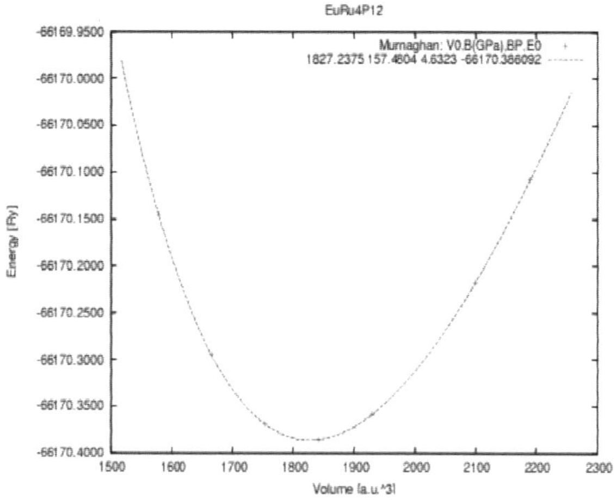

Figure 2. Plot of the total energy as a function of unit cell volume in BCC structure of $EuRU_4P_{12}$.

Table 1. Calculated equilibrium lattice parameter (a), unit cell volume (V_0), bulk modulus (B_0) and first pressure derivatives of bulk modulus (B_0') of $EuRu_4P_{12}$.

a (Å)	V_0	E_0	B_0	B'_0	Ref.
8.14	1827.2528	-66170.3864	159.8792	4.365	P.W
8.04					Exp [8]

3.2Electronic Properties

The energy distribution of the electron in the valance and conduction bands is important in determining the electronic properties of solids. The equilibrium lattice constant is used to calculate the electronic band structure and the density of states of $EuRu_4P_{12}$. The calculated band structure of $EuRu_4P_{12}$ is shown in figure 3. The energy band structure is calculated along the high symmetry points of the first Brillouin zone using PBE- GGA. From band structure, we observe that there is no band gap between conduction and valence band as a result $EuRu_4P_{12}$ has metallic character which is in good agreement with earlier result. In order to support band structure plots finding we have also plotted the total and partial density of states for atoms Eu, Ru and P. In PDOS plots, *d*-electrons have the highest contribution in Eu whereas *p*-electrons have highest contribution in atoms Ru and P. But DOS plots also do not show any band gap between valance and conduction bands. This supports that $EuRu_4P_{12}$ exhibit metallic nature in agreement with our band structure plots finding.

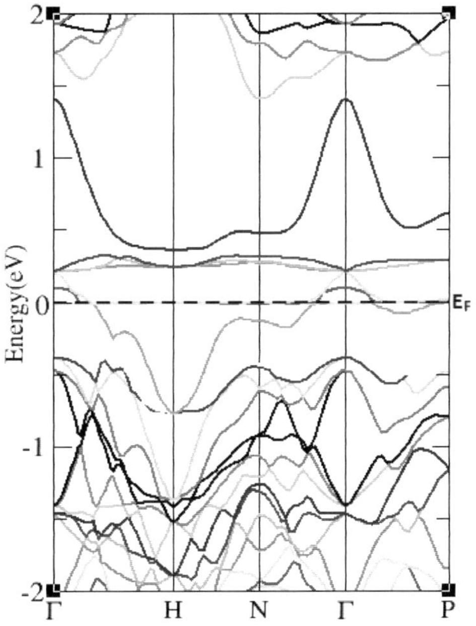

Figure 3. A plot ofelectronic bands of $EuRu_4P_{12}$ at ambient condition in bcc structure.

4. Conclusions

We have performed structural optimization and electronic structure calculation of the Rare Earth Skutterudites $EuRu_4P_{12}$, using FP-LAPW method to evaluate theoretical lattice parameter and density of electron states near Fermi energy level. The optimized lattice parameter is in good agreement with the available experimental value. There is no band gap between the valence band and the conduction band and it indicates metallic nature of this compound and likely to be the candidate for thermoelectric applications.

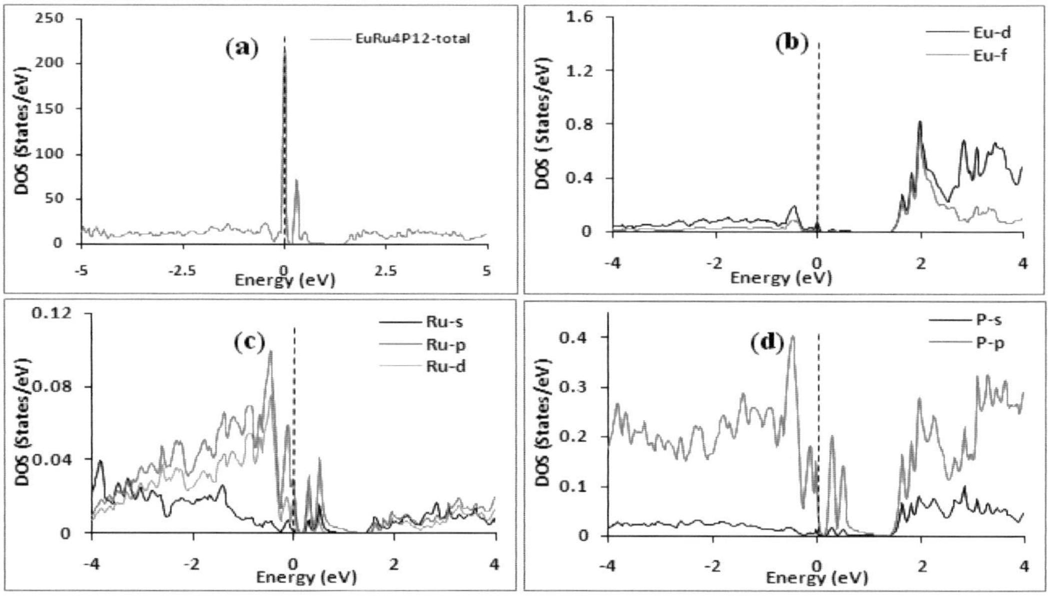

Figure 4. TDOS (a) and PDOS for (b) d and f statesof Eu (c) s, p and d states of Ru (d) s and p states of P of $EuRu_4P_{12}$.

References

[1] Zheng J C 2008 *Front. Phys. China* **3** 269
[2] Takegahar K, Harima H, Kaneta Y and Yanase A 1993 *J. Phys. Soc. Jpn.* **62** 2103
[3] Chen L, Tang X, Goto T and Hirai T 2000 *J. Matter. Res.* **15** 2276
[4] Bertini L, Stiewe C, Toprak M, Williams S, Platzek D, Mrotzek A, Zhang Y, Gatti C, Muller E, Muhammed M and Rowe M 2003 *J. Appl. Phys.* **93** 438
[5] Lue C S, Kuo Y K, Huang C L and Lai W J 2004 *Phys. Rev. B.* **69** 125111
[6] Sales B C et al. 2003 *Handbook on the Physics and Chemistry of Rare Earths* (Amsterdam: Elsevier) Chapter **211** 1
[7] Sales B C, Mandrus D, Chakoumakos B C, Keppens V, Thompson J R 1997 *Phys. Rev. B.* **56** 15081
[8] Sales B C, Mandrus D, and R. K. Williams 1996 *Science* **272** 1325
[9] Nolas G S, Cohn J L, Slack G A and Schujman S B 1998 *Appl. Phys. Lett.* **73** 178
[10] Bauere E D *et al.* 2004 *J. Phys. Condens. Matter.* **16** 5095
[11] Krishnamurthy V V *et al.* 2007 *Phys. Rev. Lett.* **98** 126403
[12] Shankar A, Rai D P , Khenata R, Maibam J, Sandeep and R K Thapa 2015 *J. Alloys Compd.* **619** 621
[13] Sekinee C *et al.* 2000 *Phys. Rev. B.* **62** 581
[14] Kihou K, Shirotani I, Shimaya Y, Sekine C, Yagi T 2004 *Mater. Res. Bull.* **39** 317
[15] Murnaghan F D 1944 *Proc. Natl. Acad. Sci.* USA **30** 244

IOP Conf. Series: Journal of Physics: Conf. Series **836** (2017) 012042 doi:10.1088/1742-6596/836/1/012042

X-ray diffraction and X-ray K-absorption studies of copper doped complex

P K Sharma[1,a], Ashutosh Mishra[2], Rashmi Kame[2,b], Varsha Malviya[2] and P K Malviya[3]

[1]Holkar Science college, Indore 452017, India
[2]School of Physics, Devi Ahilya Vishwavidyalaya, Indore 452001, India
[3]Govt. College, Nagda 456335, India

E-mail: [a)]psharma29762@yahoo.co.in, [b)]rashmikame23@gmail.com

Abstract. Sample of $Ni_{1-x}Cu_xFe_2O_4$where (x=0.05, 0.10, 0.15) has been prepared by root equation method. X-ray diffraction and K-absorption fine structural measurements have been carried out. The sample structure has been found to be cubic. X-ray diffraction measurements have been used to determine particle size and lattice parameter. EXAFS spectra has been recorded at 2.5Gev Indus -2 Synchrotron radiation source at RRCAT, Indore, India. EXAFS data have been analysed by computer software Athena.

1. Introduction

Ferrites are good dielectric materials having low conductivity and have wide application of microwave device. Ferrites are technologically important material because of their unique electric, dielectric, magnetic and an optical property which makes them suitable for many technological application like microwave device, transformer, electric generators storage device[1-6] etc. Extended Absorption fine structure refers to the oscillatory variation of the x-ray absorption as the as a function of photon energy beyond the absorption edge . Such fine structure may have amplitude of up to a few tenth of the edge jump. Synchrotron based X-ray Absorption Fine Structure (XAFS) spectroscopy becomes a powerful technique providing analytical as well as structural information with applications in a wide range of scientific field because of the rapid development of the treatment technique [7-9]. EXAFS has become the technique of choice for local structural investigations in a diverse range of material systems. This technique has also proved very useful, while studying the chemical reaction of matter under extreme condition of temperature and pressure .It is used to determine thebond distance coordination no.and species of the neighbour of the absorbing atom [9].The EXAFS oscillations in the absorption spectrum are caused by interference in the final state from backscattering of the outgoing photoelectron wave by neighbour of the exited .Hence ,it is extremely sensitive to the nearest neighbour bond length.

2. Experimental

Particles of the $Ni_{1-x}Cu_xFe_2O_4$ with x (gm) varying from x=0.05 to 0.15 were prepared by solid root method at room temperature. After weighing grind 8 to 9 hours the sample and powders were sintered at 2-2:30 hours at 550^{0}c. The XRD measurements were carried out using Bruker D8 Advance X-ray diffractometer. The X-ray were produced using a sealed tube and the wavelength of X-ray was 0.154nm (Cu K-alpha).The X-ray was detected using a fast counting detector based on silicon strip technology. Powder X-ray diffraction is perhaps the most widely used X-ray diffraction technique for characterizing materials. As the name suggest, the sample is usually in a powdery form, consisting of fine grains of single crystalline material to be studied. The technique is used also widely for studying particle in liquid suspension or polycrystalline solid .The term "powder" really means that the crystalline domains are randomly oriented in the sample. Therefore when the 2-D diffraction pattern is recorded, it shows concentric rings of scattering peaks corresponding to the various d spacing in the crystal lattice. In the powder method, the crystal to be examined is reduced to a very fine powder and placed in a beam of monochromatic X-rays. Each particle of the powder is a tiny crystal oriented at random with respect to the incident beam. Just by chance, some of the particle will be correctly oriented so that their (100) planes, for example, can reflect the incident beam. Other particles will be correctly oriented for (110) reflection, and so on. The result is that every set of lattice planes will be capable of reflection. The mass of powder is equivalent, in fact, to a single crystal rotated, not about one axis, but about all possible axes.

3. Results and discussions

Figure 1 shows the x-ray diffraction pattern of the $Ni_{1-x}Cu_xFe_2O_4$(x=0.05-0.15). It shows the formation of ferrite phase in all 3 samples. The interplanar distance d (Å) are calculated using Braggs law. All the peaks in the diffraction pattern have been indexed and refinement of the lattice parameter was done using powder X software. The crystalline size for each composition are calculate from XRD line width of the (2 1 0)(2 1 1)(2 1 1) peak using Scherrer formulla [10]. The Ferrite phase $Ni_{1-x}Cu_xFe_2O_4$ (x=0.05-0.15) is prepared by solid root technique. The phase confirmed by XRD studies. The particle size (13.3,9.1,8.55,7.84,7.4) nm by XRD calculation. Crystal structure has found cubic. In XRD, particle size is decrease as same manner. In the EXAFS calculation E_A & E_K value evaluated. The bond length of CU-Ni ferrite complexes with the help of Athena software [11]. XRD and EXAFS data shown in table 1 and table 2 and graph figure 1 and figure 2.

Table 1. Structural parameters of $Ni_{(1-x)}Cu_xFe_2O_4$ (x=0.05 to 0.15)

S.No.	Sample name	2θ	h k l	Particle size (nm)	Lattice parameter (Å)
1.	$Ni_{0.95}Cu_{0.05}Fe_2O_4$	43.30	2 1 1	13.43	4.68
2.	$Ni_{0.90}Cu_{0.10}Fe_2O_4$	35.50	2 1 1	11.64	6.32
3.	$Ni_{0.85}Cu_{0.15}Fe_2O_4$	33.11	2 1 1	11.54	6.0

Table 2. XANES parameters and bond length for the K absorption edge of copper in the complexes

S.No.	Sample name	E_K-Edge (eV)	E_A (eV)	Bond length
1.	$Ni_{0.95}Cu_{0.05}Fe_2O_4$	8985	8990.0	1.29
2.	$Ni_{0.90}Cu_{0.10}Fe_2O_4$	8996	8991.4	1.57
3.	$Ni_{0.85}Cu_{0.15}Fe_2O_4$	8997	8992.1	0.73

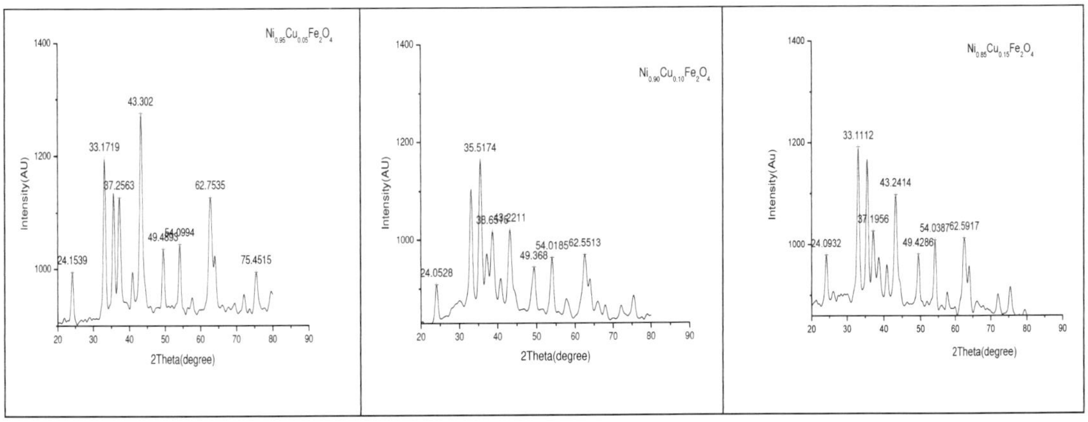

Figure 1. XRD pattern for Cu doped complexes $Ni_{(1-x)}Cu_xFe_2O_4$ (x=0.05 to 0.15)

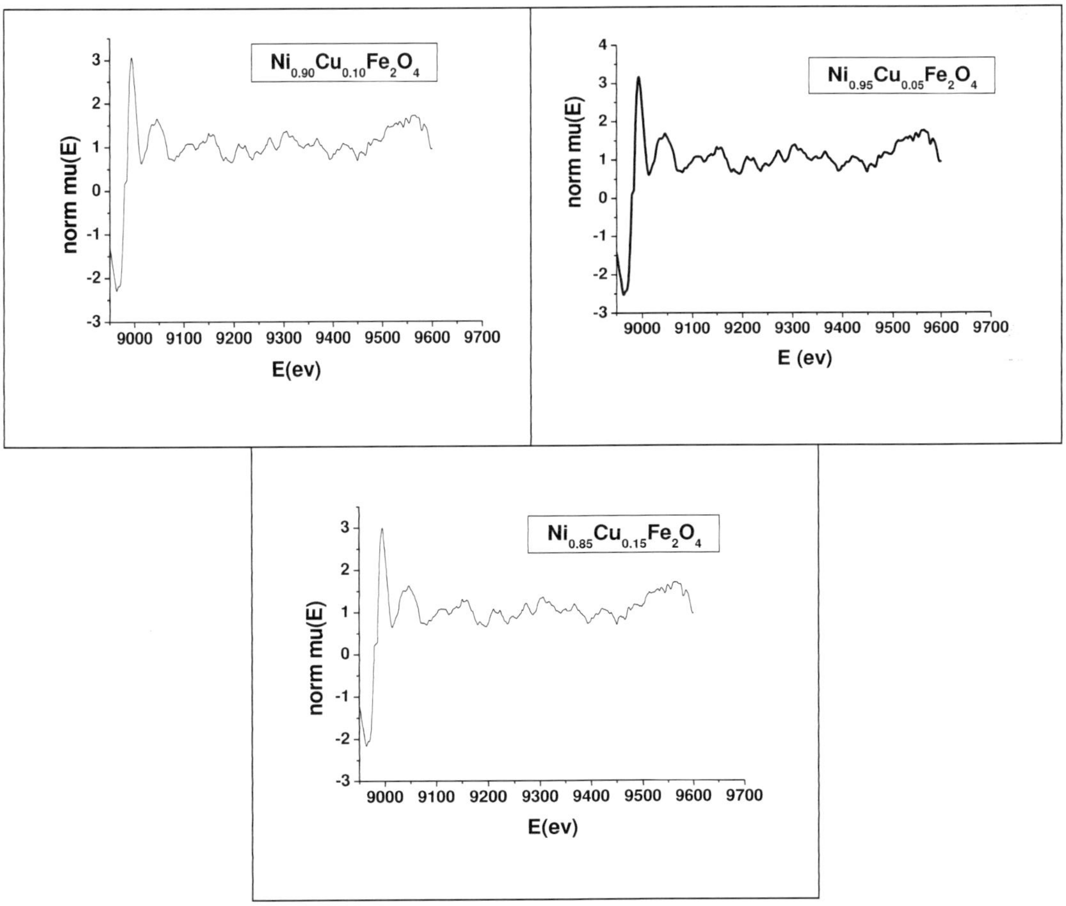

Figure 2. EXAFS spectra [μ(E) Vs E] of $Ni_{(1-x)}Cu_xFe_2O_4$ (x=0.05 to 0.15)

References

[1] Satishkumar G, Venkatraju C and Sivakumar K 2010 *Mat. Sci. Appl.* **1** 19
[2] Gubbala S, Nathani H, Koizol K and Misra R D K 2004 *Physica B: Phys. Cond. Matter* **348** 317
[3] Suryanarayana C and Norton M G 1998 *X-ray diffraction A Practical Approach* (Plenum Press)
[4] Snyder R 1999 *X-ray Diffraction in X-ray Characterization of Matter* ed E Lifshin (New York: Wiley-VCH)
[5] Speakman S A *Laboratory Module 01- Indexing X-ray Diffraction Patterns* 13-4009A
[6] Hinge V K, Joshi S K and Shrivastava B D, Prasad J and Srivastava K 2012 *J. Phys.: Conf. Series* **365** 012033
[7] Mishra A, Parsai N and Dagonkar N 2009 *Ind. J. Pure App. Phys.* **47** 337
[8] Malviya P K, Sharma P K and Mishra A 2013 *Int. J. Sci. Res. App. Phys.* **1** 2348
[9] Abbas T, Islam M U, Ashraf Ch. M 1995 *Mod. Phys. Lett. B* **9** 1419
[10] Cullity B D 1978 *Elements Of The X-ray Diffraction* (London: Addison Wesley)
[11] Rana M U, Islam M U, Ahmad I and Abbas I 1997 *Advanced materials-97: Proceeding of 5th International Symposium on Advanced Materials,* 21 - 25th September, 1997, Islamabad, Pakistan 283

Effect of ionospheric irregularities on GPS signals during declining phase of solar cycle 23 at Bhopal

Azad A Mansoori[1,a] and P K Purohit[2]

[1]Department of Electronics, Barkatullah University, Bhopal – 462026, MP, India
[2]National Institute of Technical Teachers' Training and Research, Bhopal – 462002, MP, India

E-mail: [a)]azadahmad199@gmail.com

Abstract. The present investigation is dedicated to the evaluation of GPS performance under disturbed geomagnetic conditions at the equatorial region. When GPS signals encounter the ionospheric irregularities of different size developed during high geomagnetic or solar activity, they undergo rapid changes in their phase and amplitude, known as scintillations. We have studied the occurrence characteristics of scintillation events during geomagnetic storms of different intensity at the crest of equatorial anomaly station Bhopal (23.2N, 77.6E). To accomplish this study we have used two data sets: Disturbed Storm Time (Dst) index and Amplitude Scintillation (S4) index. The geomagnetic storm activity is characterized by the Dst index and the evaluation of GPS performance during disturbed geomagnetic condition is realized through S4 index. We have selected thirty one geomagnetic storms that occurred during the year 2004 and 2005. We then classified these geomagnetic storms and scintillation events into weak, moderate and intense and weak, moderate and strong according to Dst index and S4 index respectively. During all the storm events we observed a good number of scintillation events. We then performed the correlation analysis between the Dst index and S4 index, to find out the impact of storm intensity on the occurrence of scintillation events.

1. Introduction

Ionosphere is a dispersive medium produced by ionization of gases and molecules by ionizing radiations from the sun. F-region of the ionosphere which is generally lies between 250 km to 400 km above the earth has a huge amount of free electrons. At times, F-region of the ionosphere get disturbed and small scale irregularities develop. When local electron density differs significantly from the surrounding electron density it produces localized ionospheric irregularities. Sufficiently intense irregularities scatter the radio waves and produce rapid fluctuations (Scintillations) in the amplitude and phase of the radio signals. Scintillation especially in the L-band has received a great concern of researchers over the decades because of its impact on communication and navigation systems like Global Positioning System (GPS). Scintillation has the capacity to affect the accuracy and reliability of GPS system by degrading the performances of GPS receivers. Variability of ionospheric irregularities is of serious concern to GPS system because these irregularities affect the amplitude and phase of trans-ionospheric radio (satellite) signals. Amplitude scintillation may induce signal fading and when depth of fading exceeds the fade margin of a receiving system, message errors are introduced in satellite communication systems. Amplitude scintillation leads to loss of lock by degrading the carrier-to-noise ratio (C/N_0) to below the receiver lock threshold [1], requiring the receiver to attempt reacquisition of the satellite signals. Losing signal is a major concern in GPS receiver navigation performance [2, 3]. Fortunately, many of the important characteristics of scintillation are already well known [1,3, 4, 5]. These studies revealed that scintillation activity varies with Season, local time, operating frequency, geographic location, magnetic activity and 11 year solar cycle. To determine the impact of scintillation on GPS, it is important to clearly understand the location, magnitude and frequency of occurrence of scintillation effects.Scintillation activity is most severe and frequent in and around the equatorial region, especially a few hours after sunset. Scintillation is more intense in the anomaly regions (±25^0 equator) than at the magnetic equator because of special characteristics of the equatorial ionosphere. The combination of electric and magnetic fields about the Earth produce free electrons to be lifted vertically and then diffuse northward and southward. Due to this process ionization over the magnetic equator reduces directly and increased the ionization over the anomaly regions. Low-latitude scintillation is seasonally dependent and is limited to local night-time hours. High latitude or polar scintillation is frequent but less severe in magnitude than that of the equatorial regions. The scintillation occurrence at auroral latitudes is strongly dependent on geomagnetic activity levels, but can occur in all seasons and is not limited to local night-time hours. Scintillation very rarely occurred in the mid latitudes, it occurs only during very intense magnetic storm periods [6, 7]. In low latitudes while both amplitude and phase scintillations may occur but in general amplitude scintillation is more severe than phase scintillation [8, 9]. Both amplitude and phase scintillations can degrade the GPS positioning performance by increasing the tracking error, number of cycle slips and probability of loss of lock.Forecasting of the scintillation occurrence for the local time, season, solar cycle and during the geomagnetic disturbances are very important for the radio communication.

2. Data sets and methodology

To accomplish this study we have used two data sets: Dst index and S4 index. The geomagnetic storm activity is characterized by the Dst index. The Dst index, is the global index to monitor the worldwide magnetic storm level. We have collected the data for Dst indices with 1h resolution from the Space Physics Data Facility OMNI website at http://omniweb.gsfc.nasa.gov/. On the basis of Dst index thresholds we classified the selected geomagnetic storms into three categories as intense storms (Dst ≤ -100nT), moderate storms (-100nT <Dst≤ -50nT) and weak storms (Dst ≥ -50nT). The evaluation of GPS performance during disturbed geomagnetic condition is realized through amplitude scintillation index (S4). In this study, the ionospheric scintillations has been measured using high data-rate NovAtel dual frequency GPS receiver GSV4004A GISTM receivers installed and operated at Department of Physics and Electronics, Barkatullah University, Bhopal, India (23.2°N, 77.6°E).

3. Results and discussions

To evaluate the performance of GPS under disturbed geomagnetic condition we have selected 31 geomagnetic storms that were observed during 2004 and 2005. As these years correspond to the high solar activity phase of the solar cycle 23, the geomagnetic activity during these years also remained high. The distribution of weak, moderate and weak geomagnetic storms is shown in Figure 1. From the Figure we find that among 31 storms, 11 belonged to the weak category, 13 storms were moderate and 07 intense storms were also observed.

We then identified the scintillations observed during all the three categories of geomagnetic storms viz. Weak, moderate and intense, separately for each category. The scintillations observed during a particular category e.g weak storm were then categorized into weak scintillation, moderate scintillation and strong scintillations. The Figure 2 shows the percentage occurrence of weak, moderate and strong scintillation during three different categories of geomagnetic storms separately. From the Figure during all three types of storms weak, moderate and intense the maximum percentage occurrence is that of weak scintillation followed by moderate scintillations and the least is that of strong scintillation. This shows that irrespective of the intensity of a geomagnetic storm, the highest percentage of scintillation observed during geomagnetic storms is those of weak scintillation while the lowest percentage is those of intense scintillation. The intensity of geomagnetic storm seems to have no association with the intensity of scintillation. To further investigate this we took the storm intensity index Dst and the scintillation index S4 and derived the correlation between them to access the magnitude of association between the two.

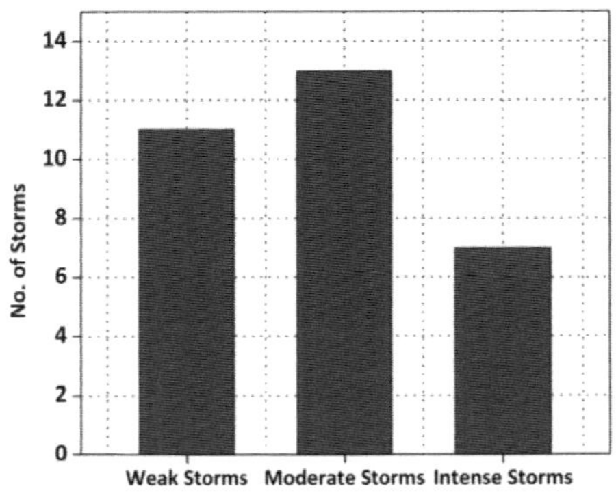

Figure 1. Distribution of geomagnetic storms into weak, moderate and intense storms observed during 2004 and 2005.

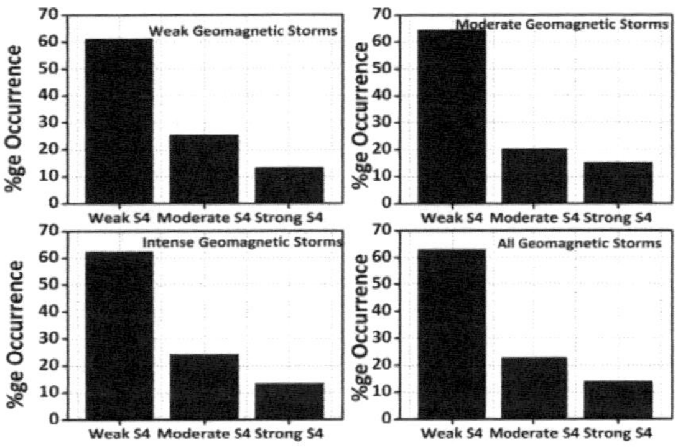

Figure 2. Percentage occurrence of weak, moderate and strong scintillations during three categories of geomagnetic storms.

The correlation analysis was performed separately for weak, moderate and intense storms. In each category of storms, we derived the correlation between the Dst index and no. of S4 occurrences separately for weak, moderate, strong and all scintillations. The scatter plot between the Dst index and no. of S4 occurrences separately for weak, moderate,strong and all scintillations observed during all the selected weak, moderate and intense geomagnetic storms separately is shown in Figure 3 to 5. To construct these plots we have taken the peak values or minimum values of the Dst index during each selected weak storms. From the Figure 3 we find that no. of S4 occurrences do not exhibit a good correlation with weak, moderate and strong scintillation separately. However if we examine the correlation between the total no. of S4 occurrences and the Dst index, we find the correlation is good.

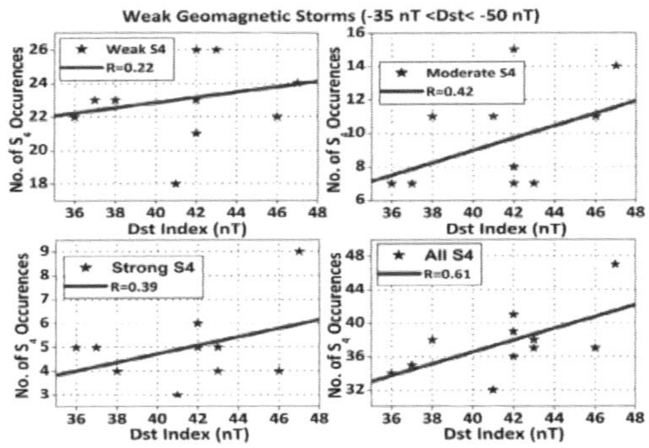

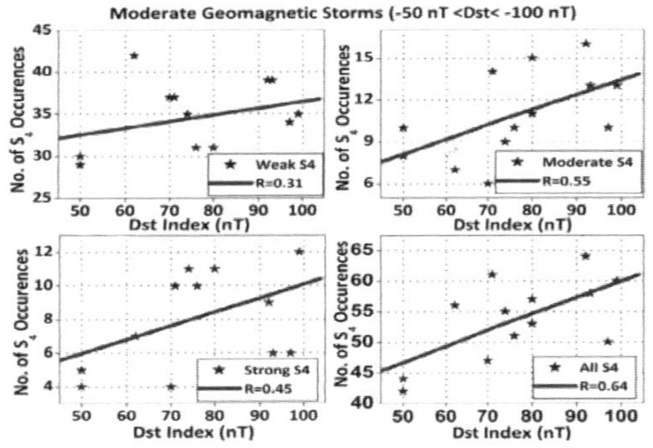

Figure 3. Scatter plot and correlation between the Peak Dst index and the No. of S4 occurrences for weak, moderate and strong scintillations during weak geomagnetic storms.

Figure 4. Scatter plot and correlation between the Peak Dst index and the No. of S4 occurrences for weak, moderate and strong scintillations during moderate geomagnetic storms.

From the figure 4 we find the no. of S4 occurrences exhibit a weak correlation with Dst index when examined separately for weak, moderate and strong scintillation. However, when the correlation between the total numbers of S4 occurrences is examined with the Dst index, the correlation is good. But the correlation is slightly better than observed between the same in case of weak geomagnetic storms.

From the figure 5 we find Dst index exhibits a good correlation with the no. of S4 occurrences for weak, moderate and strong scintillation. At the same time we also find that the total number of S4 occurrences exhibits a strong correlation with the Dst index. The correlation between the two is found to be much better than observed during the case of weak or moderate storms. The least correlation was observed during the case of weak geomagnetic storms while the strongest correlation was observed during the intense storms. Thus we find the number of S4 occurrences is highest during intense geomagnetic storms and least during weak geomagnetic storms showing that as the intensity of geomagnetic storm increases the number of scintillation occurrences increases.

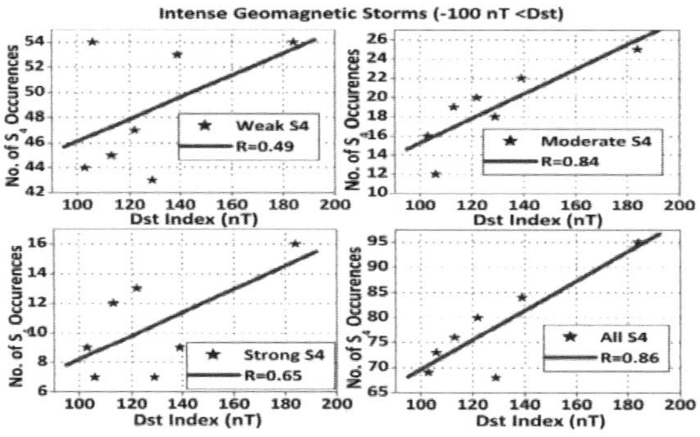

Figure 5. Scatter plot and correlation between the Peak Dst index and the No. of S4 occurrences for weak, moderate and strong scintillations during intense geomagnetic storms.

4. Conclusions

We summarize the main results of our investigation as follows:

- ★ During all types of storms the maximum percentage occurrence of scintillations was those of weakscintillations (60-70%), followed by moderate scintillations (20-30%) and the least was those of strongscintillations (10-20%).
- ★ A good correlation was observed between the number of S4 occurrences andDst index during the moderate and intense storms for all the three types of scintillationsweak, moderate and strong as well as the total number of scintillation occurrences.
- ★ A weak correlation exists between the Dst index and the number of S4 occurrences during the weak storms for all the three types of scintillations. However, a good correlation exists if the correlation is examined for total number of scintillation for the same category of storms.
- ★ The strongest correlation coefficients were found to exist for the intense storm category followed by moderate and least for weak storms for all three types of scintillation events.

References

[1] Moraes A de O, Rodrigues F da S, Perrella W J and Rodrigues de Paula E 2011 *Surv. Geophys.* **33** 1107
[2] Aquino M *et al.* 2005 *J. Nav.* **58** 241
[3] Kintner P M, Ledvina B M and De Paula E R 2007 *Space Weather* **5** S09003
[4] Aarons J 1993 *Space Sci. Rev.***63** 209
[5] Basu S, MacKenzie E and Basu S 1988 *Radio Sci.* **23** 363
[6] Doherty P, CosterA J and Murtagh W 2004 *GPS Solutions* **8** 267
[7] Basu Su *et al.* 2005 *Geophys. Res. Lett.* **32** L12S06
[8] Gwal A K *et al.* 2006 *Adv. Space Res.* **38** 2361
[9] Forte B 2012 *Radio Sci.* **47** RS4009

IOP Conf. Series: Journal of Physics: Conf. Series **836** (2017) 012044 doi:10.1088/1742-6596/836/1/012044

The full-potential study of ErMg

Meena Kumari[b], Manju Sikarwar, Priya Yadav, Priyanka Rajpoot, Shashank Nautiyal and U.P. Verma[a]

School of Studies in Physics, Jiwaji University, Gwalior-474011 M.P. India

Email: [a)]upv.udai@gmail.com, [b)]meenaranot@gmail.com

Abstract. The structural, electronic and magnetic properties of ErMg are investigated using the full potential linear augmented plane wave method based on density functional theory. The calculated structural parameters using LDA are in excellent agreement with the available experimental result. The calculated band structure is at 0 GPa. The valence and conduction bands are overlapped notable at the Fermi level (E_F = 0 eV). Consequently, there is no energy gap. The calculated band structure shows that this alloy has metallic nature. Total magnetic moment at equilibrium lattice constant of ErMg intermetallic in B2 phase along with the magnetic moment on the atomic and interstitial sites have also been presented in this paper.

1. Introduction

Magnesium is a superb metal because it is readily available and it is the lightest of all the structural metals having a density of 1.7g/cm^3. Magnesium mostly found in the earth's ocean. Magnesium has hexagonal close packed crystal structure. Therefore the property of deformation is difficultin such type of alloy. This structure restricts its ability to deform because it has fewer slip systems at lower temperatures. Magnesium has low melting point. Due to this property make easier to melt for casting. Pure magnesium is rarely used in the manufacturing of aerospace and automotive parts. In order to make use in developing purpose, magnesium is alloyed with other metals. Most common alloyed elements in commercial alloys are: aluminum, zinc, cerium, silver, and thorium. Magnesium can also be alloyed with rare earth elements, which increase the strength of magnesium especially at high temperature. The significant properties of magnesium alloys are light weight, low density (two thirds that of aluminum), and have good high temperature mechanical properties with good to excellent corrosion resistance. Magnesium alloys are good for engineering applications because they have good strength, ductility and creep properties. Magnesium alloys have replaced engineering plastics in many applications because they have a comparable density but are stiffer, more recyclable and less costly to produce. Magnesium is strong and light, making it an excellent choice for aerospace applications. Magnesium is also used in other products such as hand-held devices (chain saws, power tools, hedge clippers), in automobiles (steering wheels and columns, seat frames, transmission cases, crank case, camshaft sprocket, gearbox housings), and in audio-video-computer-communications equipment (laptop computers, camcorders, TV sets, cellular telephones.The Mg-RE based alloys serve a useful purpose in automotive industry as superior light metal-alloys in cast or wrought condition [1]. Mg-RE alloys also gained prominence in biomedical applications [2]. To recognize the magnetic properties of rare earth magnesium alloy Aleonard et al. [3] have performed neutron powder diffraction measurement. The magnetic properties of DyMg and ErMg using Mossbauer spectroscopy are also studied by Belakhovsky et al. [4]. In an early work, Buschow et al. [5] have analyzed the crystal structures and some physical properties of the intermetallics compounds of the rare earth (from La to

IOP Conf. Series: Journal of Physics: Conf. Series **836** (2017) 012044 doi:10.1088/1742-6596/836/1/012044

Lu) and non magnetic metals (B, Be, Mg, Ru, Rh, Pd) experimentally. The magnetic properties of rare earth magnesium compounds with CsCl structure have been determined in the temperature range 4.2 to 300K by Buschow et al [6] and also reported by Kirchmayr at etal. [7]. Zhang et al. [8] have predicted the enthalpy of formation of MgX (X =As, Ba, Ca, Cd, Cu, Dy, Ga, Ge, La, Ni, Lu, Pb, Si, Sn and Y) compounds from the first –principles calculation. The Brittle and elastic properties of the MgRE (RE= Sc, Y, Pr, Nd, Dy, Ho, Er) in CsCl structure have been investigated using the first principles density functional calculation by Wu and Hu [9]. The intermetallics phases in their hand book of crystallographic data are presented by Villars et al. [10]. The elastic properties of B2-MgRE (RE=Sc, Y, La–Lu) intermetallics have been calculated at T =0K by using first principles within the generalized gradient approximation (GGA) by Tao et al. [11]. Wang et al. [12] have studied lattice dynamical and thermodynamic properties such as thermal expansion, bulk modulus,heat capacities at constant volume and constant pressure as a function of temperature of the rare earth magnesium intermetallics compounds MgRE (RE=Y, Dy ,Pr, Tb) using density functional theory (DFT). They have also computed the temperature-dependent elastic properties and second and third order elastic constants for same family in CsCl type structure.In this paper, we have presented the structural, electronic and magnetic properties of ErMg alloy by using the wien2k code, theoretically.

2. Computational details

The zero-temperature energy calculations are performed using both the all-electron full-potential linear augmented-plane-wave method (FP-LAPW).All the calculations have been carried out using the WIEN2k code [13]. The sphere radii used in the calculations are 3.0 and 2.2 *a.u.* for Er and Mg, respectively. Within these spheres, the charge density and potential are expanded in terms of crystal harmonics up to angular momenta l = 10, and a plane wave expansion has been used in the interstitial region. The Brillouin zone integrations for the total energy have been carried out using 35 special k-points in the irreducible Brillouin zone. Well converged solutions were found for $R_{cut}K_{max}$= 7, where K_{max} is the plane wave cut-off and R_{cut} is the atomic sphere radii. Both the muffin-tin radius and the number of k-points are varied to ensure convergence. Core states are treated fully relativistic but for valence states relativistic effects have been included in a scalar relativistic treatment. The self-consistent calculation stops only when the total energy and the charge of the system meet the convergent limit of the order 10^{-4} Ry and $10^{-3}e^{-}$, respectively. The equilibrium lattice constant was obtained by fitting the total energy versus volume according to the Murnaghan equation of states [14].

3. Results and discussions

3.1 Structural properties

The equilibrium lattice parameters are determined by the energy minimization procedure. As a first step, a set of the unit cell volume and total energy is calculated using local density approximation (LDA). At the equilibrium volume, we set the atomic positions to obtain the equilibrium structure of the unit cell. The results thus achieved are operated to obtain a new set of the total energy and the unit cell volume, which are used to obtain equilibrium bulk modulus and its pressure derivative using the Murnaghan equation of states. The obtained results are shown in table 1. From table 1, it can be seen that the calculated lattice constants using the LDA formalism are consistent with the available experimental results.

Table1. Calculated lattice constant, a(Å), the bulk modulus B_0 (GPa), bulk modulus first pressure derivative B_0' and total energy E_0 (Ry) of ErMg along with the available theoretical and experimental data in B2 phase.

Material	Phase	a (Å)	B_0	B_0'	E_0	Ref.
ErMg	B2	3.72	35.96	3.96	-26582.03	P.W
		3.74	-	-	-	Exp.[3]
		3.75	40.06	3.47	-	The. [15]

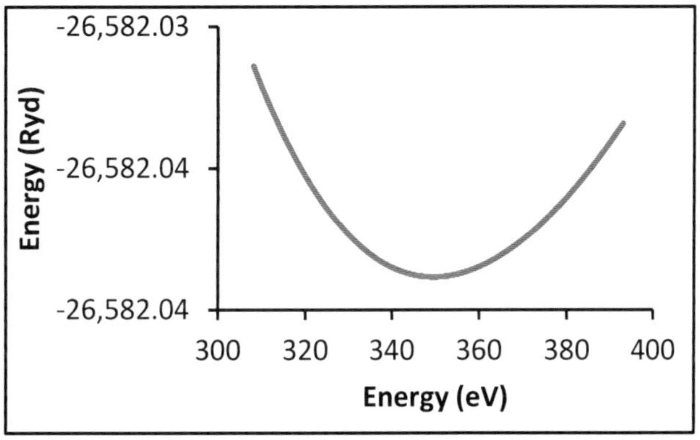

Figure 1. Total energy versus cell volume of HoMg using local density approximation (LDA).

3.2 Electronic properties

The self-consistent band structures of ErMg were calculated in B2 phase at equilibrium lattice constant within the LDA scheme in spin up and spin down case. The overall profile of band structures in spin up and spin down cases show the metallic character. The obtained band structures are good agreement with the available data. But in spin up case the band is shifted towards the Fermi level. The obtained band structures are shown in Figure 2.

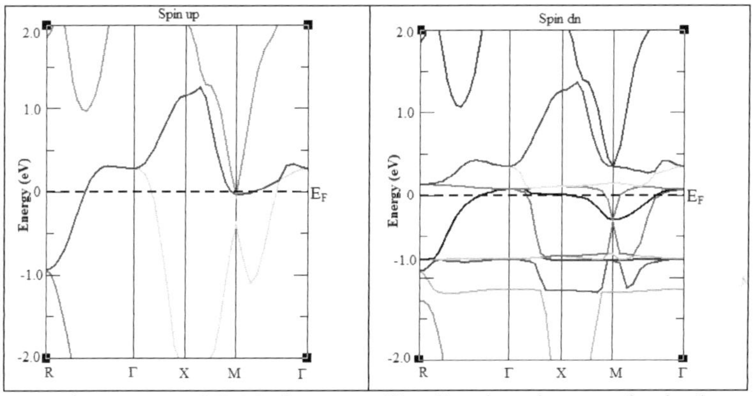

Figure 2. Band structure of ErMg intermetallic alloy in spin up and spin down channel.

3.3 Density of states

To understand the behaviour of electronic band structure, we have plotted the total density of states (TDOS) and partial density of states (PDOS) in spin and spin down case. In case of TDOS, it seems the bands lie within the region -2.0eV to -4.0eV is due to the main contribution from Er- 4f states. Peaks are concentrated in the range -2.0 to – 4.0eV in spin up cases while in spin down cases f states overlap Fermi level completely from 1.0eV to -1.0eV.The contribution of Mg-s states in band formation are also shown in figure 3.

3.3 Magnetic properties

Magnetic moment of any compound play very important role to identify the internal contribution of the spin of the electrons. The spin polarized self-consistent band structure calculation has been very successful in predicting the magnetic moment by LDA scheme. The obtained value of magnetic moments of ErMg in B2 phase is listed in Table 2. It is clear that the total and local magnetic moments (M and m) decrease with the increase in lattice parameter. The main contribution to the magnetic moment is mainly due to the 4f electrons of Er, while the contribution of Mg almost negligible. The magnetic moment Mg atom is negative, which specify that the involvement of 2p states of Mg atom is anti-parallel to that of Er.

IOP Conf. Series: Journal of Physics: Conf. Series **836** (2017) 012044 doi:10.1088/1742-6596/836/1/012044

Table 2. Obtained values of atomic and total spin magnetic moments of ErMg in stable CsCl phase

Lattice parameter	m^{Er}	m^{Mg}	m^{int}	**Total**
6.80	2.3531	-0.0203	-0.0972	2.2355
6.92	2.3533	-0.0182	-0.0922	2.2428
7.04	2.3490	-0.0165	-0.0893	2.2432
7.16	2.3428	-0.0151	-0.0875	2.2402
7.27	2.3343	-0.0142	-0.0891	2.2309

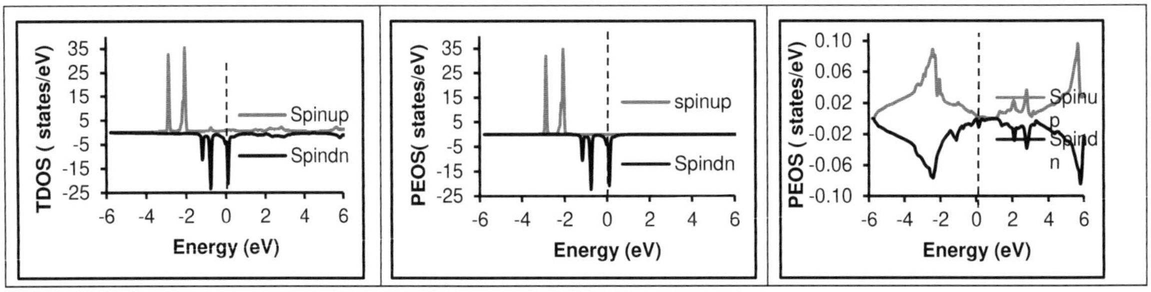

Figure 3. Spin-dependent total (a) and partial (b), (c) density of states of ErMg, using LDA scheme.

4. Conclusions

The structural, electronic and magnetic properties of ErMg intermetallic alloy are investigated by employing the first principles FP-LAPW based on DFT. The obtained equilibrium structural parameters are in excellent agreement with the experimental ones and also are served in this work to study the other properties. The electronic band structure results show that this alloy has metallic nature and the density of states also indicate that there is no gap at Fermi level. The total and local magnetic moment decrease with the increase lattice parameter. Rare earth atom i.e. Er show the main contribution in total magnetic moment due to 4f electrons. Negative sign of Mg atom specify that antiprallel spin to that of Er.

References

[1] Mordike B L and Ebert T 2001*Mat. Sci. Eng. A* **302** 37
[2] Willbold E *et al.* 2015 *Acta Biomater* **11** 554
[3] Aleonard R, Morin P, Pierre J and Schmitt D 1976 *J. Phys. F: Metal Phys.* **6**
[4] Belakhovsky M, Chappert J and Schmitt D 1977 *J. Phys. C: Solid State Phys.* **10** L493
[5] Buschow K H J 1979 *Rep. Prog. Phys.* **42** 1373
[6] Buschow K H J 1973 *J. Less Common Metal* **33** 239
[7] Kirchmayr Hans R, Poldy Carl A, Groessinger R, Haferl R, Hilscher G, Steiner W and Wiesinger G 1979 *Handbook Phys. Chem. Rare Earths* **25** 5
[8] Zhang H, Shang S, Saal J E, Saengdeejing A, Wang Y and Liu Z K 2009 *Intermetallics* **17** 878
[9] Wu Y and Hu W 2007 *Eur. Phys. J. B* **60** 75
[10] Villars P and Calvert L D 1985 *Pearson's Handbook of Crystallographic Data for Intermetallic Phases* (ASM: Metals Park, OH)
[11] Tao X, Ouyang Y, Liu H, Feng Y, Du Y and Jin Z 2008 *Solid State Commun.* **148** 314
[12] Wang R, Wang S and Wu X 2011 *Phys. Scr.* **83** 065707
[13] Bhaha P, Schwarz K, Sorantin P and Rickey S B 1990 *Comput. Phys. Commun.* **59** 399
[14] Murnaghan F D 1994 *Proc. Natl. Acad. Sci. U.S.A.* **30** 244
[15] Yasemin *et al.* 2014 *Gazi Univ. J. Sci.* **27** 761

IOP Conf. Series: Journal of Physics: Conf. Series **836** (2017) 012045 doi:10.1088/1742-6596/836/1/012045

Study of XAFS of some Fe compounds and determination of first shell radial distance

Neetu Parsai[1,a] and Ashutosh Mishra[2]

[1]Shree Vaishnav Institute of Management, Indore, India
[2]School of Physics, Devi Ahilya Vishwavidhyalaya, Indore, India

E-mail: [a)]parsai.neetu@gmail.com

Abstract. X-ray absorption fine structure (XAFS) of some Fe compounds have been studied using the latest XAFS analysis software *Demeter with Strawberry Perl*. The processed XAFS data of the Fe compounds have been taken from available model compound library. The XAFS data have been processed to plot the μ(E) verses E spectra. These spectra have been converted into K-space, R-space and q-space. R-space spectra have been used to obtain first shell radial distance in Fe compounds. Structural parameters like first shell radial distance is useful in determination of bond length in Fe compounds. Hence the study play important role in biological applications.

1. Introduction

X-ray absorption fine structure (XAFS) technique enables the measurement of fine structures in the X-ray absorption spectra, above the absorption edge of the atoms in a material. Generally an EXAFS spectrum extends from ~100 eV below the absorption edge of the particular atom, to ~700 eV above the absorption edge. The spectrum near the absorption edge (viz., the XANES part), gives information about the long range order existing in the sample and the oxidation states of the atoms involved, while the part of the spectrum well above the absorption edge (the EXAFS part) gives information regarding the short range order and local structure around the atomic species. With the advent of modern bright Synchrotron radiation sources, this technique has emerged as the most powerful technique for local structure determination, which can be applied to any type of material viz. amorphous, polycrystalline, polymers, surfaces and solutions.

A simplified EXAFS analysis method had been proposed previously for some transition metals[1,2] and copper compounds[3]. In this procedure, the theoretical EXAFS data was generated, employing computer software *Mathcad*, for the first coordination shell around the absorbing atom, using standard EXAFS equation[4]. For this calculation, the required phase shift was extracted from the experimental data itself using LSS method[5] and backscattering amplitude was taken from tables of Mackale[6] for 2.5Å radial distance from the central absorbing atom and Teo and Lee[7]. The other parameters were taken from crystallographic data[8]. First shell bond length has been calculated employing commonly used Fourier transform technique and compared with standard crystallographic value. In the present work analysis has been done on Fe foil and some Fe compounds using the latest XAFS analysis software *Demeter with Strawberry Perl*. The processed XAFS data of the Fe compounds have been taken from available model compound library. The XAFS data have been processed to plot the μ(E)

IOP Conf. Series: Journal of Physics: Conf. Series **836** (2017) 012045 doi:10.1088/1742-6596/836/1/012045

verses E spectra. These spectra have been converted into K-space, R-space and q-space. R-space spectra have been used to obtain first shell radial distance in Fe compounds.

2. Experimental

Since EXAFS spectroscopy requires an intense polychromatic X-ray source, the most suitable light source is a synchrotron storage ring. The most common EXAFS beam-line works in transmission. It collects data measuring how the beam intensity decreases as it passes through the sample while scanning energy using a crystal monochromator. Experimental spectra are usually recorded by scanning energy from about 200 eV below the explored edge to 1000 eV above it. Energy steps can be of 0.1–2.0 eV, depending on the energy interval and the experimental setup.

3. Data Analysis

The purpose of EXAFS analysis is to obtain the parameters of the local structure of the absorbing atom from its experimental cross-section. Many analysis methods have been developed and in this section we will describe one of the most general and simple ones. The first step in data analysis is signal extraction by background removing. This delicate operation can be divided
into three steps: a) choosing the threshold energy, for wave vector definition; b) pre-edge extrapolation (μ_0); c) atomic absorption modeling (μ). After these steps, it is possible to obtain the EXAFS signal as:

$$\chi(E) = \frac{\mu(E) - \mu_0(E)}{\Delta\mu_0(E)}$$

The EXAFS signal contains the contributions of all of the coordination shells. Each contribution can be approximated to a damped sinusoidal function in k-space, whose frequency is proportional to the absorber-scatterer distance.

The EXAFS signal is Fourier transformed to obtain a radial distribution function. Fourier transform (FT) is the standard tool used for frequency separation. In fact, this operation transforms each sinusoidal component in a FT modulus peak, going from the k(Å^{-1}) space to R(Å) space. To minimize distortions and to avoid the presence of ripples (peak side-lobes which arise from the truncated integration) in the FT modulus, $\chi(k)$ is multiplied by a window W(k). The height of peaks depends on the amplitude parameters of the EXAFS equation, while their position depends on phase parameters. Total or partial overlapping of more peaks often occurs. The position of a peak depends in the first approximation on the distance between absorber and scatterer and in the second approximation on the phase function. Since $\Phi(k)$ depends on k (always with a negative slope), peaks are shifted to lower R values. Once the peak associated to a single coordination shell has been recognized in the Fourier transform of the EXAFS, its contribution to the total signal can be obtained by an inverse Fourier transform carried out over the limits of the peak in R-space (otherwise the filter is calculated in a region containing more than one contribution). The outcome of this operation is the $\chi(k)$ function related to absorber-scatterer pairs whose interatomic distances belong to that specific integration interval.

Structural parameters can be obtained from $\chi(k)$ by means of a fitting procedure using known phase and amplitude functions. EXAFS analysis allows obtaining coordination numbers, interatomic distances and an estimate of the disorder around the central atom.

4. Results and Discussions

The experimental EXAFS data of the studied Fe foil and compounds have been taken from "the EXAFS model compound library" available at (http://cars9.uchicago.edu/~newville/ModelLib/). The analysis has been done using latest XAFS analysis software *Demeter with Strawberry Perl*. Figures 1(a)-(e) are showing the χ(E) verses E, $\chi(k)$ versus k, magnitude of Fourier transform versus radial distance and inverse Fourier transform versus q curves for Fe foil and its compounds. The value of first shell radial distance calculated from Fourier transform for Fe foil and its compounds have been

IOP Conf. Series: Journal of Physics: Conf. Series **836** (2017) 012045 doi:10.1088/1742-6596/836/1/012045

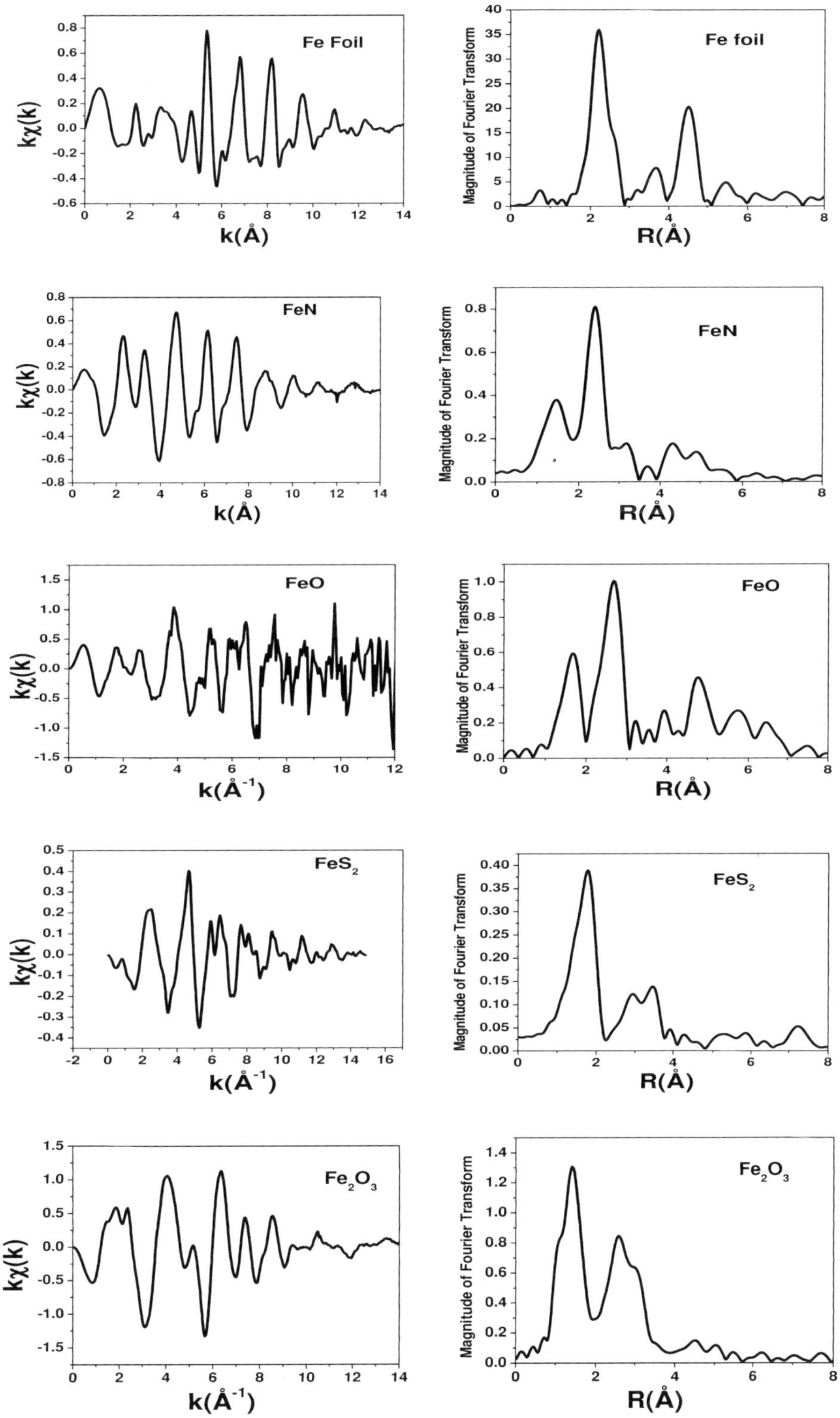

Figure 1. χ(k) versus k, magnitude of Fourier transform curves of Fe metal foil and compounds at room temperature.

IOP Conf. Series: Journal of Physics: Conf. Series **836** (2017) 012045 doi:10.1088/1742-6596/836/1/012045

shown in table 1. The values of first shell radial distance for the Fe metal foil and compounds are in good agreement with the standard crystallographic values except Fe_2O_3. Hence the proposed method of data analysis using the latest XAFS analysis software *Demeter with Strawberry Perl* is useful and efficient for the determination of structural parameters like inter atomic distances.

Table 1. First shell bond length for the Iron compounds.

S. No.	Compounds	$R_{Crystallographic}$ [Å]	$R_{experimental}$ [Å]
1	Fe	2.48	2.16
2	FeN	1.97	2.31
3	FeO	2.00	2.58
4	FeS_2	2.26	1.83
5	Fe_2O_3	2.03	1.39

5. Conclusions

EXAFS has developed into a mature technique with numerous applications in physics, chemistry, materials science, biology, and environmental science. The range of application of EXAFS will be further extended when third generation synchrotron radiation sources come on-line. The method outlined in the present work can be used for easy and quick determination of the first shell bond lengths, before one prefers to use elaborate and time consuming method of analysis of EXAFS data.

References

[1] Parsai N, Mishra A and Shrivastava B D 2012 *X-Ray Spectrum* **41** 219
[2] Mishra A, Parsai N and Shrivastava B D 2011 *Indian J. Pure App. Phys.***49** 25
[3] Mishra A, Parsai N and Shrivastava B D 2012 *Solid State Phys. AIP Conf. Proc.***1447** 173
[4] Rehr J Jand Albers R C 2000 *Rev. Mod. Phys.***72** 621
[5] Lytle F W, Sayers D E and Stern E A 1975 *Phys. Rev. B* **11** 4825
[6] McKale A G, Knapp G S and Chan S K 1986 *Phys. Rev. B* **33** 841
[7] Teo B K and Lee P A 1979 *J Amer. Chem. Soc.* **101** 2815
[8] 1962*International tables for X-ray crystallography* vol III ed Caroline H MacGiilavry and Gerard D. Rieck (Kynoch press, Birmingham, UK) p 277

IOP Conf. Series: Journal of Physics: Conf. Series **836** (2017) 012046 doi:10.1088/1742-6596/836/1/012046

Raman and X-ray diffraction studies of superconducting $FeTe_{1-x}Se_x$ compounds

Pavitra Devi Lodhi[1,a], V P S Awana[2] and Netram Kaurav[1,b]

[1] Department of Physics, Government Holkar Science College, A. B. Road, Indore 452017, MP, India
[2] Superconductivity and Cryogenics Division, National Physical Laboratory, New Delhi-110012, India

E-mail: [a)]pavitralodhi14@gmail.com, [b)]netramkaurav@yahoo.co.uk

Abstract. Iron-based superconducting layered compounds have the second highest transition temperature after cuprate superconductors. Their discovery is a milestone in the history of high-temperature superconductivity and will have profound implications for high-temperature superconducting mechanism as well as industrial applications. We have measured X-ray diffraction and Raman spectra of poly-crystalline $FeTe_{1-x}Se_x$ samples with $x = 0.0$, 0.5 and 1 at room temperature. X-ray profile of FeTe and $FeTe_{0.5}Se_{0.5}$ confirmed the primitive tetragonal unit cell. FeSe is found to be in double phase with tetragonal phase (about 66.62%) and orthorhombic (33.38%). A_{1g} and B_{1g} phonon modes were observed in Raman spectra. At room temperature, these lines are located at 95 and 130 cm^{-1} for FeTe, 93 and 129 cm^{-1} for $FeTe_{0.5}Se_{0.5}$ and 93 and 129 cm^{-1} for FeSe samples.

1. Introduction

The recent discovery of superconductivity in quaternary, rare-earth transition-metal oxypnictides, and especially the subsequent raising of the superconducting transition temperature (T_c) has drawn great interest in the condensed matter community [1-2]. $REFeAsO_{1-x}F_x$, which was abbreviated as FeAs-1111, is the first series of superconductors showing such high T_c values without copper-oxide planes as traditionally observed in cuprate superconductors [3]. As such, it provides a new system, quite different from the cuprate superconductors, in which the mechanisms of high-temperature superconductivity were understand in recent years. It has been thought that superconductivity in the FeAs-based series may have a direct connection to a spin-density-wave (SDW) anomaly occurring in the FeAs layer [4]. In particular, superconductivity emerges when such SDW order is suppressed by chemical doping [5] or by high pressures [6]. However, all these series of iron-based superconductors contain the element As, which is toxic on its own and would be even more so when oxidized to As_2O_3 and/or related compounds [7]. Therefore, there was a quest for search of alternate compound s/elements to serve As role and make these systems user and environment friendly as well. In continuation, as a substitute, FeSe with some Se deficiency, which is less toxic and easier to handle than arsenides, was found to exhibit superconductivity [3]. Among the iron-based superconductors, these systems, called the 11 system. The FeSe/Te superconductor is a remarkable material. The structural simplicity of FeCh4 (Ch = S, Se, Te), in which the conducting layers are not separated by any third-atom layers, offers the best tool for investigating the nature of superconductivity and magnetism in Fe-based compounds [8]. Either the pressure or the chemical composition distort the FeCh4 tetrahedron and tune the electronic properties.

However, pure FeTe is not superconducting and the two end compounds FeSe and FeTe are structurally isomorphic but revels different physical and magnetic properties [9]. Density functional calculations on FeS, FeSe and FeTe indicated that the strength of spin density wave (SDW) in FeTe and the possibility of higher T_c in doped FeTe alloy compared to FeSe [10]. Further, the enhancement of T_c in Te substituted FeSe explained on the basis of the density functional calculations. When partially substituting Se for Te in the antiferromagnetic FeTe, the Se substitution affects the structure and must be kept under control for tuning the structure deformation and the electronic properties. In fact, the superconducting state exists over quite a wide range of doping in the Fe (Se, Te) system (up to 90% Te substitution for Se in polycrystalline samples) with a maximum of 15 K [11]. FeSe has been studied quite extensively [9], a key observation is that the phase pure superconducting sample exists only for those samples prepared with intentional Se deficiency. In this paper, we report Raman-scattering results obtained on $FeTe_{1-x}Se_x$ bulk samples. The zone-center optical modes were classified by a group-theoretical analysis and the Raman-active phonons assigned accordingly.

2. Experimental

The bulk polycrystalline $FeTe_{1-x}Se_x$ samples with x = 0.0, 0.5 and 1, were synthesized through a standard solid state reaction route via vacuum encapsulation. The high purity chemicals Fe, Se, and Te were weighed in the stoichiometric ratio and ground thoroughly in a glove box having pure Argon atmosphere. The mixed powder was subsequently pelletized and then encapsulated in an evacuated (10^{-3} Torr) quartz tube. The encapsulated tube was then heated at 750 °C for 12 hours and slowly cooled to room temperature. The heating schedule was repeated couple of times with intermediate grinding. The x-ray diffraction (XRD) pattern was recorded at room temperature in the scattering angular (2θ) range of 10° to 60° in equal 2θ steps of 0.02° using a Rigaku Diffractometer with Cu Kα (λ = 1.54 Å). Raman spectra were recorded with Horiba Jobin Yvon T6400 Micro Raman using a He–Ne Laser operating at λ = 632.8 nm.

3. Results and discussions

Figure 1 shows the x-ray diffraction profile and results of Rietveld refinement. The x-ray profile of FeTe, $FeTe_{0.5}Se_{0.5}$ and FeSe confirmed the primitive tetragonal unit cell and the space group P4/nmm, z-coordination parameter = 0.2850). The FeTe and $FeTe_{0.5}Se_{0.5}$ samples are in single phase as confirm from the Rietveld refinement. However, FeSe is found to be in double phase, were in, tetragonal phase (about 66.62%) and orthorhombic with space group *Cmcm* (33.38%). The estimated values of lattice parameter a and c from the refinement results are listed in table 1 and both were found to be decreasing with the selenium concentration. Similar trends also true for volume concentration. Further, the bond length of Fe–Te = 2.648 Å, Fe–(Se,Te) = 2.505 Å and Fe-Se = 2.33 Å were calculated using the Vesta software and found that the bond length to be decreasing as the concentration of Se increases. Observation of such trend in bond length is quite consistent as the smaller ionic radius of Se over the Te, presumably creates the positive chemical pressure.

Figure 2 shows the comparison of phonon Raman spectra between the parent compound FeTe and the superconducting $FeTe_{0.5}Se_{0.5}$ and FeSe. From the group theory considerations [12], Γ- point phonon modes of the tetragonal Fe(Te,Se) can be expressed as $\Gamma = A_{1g} + 2A_{2u} + B1g + 2E_g + 2E_u$. Fe and Te ions occupy 2a and 2c Wyckoff positions, respectively. Symmetry analysis shows that there are four Raman-active modes [A_{1g}(Te) + B_{1g}(Fe) + $2E_g$(Te, Fe)] and two infrared active modes [A_{2u}(Te, Fe) + E_u(Te, Fe)] [13]. Acordingly, the Raman tensors take the form:

$$A_{1g} = \begin{pmatrix} a & 0 & 0 \\ 0 & a & 0 \\ 0 & 0 & b \end{pmatrix}, \tag{1}$$

IOP Conf. Series: Journal of Physics: Conf. Series **836** (2017) 012046 doi:10.1088/1742-6596/836/1/012046

$$B_{1g} = \begin{pmatrix} c & 0 & 0 \\ 0 & -c & 0 \\ 0 & 0 & 0 \end{pmatrix}, \quad (2)$$

$$E_g = \begin{pmatrix} 0 & 0 & -e \\ 0 & 0 & e \\ -e & e & 0 \end{pmatrix}, \quad (3)$$

are the Raman active modes and

$$B_{2g} = \begin{pmatrix} 0 & d & 0 \\ d & 0 & 0 \\ 0 & 0 & 0 \end{pmatrix} \quad (4)$$

is the Raman inactive mode.

Table 1. Structural parameters obtained from the Rietveld refinement analysis of $FeTe_{1-x}Se_x$ at 300 K. Space group: *P4/nmm* (# 129) and Space group: *Cmcm* (# 63) . Atomic positions: Fe: 2a (3/4, 1/2, 0); Te: 2c (1/2,1/2, 0.28); respectively.

X	0.0	0.5	1 (*P4/nmm*)	1 (*Cmcm*)
a (Å)	3.822	3.794	3.809	5.241
b (Å)	3.822	3.794	3.809	4.572
c (Å)	6.281	6.017	5.389	1.292
V (Å^3)	91.757	86.624	66.62	33.38

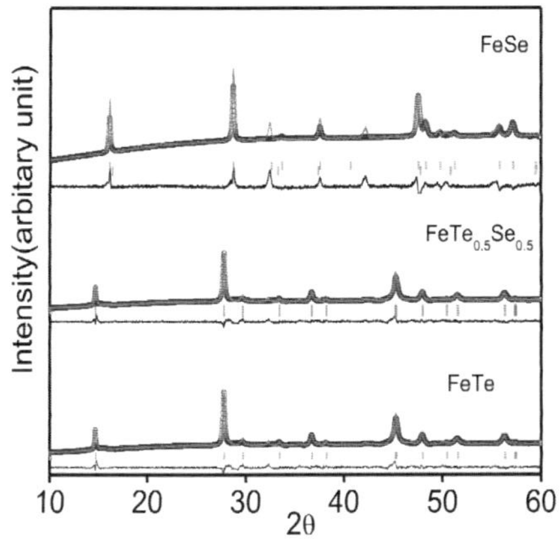

Figure 1. X-ray diffraction data along with the Rietveld refinement results of $FeTe_{1-x}Se_x$ samples.

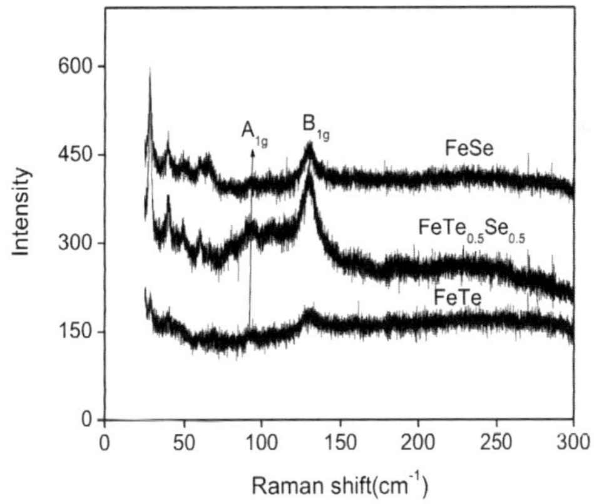

Figure 2. Comparison of phonon Raman spectra between the parent sample FeTe and superconducting $FeTe_{0.5}Se_{0.5}$ and FeSe samples. Vertical line shows the shifting of A_{1g} mode.

Herein, the Raman-active modes, $\Gamma_{Raman} = A_{1g} + B_{1g} + 2E_g$ can be assigned to the observed frequencies. However, we have observed A_{1g} and B_{1g} modes in our measurements. At room temperature, these lines are located at 95 and 130 cm^{-1} for FeTe, 93 and 129 cm $^{-1}$ for $FeTe_{0.5}Se_{0.5}$ and 93 and 129 cm $^{-1}$ for FeSe samples. The peak position of A_{1g} is found to be decreasing as the Se concentration increases. Such observation could be due the Fe magnetic moment and anharmonicity [13]. As mentioned above, the A_{1g} type of Te vibrations plays the important role for the mode vibration in the chalcogen's tetrahedral formation. Only chalcogen's antiphase vibrations that occurred perpendicular to the Fe layer has an internal source of anharmonicity. While the peak position of B_{1g} has been shifted slightly to lower energy with increasing Se concentration. It is instructive to mention that Fe magnetic moment and anharmonicity may not alter the B_{1g} modes. Therefore, the B_{1g} vibration of Fe ions should not be affected by this structure due to the Fe layer topology. In other phonon Raman-scattering measurements, Okazaki et al. [14] have observed these modes at 158 and 202 cm^{-1} on single crystal $Fe_{1.074}Te$ sample. Similar results were also reported by Xia et al. [15] at 159.1 and 196.3 cm^{-1} for single crystal $FeTe_{0.92}$ and Kumar et al. [16] at 160 and 224 cm^{-1} on polycrystalline $FeSe_{0.82}$ sample.

4. Conclusions

We have measured X-ray diffraction and Raman spectra of poly-crystalline $FeTe_{1-x}Se_x$ samples with x = 0.0, 0.5 and 1 at room temperature. X-ray profile of FeTe and $FeTe_{0.5}Se_{0.5}$ confirmed the primitive tetragonal unit cell. FeSe is found to be in double phase with tetragonal phase (about 66.62%) and orthorhombic (33.38%). A_{1g} and B_{1g} phonon modes were observed in Raman spectra. The peak position of A_{1g} is found to be decreasing as the Se concentration increases. While the peak position of B_{1g} has been shifted slightly to lower energy with increasing Se concentration.

Acknowledgements

Authors are thankful to Dr. Mukul Gupta and Dr. V. Ganesan, Centre Director, UGC-DAE-CSR, Indore for XRD and Director, Raman Research Institute, Bangaluru for Raman data for providing experimental faculties. Financial support from SERB, DST, New Delhi are gratefully acknowledged.

References

[1] Kamihara Y, M. Hirano M and Hosono H 2008 *J. Am. Chem. Soc.* **130** 3296
[2] Chen XH, Wu T, Wu G, Chen H and Fang DF 2008 *Nature* **453** 761
[3] Yeh K W, Huang T W, Yan Y D C and Wu M K 2008 *Eur. Phys. Lett.* **84** 37002
[4] Cruz C de la, Huang Q, Mook H A,Wang N L and Dai P C 2008 *Nature* **453** 899
[5] Mizuguchi Y, Tomioka F and Takano Y 2009 *App. Phys. Lett.* **94** 012503
[6] Subedi A, Zhang L, Singh D J and Du M H 2008 *Phys. Rev. B* **78** 134514
[7] Fang M H, Pham H M, Liu Y, Spinu L and Mao Z Q 2008 *Phys. Rev. B* **78** 224503
[8] Ishida K, Nakai Y and Hosono H 2009 *J. Phys. Soc. Jpn.* **78** 062001
[9] Terzieff P and Komarek K L 1978 *Monats Chem.* **109** 651
[10] Okada H, Takahashi H and H. Takahashi 2009 *J. Phys. Soc. Jpn.* **78** 083709
[11] Mizuguchi Y, Tomioka F, Tsuda S and Takano Y 2008 *App. Phys. Lett.* **93** 012505
[12] Bao W, Qiu Y, Huang Q, Green M A and Mao Q 2009 *Phys. Rev. Lett.* **102** 247001
[13] Gnezdilov V, Shevtsova T, Tsurkan V and Loidl A 2011 *Phys. Rev. B* **83** 245127
[14] Okazaki K, Sugai S, Niitaka S and Takagi H 2011 *Phys. Rev. B* **83** 035103
[15] Xia T L, Hou D, Wei J H, Lu Z Y and Zhang Q M 2009 *Phys. Rev. B* **79** 140510
[16] Kumar P, Kumar A, Saha S, Ganguli A and Sood A 2010 *Solid State Comm.* **150** 557

IOP Conf. Series: Journal of Physics: Conf. Series **836** (2017) 012047 doi:10.1088/1742-6596/836/1/012047

The thermal expansion and specific heat of solid oxide fuel cell cathode material $Nd_{1-x}Sr_xCoO_{3-\delta}$

R Thakur[1,a], A Srivastava[2], R K Thakur[1] and N K Gaur[1]

[1]Department of Physics, Barkatullah University, Bhopal - 462026, India
[2]Department of Physics, Sri Satya Sai College for Women, Bhopal 462024, India

E-mail: [a)]rasnathakur@yahoo.com

Abstract. The thermal and allied properties of $Nd_{1-x}Sr_xCoO_{3-\delta}$ ($0\leq x\leq 0.8$) were investigated in this work by means of Modified Rigid Ion Model (MRIM). Theoretically, MRIM provides arguably the most realistic interaction potential to treat these properties. We have computed the variation of specific heat and volume thermal expansion coefficient for these cobaltates. Here, Bulk modulus obtained from AIM theory is observed to be decreasing with increasing doping concentration x. The lattice specific heat is related to Debye temperature and hence this property shows a increasing trend with x and thermal expansion also follows the same increasing trend. The trend of variation of Debye temperature (θ_D), thermal expansion (α), bulk modulus (B) and cohesive energy (ϕ) with A-site cationic radius is predicted probably for the first time for these technologically important doped rare earth cobaltates.

1. Introducton

Solid oxide fuel cells (SOFCs) are a considerable promising energy conversion system with very high efficiency, low emissions, excellent fuel flexibility, and environmental friendliness at high operating temperature (~1000 °C). The high operating temperature, however, induces electrode sintering, high material and manufacturing costs, and interface reactions between cell components. Recently, significant efforts have been focused on the development of intermediate-temperature SOFCs operating at 500–700 °C in order to reduce fabrication costs, improve long-term performance, and effectively extend the choice of materials. The reduction of SOFC operating temperature brings critical problems such as slow kinetics of the oxygen reduction reaction, which leads to large over potential at the electrode–electrolyte interface. Therefore, developing new cathode materials with good performance at low operating temperature is a key area of research for SOFC applications. A desirable cathode material for intermediate temperature SOFCs should have high electronic and oxide ion conductivities, low thermal expansion to be compatible with the electrolyte, as well as high catalytic activity for the oxygen reduction reaction. Mixed ionic-electronic conductors (MIECs) containing Mn, Fe, Co, and/or Ni with the capability to conduct oxygen ions and electrons simultaneously have been identified as strong potential IT-SOFC cathode candidates in recent years.

Among the various MIEC oxides, Sr-doped lanthanide cobaltates, $Ln_{1-x}Sr_xCoO_{3-\delta}$ (Ln=Pr, Nd, Sm, and Gd) have been investigated as cathode materials for intermediate temperature application. It was reported that Ln = Sm and Gd exhibit low electronic conductivity due to bending of the Co-O-Co bonds from 180° and a consequent decrease in the bandwidth and the covalency of the Co-O bonds. Most of the $Nd_{1-x}Sr_xCoO_{3-\delta}$ cathode researches focus on the characterization and electrochemical performance based on $La_{0.8}Sr_{0.2}Ga_{0.8}Mg_{0.2}O_{2.8}$ (LSGM) electrolyte. The LSGM electrolyte and yttria-stabilized zirconia (YSZ) electrolyte, however, react with Sr-doped lanthanide cobaltates. For example, the Manthiram et al. [1] group reported that $Nd_{1-x}Sr_xCoO_{3-\delta}$ cathode forms a secondary phase with LSGM electrolyte at high Sr content (x=0.5). Taking into consideration these results and the fact that strong electron-phonon coupling is present in these compounds, the $Nd_{1-x}Sr_xCoO_{3-\delta}$ have been investigated to study their physical properties like cohesive energy (ϕ), molecular force constant (*f*), Reststrahlen frequency (υ), Debye temperature (θ_D) Gruneisen parameter (γ) and Thermal Expansion (α). Also, the temperature dependent lattice specific heat of $Nd_{1-x}Sr_xCoO_{3-\delta}$ cobaltates is explored. The interaction potential within the framework of modified rigid ion model is given in the next section and the results obtained from its application are presented in subsequent sections.

2. Present potential model and method of computation

The formalism of Modified Rigid Ion Model (MRIM) has been derived for the following effective interatomic potential [2-4].

$$\phi(r) = \frac{-e^2}{2}\sum_{kk'} Z_k Z_{k'} r_{kk'}^{-1} - \sum_{kk'} C_{kk'} r_{kk'}^{-6} + \sum_{i,kk'} \left[\begin{array}{l} n_i b_i \beta_{kk'} \exp\{(r_k + r_{k'} - r_{kk'})/\rho_i\} \\ + \frac{n_i'}{2} b_i \left(\beta_{kk} \exp\{(2r_k - r_{kk})/\rho_i\} + \beta_{k'k'} \exp\{(2r_{k'} - r_{k'k'})/\rho_i\}\right) \end{array} \right] \quad (1)$$

The symbols involved are in it are the same as those defined in our earlier papers [2-4]. Here, k (k') denotes the positive (negative) ions and the sum is taken over all the ions (k, k') and the $\beta_i^{kk'}$ is the Pauling coefficient [5] given by:

$$\beta_i^{kk'} = 1 + (Z_k / N_k) + (Z_{k'} / N_{k'}) \quad (2)$$

where Z_k ($Z_{k'}$) and N_k ($N_{k'}$) are the valence and the number of electrons in the outermost orbit of the k (k′) ion, respectively. The contributions of van der Waals (vdW) attraction for the dipole-dipole interaction are determined by using Slater- Kirkwood Variational (SKV) method [6].The model parameters, hardness (b) and range (ρ) are determined from the equilibrium condition:

$$[d\phi/dr]_{r=r_0} = 0 \quad (3)$$

and the bulk modulus,

$$B = 1/9Kr_0 \left[d^2\phi/dr^2\right]_{r=r_0} \quad (4)$$

where K is the crystal structure-dependent constant and r_0 is the equilibrium nearest neighbour distance. The model parameters (b and ρ) obtained from the equations (3) and (4) have been used to compute the thermodynamic properties of $Nd_{1-x}Sr_xCoO_{3-\delta}$ perovskites. The lattice specific heat is calculated using the formula

$$C_{V(lattice)} = 9R\left(\frac{T}{\theta_D}\right)^3 \int_0^{\frac{\theta_D}{T}} \frac{e^x x^4}{\left(e^x - 1\right)^2} dx \quad (5)$$

Here R is the universal gas constant and θ_D is Debye temperature. The cohesive energy for Sr doped $NdCoO_{3-\delta}$ is calculated using equation (1) and other thermal properties such as the Debye temperature (θ_D), Reststrahlen frequency (υ), molecular force constant (*f*), Gruneisen parameter (γ), and thermal expansion (α) are computed using the expression given in our earlier publication [2-4]. The results are thus obtained and discussed below.

3. Results and discussions

To unravel the elastic and thermal behavior of $Nd_{1-x}Sr_xCoO_{3-\delta}$, the unit cell parameters (a, b, c) are taken from refs. [7] are essential in finding the model parameters which are in turn used in finding the thermodynamic properties of the smaller ionic radii compounds. The values of average cation radius at A/B-site are listed in Table 1. The model parameters, bulk modulus and thermal properties are listed in Table 1-3. Here, Bulk modulus obtained from AIM theory is observed to be decreasing with increasing doping concentration x. The value of the lattice energy of the compounds as estimated using the empirical Kapustinskii equation [8] is found to be in good agreement with our calculated MRIM values, giving further confidence in the validity of our model. The molecular force constant is observed to be decreasing monotonously with increasing doping level x and similar decreasing trend is observed for the other dependent properties like Reststrahlen frequency (υ), cohesive energy (ϕ), Debye temperature (θ_D) and Gruneisen parameter (γ) (Table 3).

Table 1. The average cation radius at A and B-site, critical radius (r_C), tolerance factor and variance for $Nd_{1-x}Sr_xCoO_{3-\delta}$ compounds. O indicates the orthorhombic structure.

Doping conc. x	Structure	Spin	r_A (Å)	r_B (Å)	r_C (Å)	Tolerance factor t	Variance σ^2 (Å^2)
0.0	O	IS	1.163	0.560	1.020	0.9247	0
0.1	O	IS	1.177	0.557	1.017	0.9315	0.001945
0.2	O	IS	1.192	0.554	1.012	0.9382	0.00345
0.3	O	IS	1.207	0.551	1.011	0.9450	0.004538
0.4	O	IS	1.221	0.548	1.005	0.9518	0.005186
0.5	O	IS	1.236	0.545	0.998	0.9586	0.005402
0.6	O	IS	1.251	0.542	0.990	0.9654	0.005186
0.7	O	IS	1.265	0.539	0.983	0.9723	0.004538
0.8	O	IS	1.280	0.536	0.984	0.9792	0.003457

Table 2. Values of A-site charge mismatch, size mismatch and model parameters of $Nd_{1-x}Sr_xCoO_{3-\delta}$ system.

Doping conc. x	A-site charge mismatch σ_c	A-site size mismatch σ_m	Model Parameters			
			$b_1\times10^{-19}$ (J) (Co-O)	$b_2\times10^{-19}$ (J) (Nd/Sr-O)	ρ_1 (Å) (Co-O)	ρ_2 (Å) (Nd/Sr-O)
0.0	1	2.07	1.184	1.85	0.195	0.359
0.1	0.935	2.11	1.348	1.99	0.207	0.380
0.2	0.875	2.15	1.447	2.06	0.213	0.390
0.3	0.818	2.19	1.579	2.17	0.219	0.400
0.4	0.765	2.22	1.719	2.27	0.230	0.418
0.5	0.714	2.26	1.888	2.39	0.244	0.441
0.6	0.667	2.30	2.070	2.51	0.259	0.466
0.7	0.622	2.34	2.234	2.62	0.272	0.486
0.8	0.579	2.38	2.422	2.74	0.281	0.502

Table 3. Values of bulk modulus, cohesive and thermal properties of $Nd_{1-x}Sr_xCoO_{3-\delta}$ compounds.

Doping conc. x	Bulk modulus B (GPa)	Cohesive properties		Thermal properties			
		ϕ (eV) (MRIM)	ϕ (eV) (Kap.eqn.)	f (N/m)	υ (THz)	θ_D (K)	γ
0.0	219.4	-152.7	-151.8	42.47	10.2	492.3	3.41
0.1	203.1	-152.3	-151.5	39.44	9.96	478.5	3.25
0.2	196.4	-152.6	-151.9	38.20	9.89	475.3	3.17
0.3	188.6	-152.6	-152.3	36.85	9.81	471.4	3.11
0.4	178.1	-152.5	-152.1	34.84	9.64	463.3	2.99
0.5	165.6	-152.1	-151.6	32.41	9.40	451.8	2.85
0.6	153.5	-151.6	-151.0	30.06	9.17	440.5	2.71
0.7	144.3	-151.3	-150.5	28.28	9.01	432.9	2.60
0.8	136.9	-150.9	-150.3	26.95	8.92	428.7	2.54

The lattice specific heat is related to Debye temperature and hence this property shows a increasing trend with x and thermal expansion also follows the same increasing trend (Fig. 1). The values of room temperature volume thermal expansion is high compared to its low temperature values at 10K.

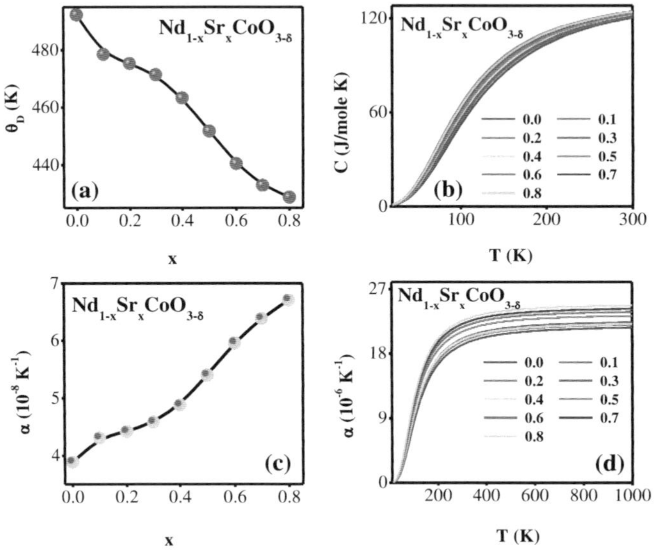

Figure 1. **(a)** Debye temperature of $Nd_{1-x}Sr_xCoO_{3-\delta}$ ($0.0 \leq x \leq 0.8$) with doping level x of strontium at A-site, **(b)** Lattice specific heat of $Nd_{1-x}Sr_xCoO_{3-\delta}$ in 1K to 300K temperatures. All the curves have been shifted by 5 J each from the preceding curve for clarity. The curve for x=0.02 is undisplaced. Variation of thermal expansion of $Nd_{1-x}Sr_xCoO_{3-\delta}$ against doping level (x) **(c)** at 10K temperature **(d)** in 10K to 1000K temperatures.

4. Conclusions

Conclusively, we have made an attempt to analyse the specific heat behavior in doped cobaltate. We found that a simple analysis of the specific heat into a T^3 contribution from the lattice yields the correct interpretation of $Nd_{1-x}Sr_xCoO_{3-\delta}$ cobaltates. Our results are probably the first reports of the bulk modulus and thermal expansion for these perovskites at various temperatures and compositions. Our results demonstrate that this method can provide amply reliable predictions for the temperature dependence of specific heat, thermal expansion, bulk modulus, cohesive energy (ϕ), molecular force constatnt (f), Reststrahlen frequency (υ), Debye temperature (θ_D) and Gruneisen parameter (γ). To the best of our knowledge, the values of the thermodynamic properties of SOFC cathode material $Nd_{1-x}Sr_xCoO_{3-\delta}$ have not yet been measured or calculated, hence our results will serve as a guide for future investigators.

Acknowledgement

The authors are thankful to the University Grant Commission (UGC), New Delhi for the award of Post Doctoral Fellowship.

References

[1] Lee K T and Manthiram A 2005 *J. Electrochem. Soc.* **152** A197
[2] Gaur N K, Thakur R and Thakur R K 2017 *J. Alloys and Compd.* **691** 866
[3] Thakur R and Gaur N K 2014 *Thermochim. Acta* **584** 79
[4] Thakur R K, Thakur R, Samatham S S, Kaurav N, Ganesan V, Gaur N K and Okram G S 2012 *J. App. Phys.* **112** 104115
[5] Pauling L 1945 *Nature of the Chemical Bond* (Ithaca: Cornell University Press)
[6] Slater J C and Kirkwood J G 1931 *Phys. Rev.* **37** 682
[7] Kakinuma K, Arisaka T and Yamamura H 2004 *J. Ceram. Soc. Japan* **112** 342
[8] Glasser L 1995 *Inorg. Chem.* **34** 4935

UV-shielding efficiency of TiO_2-polystyrene thin films prepared by solution cast method

Aman Deep Acharya[1,2, a], Bhawna Sarwan[1,2], R Sharma[1], S Moghe[1], S B Shrivastava[1] and V Ganesan[3]

[1]Vikram University, Ujjain 456010, MP, India
[2]Lovely Professional University Phagwara 144411, Punjab, India
[3]UGC-DAE Consortium for Scientific Research, Khandwa Road, Indore, MP 452001, India

Email: [a)]acharyaphysics2011@gmail.com

Abstract. In order to evaluate the UV-protection efficacy of TiO_2 doped polystyrene (PS/TiO_2) films with concentrations 0.5%, 0.75% and 1% have been fabricated by using solution cast technique. The effect of doping on the microstructure and morphological properties was studied. The obtained results reveal that the better the dispersion of filler particles, larger the UV-protection efficiency of the composite material.

1. Introduction

Incorporation of small amount of inorganic nanoparticles can considerably improve the bulk performance of polymer matrix in optical, electrical, mechanical, thermal properties etc. [1-4]. Most promising inorganic material is TiO_2, which have wide band-gap energies of around 3 eV have been widely used as inorganic UV absorbers due to its significant optical properties [5]. Consequently, TiO_2 doped polystyrene thin film nanocomposite has been considered as an excellent candidate for UV shielding applications. In present work, we have prepared TiO_2/PS thin films having different concentration of the dopant by using solution cast method. The structural and morphological properties of films were studied by using X-ray diffraction (XRD), atomic force microscopy (AFM) measurements. Transmittanceof the films has been recorded by using UV-Vis spectroscopy.

2. Experimental

The polystyrene was dissolved in dichloromethane in a 25 ml borosil glass beaker to cast a defect free film. The viscous solution was then poured over a smooth casting surface of a flat-bottomed petrie dish. In order to get the uniformity in the membrane thickness, the dish was made to float on the mercury. The solvent was allowed to evaporate slowly over a period of 12 - 24 hrs in a dry atmosphere and the membrane was peeled off the surface of petrie dish. For preparing doped polystyrene thin films, the dopant concentration was calculated from the following equation.

$$W(wt\%) = \frac{w_f}{w_p + w_f} \times 100 \quad (1)$$

wherew_f and w_p represent the weight of dopant and the polymer , respectively.

IOP Conf. Series: Journal of Physics: Conf. Series **836** (2017) 012048 doi:10.1088/1742-6596/836/1/012048

3. Structural and morphological analysis

It can be seen in figure1, XRD pattern of pure polystyrene film shows an amorphous polymeric structure without any peaks. The patterns of the 0.5 wt% TiO_2/PS thin film indicates low intensity diffraction peaks attributed to the crystalline domains of the PS chains. In case of 1wt%, the peak was slightly shifted towards lower diffraction suggesting an interaction between the filler particle and the polymer structure, leading to a rearrangement of the PS chains. The intensity increased of these peaks with higher amount of filler loading implies that the slowing of the crystallization rate caused due to inclusion of filler particles.

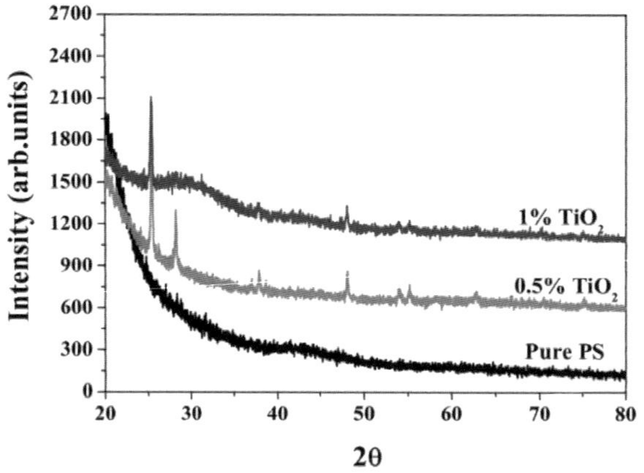

Figure 1. X-ray diffraction pattern of PS/TiO_2 thin films

Figure 2 shows AFM images of PS-TiO_2 at content of 0.5, 0.75 and 1.0 wt %. The addition of 0.5 wt% TiO_2, some rods were seen(see figure 2b) on the surface of pure PS. The formation of rods assumes the random aggregation of nucleates or acting as efficient nucleating agents.

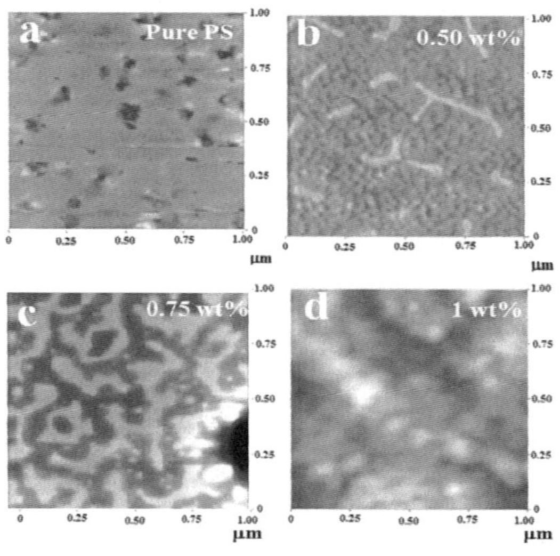

Figure 2. AFM images of PS/TiO_2 film

Additionally, as can be seen from figure 2c, TiO_2 lead to fractal type of aggregationon addition of 0.75 wt% filler content. Moreover in figure 2d the nucleates randomly agglomerate in the continuous phase and increase of titania particle volume fraction made the area larger and causes their overlap which led to opaque PS/TiO_2 films [6-7]. This suggests that the 0.75 wt% PS/TiO_2films have large particle agglomerates, while the 0.5 wt % PS/TiO_2films have improved dispersion as well as higher interface area and therefore possess higher UV- absorption efficiency.

4. PS/TiO_2 UV–Vis Shielding

In figure 3, pure PS film seems rather transparent in the visible region (400-800 nm), whereas films with increasing TiO_2 content continuously adsorb more UV light as a result of empty conduction band and filled valence band. However, UV protection behaviour was observed with 0.5 wt % content of TiO_2 and extensive UV-blocking effect with visible transmittance more than 70% in the region bellow 355 nm was seen. With the further addition of 0.75 wt% TiO_2, the material appeared opaque to eye observation and had significant absorbance in the visible region as well as in the UV region. In addition, the cluster sizes of films became more nonuniform, and irregular with increasing TiO_2 content up to 1 wt% which causes a reduction in transmittance.

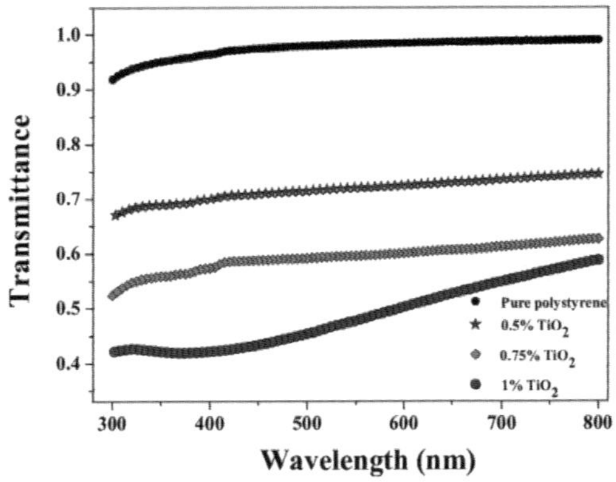

Figure 3. The transmittance spectra of PS/TiO_2 films

Figure 4 shows that the bandgap energy of samples was found to be 3.00 and 2.61 eV, for pure PS and 0.5 wt% TiO_2/PS respectively. This variation in the calculated optical band gap may reflect the role of dopant particles in modifying the electronic structure of the PS matrix due to appearance of various polaronic and defect levels [8-9]. The decrease in bandgap also reflects the increase in degree of disorder in the films which arises due to the change in polymer structure.

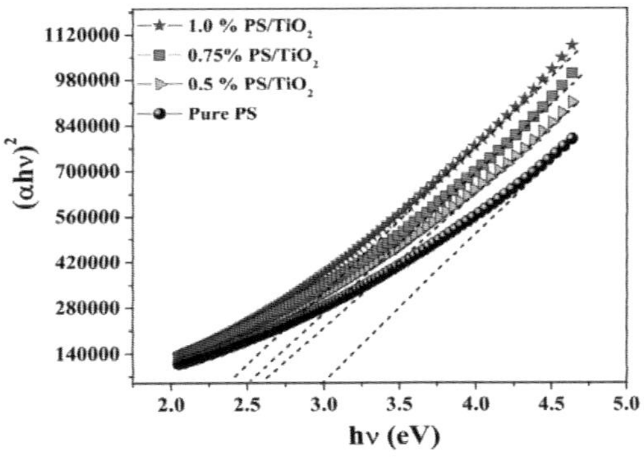

Figure 4. Plot of $(\alpha h\upsilon)^2$ v/s photon energy ($h\upsilon$) of PS/TiO_2 thin films

5. Conclusions

XRD results revels that low intensity diffraction peaks showed crystalline domains of the PS chains. Morphological studies show that the titania particles were aggregates of loosely associated primary particles. With 0.5 wt % content of TiO_2, extensive UV-blocking effect in the region bellow 355 nm was observed. This confirms the introduction of matrix compatible polymer chain doped with TiO_2 particles increased the UV shielding ability of PS film.

Acknowledgement

The authors would like to thank the Director and Center Director of UGC-DAE CSR, Indore for their concern to avail the characterization facilities and Dr. Y.K. Vijay (Prof. University of Rajasthan, Jaipur) for providing the experimental facilities to study positron lifetime.

References

[1] Pavlidou S, Papaspyrides C D 2008 *Prog. Polym. Sci.* **33** 1119
[2] Yoon K B, Hwang Y Y, Noh S K and Lee D H 2008 *Polym. J.* **40** 50
[3] Jordan J, Jacob K I, Tannenbaum R, Sharaf M A, Jasiuk I 2005 *Mater. Sci. Engg. A* **393** 1
[4] Zaragoza contreras E A, Hernandez escobar C A, Mendoza duarte M E, Flores-Gallardo S G, Ibarra gomez R, Marquez lucero A 2009 *Polym. J.* **41** 816
[5] Acharya A D, Moghe S, Panda R, Shrivastava S B, Gangrade M, Shripathi T, Phase D M and Ganesan V 2012 *Thin Solid Films* **525** 49
[6] Menga X, Zhanga Z, Luoa N, Caoa S, Yangb M 2011 *Poly. Sci. A* **53** 977
[7] Scrosati B 1988 in *Solid States ionic Devices* ed Chowdari B V R and Radhakrishna S (World Scientific Publishing Co.) **133**
[8] Devi C U, Sharma A K and Rao V V R N 2002 *Mater. Lett.* **56**167
[9] Rozra J, Saini I, Sharma A, Chandak N, Aggarwal S, Dhiman R and Sharma P K 2012 *Mater. Chem. Phys.* **134** 1121

Structure and morphology of $Sn_{1-x}Al_xO_2$ thin films prepared by spray pyrolysis technique

G Bhatia[1,a], V K Gupta[1], S B Shrivastava[1], M M Patidar[2], V Ganesan[2]

[1] School of Studies in Physics, Vikram University, Ujjain- 456010, MP (India)
[2] UGC-DAE CSR, Khandwa Road, Indore - 452001, MP (India)

E-mail: [a)]geetabhatia1@gmail.com

Abstract. Al-doped tin oxide [Sn_{1-x} Al_xO_2(x=0, 0.03 and 0.09)] thin films were prepared by spray pyrolysis technique (SPT) on glass substrate. XRD and AFM techniques have been used for the investigation of structural and morphological properties of the films. XRD spectra reveal the polycrystalline nature of the films having tetragonal structure. AFM images show an evolution in the morphology in the sense that larger grains of SnO_2 are converted in some smaller grains upon Al inclusion. This may be due to the strain developed due to the Al substitution. The roughness has been found to decrease with the increase in Al concentration.

1. Introduction

The tin oxide is a wide band gap semiconductor (energy band gap 3.6 eV). SnO_2 has a low electrical resistance with high optical transparency in the visible range. Due to these properties, apart from gas sensors, SnO_2 is being used in many other applications, such as electrode materials in solar cells, light-emitting diodes, flat panel displays, and other optoelectronics devices. In the present work, structural and morphological studies of undoped and Al doped SnO_2 thin films prepared by spray pyrolysis technique is reported. This is a chemical route, cost effective method and hence useful to cover a large area of substrate.

2. Experimental

Tin oxide (SnO_2) and Al doped SnO_2 films have been prepared by employing conventional spray pyrolysis technique on to microscopic glass substrate, which were chemically and ultrasonically cleaned before coating. Al doped SnO_2 films were prepared from dihydrated stannous chloride ($SnCl_2.2H_2O$) dissolved in 5ml concentrate HCl. Varying amounts of $AlCl_3$ was added to starting solution for Al doping in a concentration of 0.03, 0.09. The above mixture was heated up to 90°C and after 10 min methanol was added. The solution was further mixed for 15 min. 43ml precursor solution was sprayed with the maintained flow rate of6 ml/min over the cleaned glass substrate at 400°C. After the deposition the sample was kept at the same temperature for 30 min. The structural studies were done by using XRD technique and the morphology was done by using AFM technique.

3. Results and discussions

Figure 1 shows the XRD patterns of the undoped and Al doped tin oxide films on glass substrates. The peaks in the spectra are identified as originating from reflection from the (110), (101), (200), (211) planes showing a polycrystalline nature with tetragonal structure of tin oxide.

The lattice constants (a, c), crystallite size (D) and micro strain (ε) were calculated using the following equations for the pure and Al doped SnO_2 films [2-4];

$$\frac{1}{d^2} = \frac{(h^2 + k^2)}{a^2} + \frac{l^2}{c^2} \quad (1)$$

$$D = \frac{0.9\lambda}{\beta cos\theta} \quad (2)$$

$$\varepsilon = \frac{\lambda}{Dsin\theta} - \frac{\beta}{tan\theta} \quad (3)$$

where d is the interplaner distance and h, k, l are bragg constants and D is the mean crystallite size, λ (= 1.540A^0) is the wavelength of X-ray – beam, β is full width at half maxima of XRD peaks and **θ** is the diffraction angle.

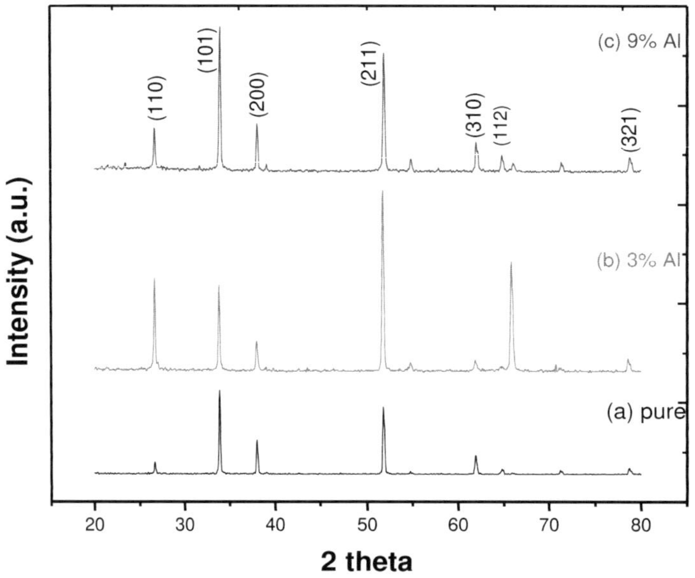

Figure 1. X-ray diffraction patterns of Al Doped SnO_2 thin film (a) pure SnO_2 (b) 3wt. % Al doped (c) 9wt. % Al doped.

The variation of the unit cell parametersa and c with Al concentration is shown in figure 2. The parametersa and c increases at 3 wt. % of Al and tends to decreases at 9 wt. % monotonously. Variation of mean crystallite size and mean microstrain with Al concentration of pure and Al doped SnO_2 thin films is shown in figure 3. From the figure it is observed that the mean crystallite size decreases and mean micro strain increases with increase in Al concentration. This increase in strain could be due to the changes in grain morphology.

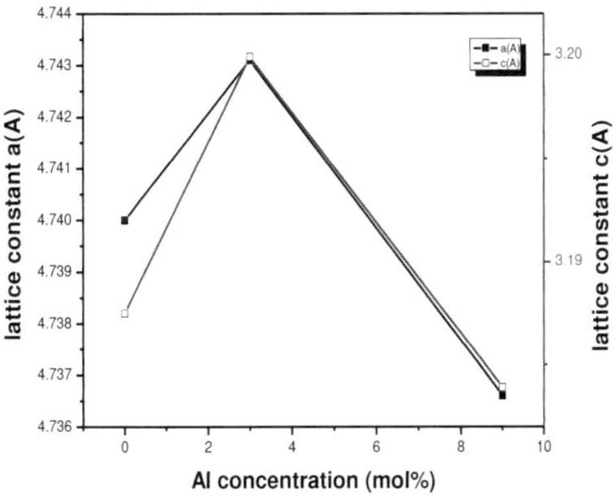

Figure2. Variation of lattice constant (a, c) with Al concentration of pure andAl doped SnO_2 films.

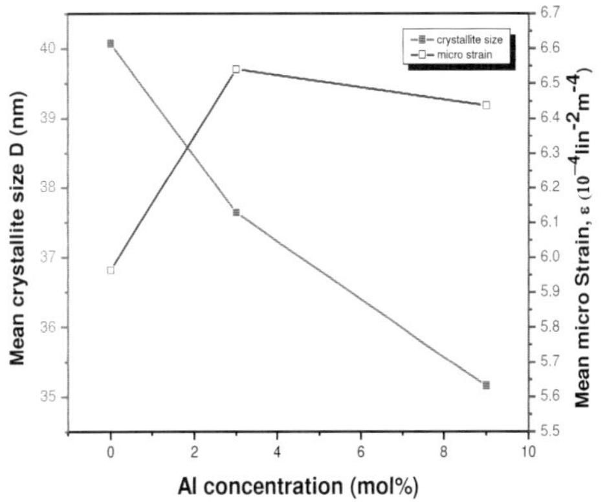

Figure 3. Variation of mean crystallite size and mean micro strain with Al concentration of pure and Al doped SnO_2 thin films.

Figure 4 (a-c) shows the AFM images of pure and Al doped SnO_2 thin films and corresponding line profile are shown in Figure 5(a-c). The Root Mean Square (RMS) roughness of pure, 3% and 9% of $Sn_{1-x}Al_xO_2$ is 142.798 nm, 200.2 nm and 127.0 nm respectively. AFM images show that the grains size decreases with the increase in Al concentration.

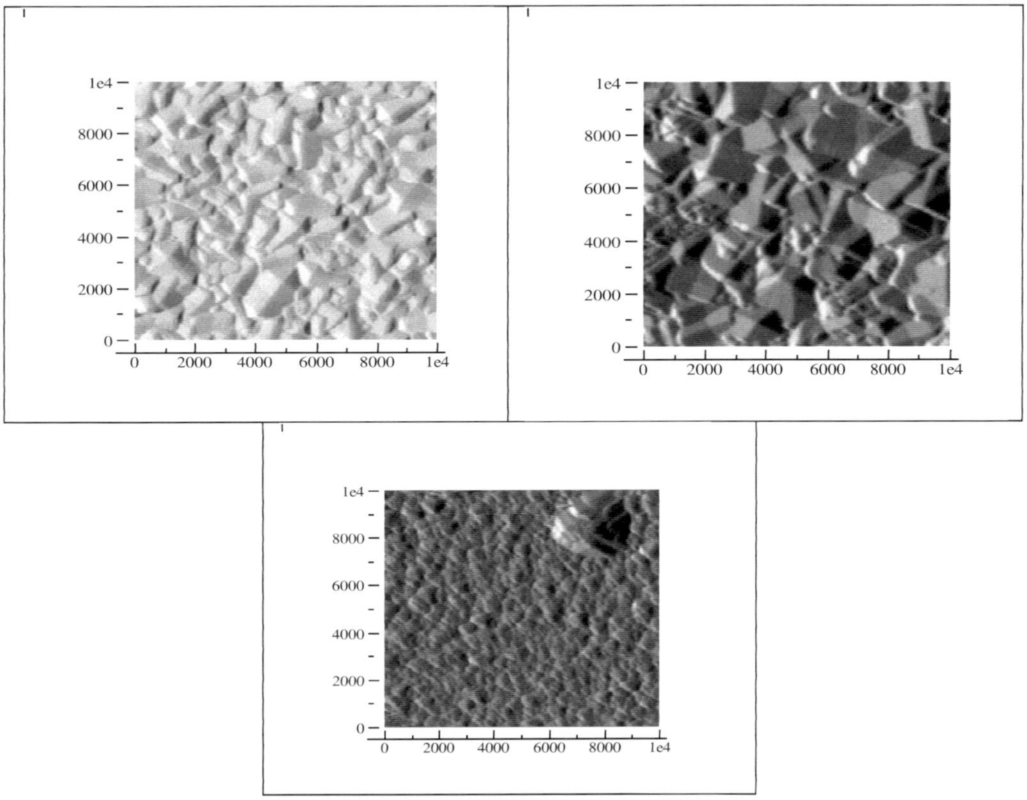

Figure4(a-c). 2D AFM images of Sn_{1-x} Al_xO_2(x=0, 0.03 and 0.09)]

IOP Conf. Series: Journal of Physics: Conf. Series **836** (2017) 012049 doi:10.1088/1742-6596/836/1/012049

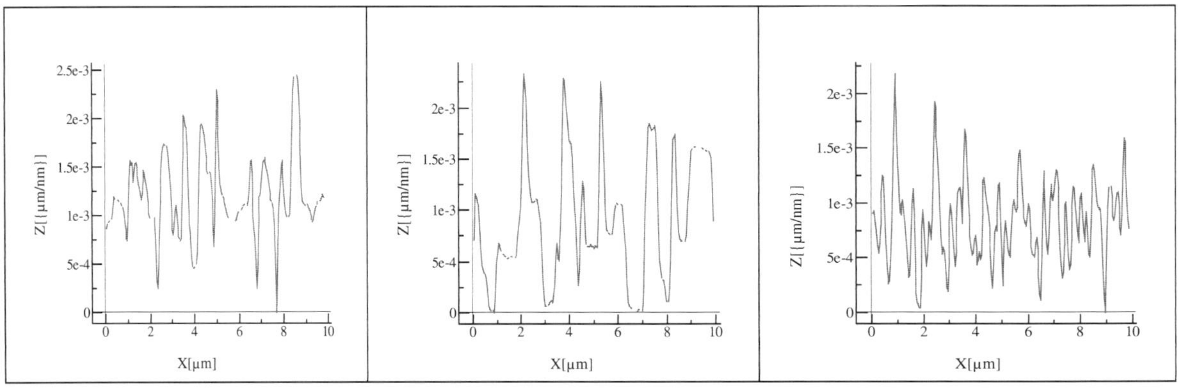

Figure 5 (a-c). Variation of height for a given line profile of Sn_{1-x} Al_xO_2(x=0, 0.03 and 0.09)]

4. Conclusions

Al-doped tin oxide [Sn_{1-x} Al_xO_2(x=0, 0.03 and 0.06)] thin films have been prepared by spray pyrolysis technique (SPT) on glass substrates at a temperature of 400^0C using $AlCl_3$ for Al doping. XRD and AFM techniques were used for structural and morphological studies of the films. XRD spectra reveal the polycrystalline nature of the films with tetragonal structure. An AFM image shows that the roughness first increases for x = 0.03 concentration and decrease at higher concentration i.e. for x = 0.09.

Acknowledgements

Authors would like to acknowledge the UGC-DAE-CSR, the members of LTL, CSR Indore especially Mr. M. Gangrade, Mr. V. Gorli and Mr. Durgesh Singh for their support.

References

[1] Kesar R R, Deshpande N G, Gudage Y G, Vyas J C and Sharma R 2008 *Physica B* **403** 3724
[2] Mairappam R, Ragavendar M and Ponnuswamy V 2011 *J. Alloys Comp.* **509** 7337
[3] Mahalingam T, John V S and Hsu L S 2007 *J. New Mat. Electroch. Syst.* **10** 9

IOP Conf. Series: Journal of Physics: Conf. Series **836** (2017) 012050 doi:10.1088/1742-6596/836/1/012050

Green synthesis of silver nanoparticles and their characterization by XRD

B K Mehta[1], Meenal Chhajlani[1,a], B D Shrivastava[2]

[1] School of Studies in Chemistry and Biochemistry,Vikram University,Ujjain,456010, India
[2] School of Studies in Physics,Vikram University,Ujjain,456010, India

Email: [a)]meenalindore@yahoo.com

Abstract. A cost effective and environment friendly technique for green synthesis of silver nanoparticles has been reported. Silver nanoparticles have been synthesized using ethanol extract of fruits of *Santalum album* (Family Santalaceae), commonly known as East Indian sandalwood. Fruits of *S.album* were collected and crushed. Ethanol was added to the crushed fruits and mixture was exposed to microwave for few minutes. Extract was concentrated by Buchi rotavaporator. To this extract, 1mM aqueous solution of silver nitrate ($AgNO_3$) was added. After about 24 hr incubation Ag^+ ions in $AgNO_3$ solution were reduced to Ag atoms by the extract. Silver nanoparticles were obtained in powder form. X-ray diffraction (XRD) pattern of the prepared sample of silver nanoparticles was recorded The diffractogram has been compared with the standard powder diffraction card of JCPDS silver file. Four peaks have been identified corresponding to (hkl) values of silver. The XRD study confirms that the resultant particles are silver nanoparticles having FCC structure. The average crystalline size D, the value of the interplanar spacing between the atoms, d, lattice constant and cell volume have been estimated. Thus, silver nanoparticles with well-defined dimensions could be synthesized by reduction of metal ions due to fruit extract of *S.album.*

1. Introduction

The present paper deals with the synthesis of nanoparticles of silver using ethanol extract of the fruits of *Santalum album,* which is commonly known as East Indian sandalwood and belongs to the family Santalaceae. It is a mid-sized evergreen parasitic tree easily grown in India. The essential oil of sandalwood is usually prepared by steam distillation from the heartwood and is used in perfumes, cosmetics, and various medicines [1]. Though different parts of this plant can be used to prepare their extracts, only the extract of heartwood have been investigated and reported extensively in literature. There are hardly any reports about other extracts. In the present work we have synthesized the extract of the fruits and utilized it for the reduction of $AgNO_3$ to obtain silver nanoparticles.

Nanoparticles are atomic or molecular aggregates with at least one dimension between 1 and 100 nm, that can drastically modify their physico-chemical properties compared to the bulk material. Nanotechnology has opened up novel fundamental and applied frontiers in materials science. Nanoparticle production and applications have been extensively studied. Nanoparticles can be made from a variety of bulk materials and that they can explicate their actions depending on both the chemical composition and on the size and/or shape of the particles. In the last decade, biosynthesis of nanoparticles has received increasing attention due to a growing need to develop environmentally benign technologies in material syntheses. The significance of such a synthetic protocol has been well demonstrated. For instance, a great deal of effort has been put into the biosynthesis of inorganic materials, especially metal nanoparticles, using microorganisms and plants**.** The formation of nanoparticles mediated by biological route, i.e., green process, is considered as better method than any other method [2].

In nanotechnology, silver nanoparticles are the most promising ones. Silver nanoparticles are nanoparticles of silver, i.e., particles size in range of between 1 and 100 nm and because of its nano size they have attracted intensive research interest. Silver nanoparticles are of interest because of the unique properties (*e.g.,* size and shape depending optical, electrical, and magnetic properties) which can be incorporated into antimicrobial applications, biosensor materials, composite fibers, cryogenic superconducting materials, cosmetic products, and electronic components. A variety of biological sources are able to produce silver nanoparticles of different shapes and nature. The biosynthetic method employing plant extracts has received some attention as a simple and reliable alternative to chemical procedures and physical methods synthesizing metal nanoparticles only in recent years [3]. Jose-Yacaman *et al* have reported the formation of gold and silver nanoparticles by living plants [4,5,6]. Sastry *et al* have extensively used the biosynthesis method of obtaining metal nanoparticles by plant leaf extracts and their potential applications [7-15]. They studied bioreduction of silver ions by a broth of geranium leaf [7,8] or Neem leaf. Very recently, they have demonstrated synthesis of silver nanoparticles using *Aloe vera* plant extracts.

Silver nanoparticles can be produced either intra or extra cellularly by using living organism like diatoms, fungi, bacteria, viruses etc. or mimicking biological systems using biomolecules isolated from them. While micro-organisms continue to be investigated for metal nanoparticle synthesis, the use of plant extracts in similar nanoparticle biosynthesis methodologies is an exciting possibility and is relatively unexplored and under-exploited. In the present paper, an attempt has been made to use plants for the extra-cellular synthesis of silver nanoparticles and demonstrate their capabilities as an alternative to chemical or micro-biological synthetic processes. To circumvent the problem of internalization of nanoparticles an attempt has been made to use the extract of a part of the plant, namely, fruit of *S.album*, instead of the live part.

2. Experimental

Extracts of different parts of *S.album* are commonly obtained by conventional distillation process which requires great energy, significant amount of solvents, and quite a long process time. Therefore, the use of new 'green technique' for extraction with minimum/low energy, solvents, and time is preferred. Hence, in the present work we have used microwave assisted extraction method for getting ethanol extract of fruits of *S.album.* All the reagents used for reaction were analytical grade chemicals and used without further purification. Redistilled deionized water was used for sample preparation.

The fruits of *S.album* are globose, fleshy drupe; red, purple to black when ripe, about 1 cm in diameter, with hard ribbed endocarp and crowned with a scar, almost stalkless, smooth, single seeded. Fruits were collected from the local area. They were washed and shed dried. These dried fruits were then crushed using a pestle and mortar. A measured amount of ethanol was added to the crushed sample. The mixture was kept in microwave for few minutes. Microwave ovens are generally operated at 2.45 GHz. Owing to their electromagnetic nature, microwaves possess electric and magnetic fields which are perpendicular to each other. The electric field causes heating via two simultaneous mechanisms, namely, dipolar rotation and ionic conduction. Dipolar rotation is due to the alignment on the electric field of the molecules possessing a dipole moment in both the solvent and the solid sample. This oscillation produces collisions with surrounding molecules and thus the liberation of thermal energy into the medium.

Extract was concentrated by Buchi rotavaporator. To 10 ml of this concentrated extract 90 ml of 1mM aqueous solution of silver nitrate ($AgNO_3$) was added. The mixture was kept in dark at room temperature. After about 24 hr incubation Ag^+ ions in silver nitrate solution were reduced to Ag atoms by the extract and the colour of the solution was observed to change from pale yellow to dark brown. A sufficient amount of dark brown precipitate was observed and it was separated by high speed centrifugation. The separated solid mass was washed with double distilled water for few times to remove impurities. After complete washing the solid mass was kept open to dry. The complete drying

IOP Conf. Series: Journal of Physics: Conf. Series **836** (2017) 012050 doi:10.1088/1742-6596/836/1/012050

of this solid mass resulted in a black colored material which was collected in powder form and sampled for characterization purpose.

3. Results and discussions

The X-ray diffraction (XRD) pattern of the prepared sample of silver nanoparticles was recorded at UGC-DAE Consortium for Scientific Research, Indore, by employing Bruker d8 Advance X-ray diffractometer, using CuKα radiation (λ = 1.5406 Å), 40 kV- 40mA, $2\theta/\theta$ scanning mode. Data was taken for the 2θ range of 30 to 80 degrees with a step of 0.0202 degree. The XRD data and it's analysis are given in table 1. The diffractogram (Fig. 1) has been compared with the standard powder diffraction card of JCPDS, silver file No. 04-0783. Four peaks at 2θ values of 38.2901, 44.5583, 64.8185, and 77.4383 degree in the experimental diffractogram have been identified to be due to silver metal and corresponding to (hkl) values - (111), (200), (220) and (311) planes of silver and are labeled as 1-4 in fig.1. The details of these peaks are given in table 1. The XRD study has thus confirmed that the resultant particles in the prepared sample are silver nanoparticles having face centered cubic crystal structure. There are four more peaks in the diffractogram at 32.35, 46.38, 54.03 and 57.66 degrees. These peaks have been identified to be due to $AgNO_3$, which might have not been reduced and hence remained in the sample in minute quantity. These peaks are not included in table 1.

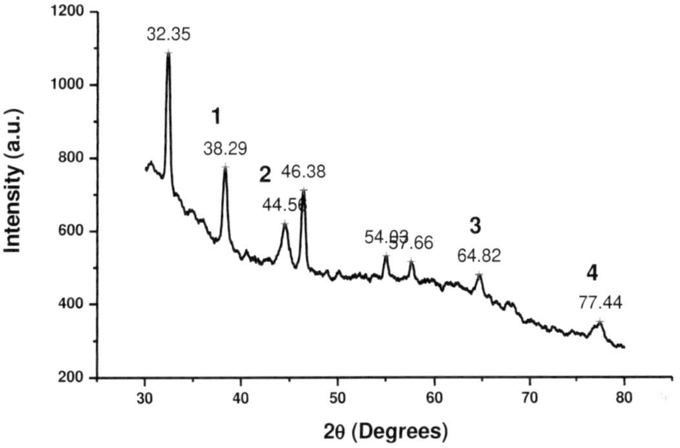

Figure 1. XRD of silver nanoparticles. Ag peaks are marked 1-4 and 2θ values are given.

The average crystalline size D of the silver nanoparticles have been estimated from the diffractogram by using Debye-Scherrer formula, $D = 0.9\lambda/\ \beta \cos\theta$, where λ is the wavelength of the X-rays used for diffraction and β is full width at half maximum (FWHM) of a peak [16]. To estimate FWHM, each of the four peaks was fitted with a Gaussian function. The FWHM of the fitted gaussian curve is taken as FWHM of the peak. This could be done in the software *origin 6.1*. Four values of 'D'

Table 1. XRD results of silver nanoparticles.

P. No	2θ	Cos θ	Sin θ	FWHM degree	β radian	Cryst--alline size 'D' nm	Inter-planar spacing 'd' Å	h k l Identi-fied from peak	$h^2+k^2+l^2$ from identi-fied h k l	Lattice const. 'a' from d Å	Cell vol--ume $Å^3$
1	38.2901	0.9447	0.328	0.3407	0.0059	24.68	2.3485	1 1 1	3	4.0677	67.30
2	44.5583	0.9253	0.3791	0.6397	0.0111	13.41	2.0319	2 0 0	4	4.0638	67.11
3	64.8185	0.8442	0.536	0.6033	0.0105	15.59	1.4371	2 2 0	8	4.0648	67.16
4	77.4383	0.7789	0.6271	0.3607	0.0063	28.27	1.2284	3 1 1	11	4.0740	67.62

have thus been obtained from the data for the four peaks. It is seen from table 1 that the four values are nearly the same.

The average crystalline size of the silver nanoparticles is found to be 20 nm. The value of the interplanar spacing between the atoms, d, has been calculated using Bragg's Law, $2d\text{Sin}\theta = n\lambda$, where n is the order of diffraction pattern. In the present case n is equal to 1. From the four data sets for the four peaks, four values of 'd' are obtained and are given in table 1. It is seen that the four values are nearly the same.

Lattice constant has been estimated using the formula, $a = d*\sqrt{(h^2+k^2+l^2)}$. Average of the four values of 'a' calculated from the four values of 'd' as obtained from the data for the four peaks is found to be 4.0676 Å. This is in fair agreement with the standard value for silver that is 4.0857 Å.
Cell volume has also been estimated from the value of lattice constant and the values are given in table 1.

Thus, the XRD analysis has shown that silver nanoparticles with well-defined dimensions could be synthesized by reduction of metal ions due to fruit extract of *S.album.*

4. Conclusions

The present work shows that the reduction of silver ions from $AgNO_3$ to Ag occurs due to fruit extract of *S.album.* The synthesized silver particles have been characterized by XRD. The obtained XRD data for 2θ positions identifies the sample as silver crysatlline particles having (hkl) values corresponding to FCC silver. The d-spacing values, lattice constant and cell volume, all comfirm the sample to be silver crysatlline particles. The average crystalline size was estimated to be 20 nm. confirming the nanoparticle nature of the obtained sample. Thus, the present method leads to the formation of silver nanoparticles with well-defined dimensions.

References

[1] Kim T H, Ito H, Hayashi K, Hasegawa T, Machiguchi T and Yoshida T 2005 *Chem. Pharm. Bull.* **53** 641
[2] Mandal D, Bolander M E, Mukhopadhyay D, Sarkar G and Mukherjee P 2006 *Appl. Microbiol. Biotechnol.* **69** 485
[3] Huang J, Li Q, Sun D, Lu Y, Su Y, et al., 2007 *Nanotechnology* **18** 11
[4] Gardea-Torresdey J L, Tiemann K J, Gomez E, Dokken K, Tehuacanero S and Jose-Yacaman M 1999 *J. Nanopart. Res.* **1** 397
[5] Gardea-Torresdey J L, Parsons J G, Dokken K, Peralta-Videa J, Troiani H E, Santiago P and Jose-Yacaman M 2002 *Nano Lett.* **2** 397
[6] Gardea-Torresdey J L, Gomez E, Peralta-Videa J, Parsons J G, Troiani H and Jose-Yacaman M 2003 *Langmuir* **19** 1357
[7] Shankar S S, Ahmad A, Pasricha R and Sastry M 2003 *J. Mater. Chem.* **13** 1822
[8] Shankar S S, Ahmad A and Sastry M 2003 *Biotechnol. Prog.* **19** 1627
[9] Shankar S S, Rai A, Ahmad A and Sastry M 2004 *J. Colloid Interface Sci.* 275 496
[10] Shankar S S, Rai A, Ankamwar B, Singh A, Ahmad A and Sastry M 2004 *Nat. Mater.* **3** 482
[11] Ankamwar B, Damle C, Ahmad A and Sastry M 2005 *J. Nanosci. Nanotechnol.* **5** 1665
[12] Ankamwar B, Chaudhary M and Sastry M 2005 *Synth. React. Inorg. Met.-Org. Nano-Metal Chem.* **35** 19
[13] Shankar S S, Rai A, Ahmad A and Sastry M 2005 *Chem. Mater.* **17** 566
[14] Rai A, Singh A, Ahmad A and Sastry M 2006 *Langmuir* **22** 736
[15] Chandran S P, Chaudhary M, Pasricha R, Ahmad A and Sastry M 2006 *Biotechnol. Prog.* **22** 577
[16] Sharma R, Acharya A D, Moghe S, Shrivastava S B, Gangrade M, Shripathi T, and Ganesan V 2014 *Material Science in Semiconductor Processing* **23** 42

IOP Conf. Series: Journal of Physics: Conf. Series **836** (2017) 012051 doi:10.1088/1742-6596/836/1/012051

Origin of $L\beta_2^0$ satellite in higher Z elements

Rajeev K Trivedi[1], **Renuka Kendurkar**[2,a], **B D Shrivastava**[2]

[1] Alpine Institute of Technology, Ujjain, 456010, India
[2]School of Studies in Physics, Vikram University, Ujjain, 456010, India

Email:[a)] renukakendurkar@gmail.com

Abstract. One of the satellite lines accompanied with the intense diagram line $L\beta_2$ (L_3-N_5) on the higher energy side, is the satellite β_2^0 in the elements from $_{71}$Lu to $_{84}$Po, $_{88}$Ra, $_{90}$Th and $_{92}$U. Shahlot and Soni have theoretically investigated this satellite and have found all the possible transitions using jj coupling scheme using Hartree-Fock-Slater formulae. A perusal of their results shows that in some cases the agreement between theoretical and experimental values is not so good. Hence, in the present investigation we have tried alternative calculations by using the tables of Parente et al. While these calculations are relativistic ab initio calculations, those of Shahlot and Soni are non-relativistic semi-empirical calculations. Considering the same grouping of transition schemes as assigned by Shahlot and Soni, calculations have been done by us using the tables of Parente et al, which gives the values of transition energies only for the 11 elements. The transition energies for intermediate elements have been calculated by us by linear interpolation method. Our calculations show better agreement with the experimental values than that obtained from the values of Shahlot and Soni. However, in some cases, our calculations also do not yield good results and this has been discussed.

1. Introduction

The X-ray emission spectra consists of diagram lines and non-diagram lines or satellites. The diagram lines result from transitions between atomic states, involving single vacancy. Energy of such lines can be expressed as the difference of two terms in the 'single vacancy' energy level diagram. The diagram lines are found to be accompanied by groups of lines of slightly different energies and usually much smaller intensities called satellites. Such lines have energies, which do not correspond to the energy difference between any two states of the normal single vacancy energy level diagram of the element concerned. A survey of the theories reveals that the most widely accepted theory of X-ray satellites is the multiply ionization theory. In a multiple ionized atom, the energy levels of the atom are different than those in a singly ionized atom because the electrostatic attraction of the nucleus for the remaining electrons is increased by the absence of a second or more electrons. The X-ray satellite emitted due to transition in the multiply ionized atom has a different energy than the parent diagram line emitted by a transition in a singly ionized atom. Several methods have been used to calculate the energy difference between the double vacancy initial and final states, which could explain the origin of X-ray satellites. In the present work we have attempted to explain the origin of one of the L-emission satellites.

The intense diagram line $L\beta_2$ in the L-emission spectra arises because of strong dipole transition L_3-N_5 ($2p_{3/2}$-$4d_{5/2}$) between singly ionized states. In the elements with $Z \geq 71$, the line $L\beta_2$ is accompanied with its satellites on the higher energy side. One of these satellite lines is the satellite β_2^0 [1]. In the present investigation, we have theoretically investigated the origin of $L\beta_2^0$ satellite in the elements, in which this satellite has been observed. Shahlot [2] and Soni have theoretically investigated the $L\beta_2^0$

satellite and have found all the possible transitions using jj coupling scheme for the transition arrays L_3N_x- N_xN_5 (x = 1 - 5). They have used Hartree-Fock-Slater (HFS) formulae applicable in jj coupling for two hole states and have devised a new method to find adiabatic relaxation energy which should be subtracted from the value of energy found by HFS method to arrive at the corrected value of the theoretical energy for a satellite. A perusal of their results shows that in some cases the agreement between theoretical and experimental values is not so good. Hence, in the present investigation we have tried alternative calculations for calculating the energy of the satellite using the theoretical transition energies by Parente *et al.* and assuming the same transition array as the origin of this satellite, as assumed by Shahlot and Soni. The experimental values have been compared with our theoretically calculated values as well as with those of Shahlot and Soni.

2. Calculations

In the present investigation, we are concerned with the theoretical investigations of $L\beta_2^0$ satellite in the elements from $_{71}$Lu to $_{84}$Po, $_{86}$Rn, $_{88}$Ra, $_{90}$Th and $_{92}$U, as in most of these elements, the satellites have been observed. We have tried to assign transitions to the $L\beta_2^0$ satellite using energies of transitions in the doubly ionized atom as theoretically calculated by Parente *et al.* [3] and given in the form of tables. For our present calculations of the energies of $L\beta_2^0$ satellite, we have assumed the transition arrays responsible for the origin of the satellite $L\beta_2^0$ as $2p_{3/2}^{-1}4x$ - $4x\ 4d_{5/2}^{-1}$(x = s, p, d) i.e. L_3N_x - N_xN_5 (x = 1-5), i.e., the same as assumed by Shahlot and Soni. Our method of calculation is given below.

2.1. Method of Calculation of Energies of Satellites using tables of Parente et al.

As pointed out above, the available theoretical calculations by Parente *et al.*, which give energies of transitions in doubly ionized atoms that arise from the presence of one 'spectator hole' in the M- or N-shells of the emitting atom which can be used to explain the origin of L- emission satellites. They have listed for 11 elements, namely , $_{65}$Tb, $_{67}$Ho, $_{70}$Yb, $_{74}$W, $_{78}$Pt, $_{80}$Hg, $_{85}$At, $_{88}$Ra, $_{90}$Th, $_{92}$U, $_{95}$Am, the energies of all electric dipole and electric quadrupole L X-rays for which the final state comprises two vacancies in the M- and N-Shells.

Energies of initial and final atomic states have been computed separately to allow for complete relaxation. With a local approximation to the atomic potential, Dirac-Hartree-Slater (DHS) wave functions have been computed. A vacuum polarization correction has been included by means of first order perturbation according to the method of Huang [4] The self-energy shift has been calculated from Mohr's coulomb shifts [5]. The effect of screening, relaxation and finite nuclear extent on self - energy has been taken into account approximately. The computations have been carried out in j-j coupling appropriate for the high Z-region. The electrostatic interactions between the two inner-shell holes in the initial and final states have been included, but coupling with unfilled sub-shells, if any, has been neglected. No electron-electron coulomb correlation effects have been included.

In these tables have been listed all the X-ray satellite theoretical energies for the selected 11 elements. The tables do not give which observed satellites correspond to which transitions. One has to identify the transitions corresponding to the observed satellites.

In the present investigation, we have used the tables of Parente et al. not only for the 11 elements for which Parente et al have listed the transition energies, but also for nearly all the elements from Z = 60 onwards. This has been done by us using interpolation method. The transition energies for intermediate elements have been calculated by interpolation from transition energies for nearby elements among the 11 elements. Our calculated values for the energies of $L\beta_2^0$ satellite using the tables of Parente et al. are given in table 1 along with the experimental values.

As already pointed out in the introduction, Shahlot and Soni have also theoretically calculated the energies of the satellite $L\beta_2^0$. In order to compare our method with their method of calculation and also to compare our theoretical values with their theoretical values, their method of calculation and the results obtained by them are described below in brief.

IOP Conf. Series: Journal of Physics: Conf. Series **836** (2017) 012051 doi:10.1088/1742-6596/836/1/012051

Table 1: Theoretical values of energy (in eV) $L\beta_2^0$ satellite and their comparison with experimental values

S. No.	Elements	Present calculated value using tables of Parente *et al.* as mean of transition schemes $L_3N_x - N_xN_5$ (x = 1-5)	Shilpa Shahlot's theoretical value as corrected value of mean of transition schemes $L_3N_x - N_xN_5$ (x =1-5)	Experimental value
1.	$_{71}$Lu	9061.88[a]	9064.76[a]	9055.64
2.	$_{72}$Hf	9363.55[a]	9366.85	9354.84
3.	$_{73}$Ta	9665.22[a]	9668.17[a]	9659.39
4.	$_{74}$W	9966.89	9971.14	9969.96
5.	$_{75}$Re	10289.42[a]	10290.12[a]	10282.32
6.	$_{78}$Pt	11257.00	11247.48	11259.18
7.	$_{79}$Au	11612.30[a]	11589.52[a]	11594.71
8.	$_{81}$Tl	12328.22[a]	12273.61[a]	12283.05
9.	$_{82}$Pb	12688.85[a]	12618.72	12632.96
10.	$_{83}$Bi	13049.51[a]	12991.56[a]	12991.09
11.	$_{88}$Ra	14852.60	14871.23	14858.55
12.	$_{90}$Th	15636.90	15623.36[a]	-
13.	$_{92}$U	16443.84	16375.29[a]	16444.34

Note - [a] Denotes interpolated values

2.2 *Shilpa Shahlot and Soni's theoretical method of calculation of energies of the satellite $L\beta_2^0$*

Shahlot and Soni have calculated L_3N_x-N_xN_5 transition energies in elements with $Z \geq 72$ in which jj - coupling approximation is most suitable. The HFS formulas for the energies of two hole states like, L_3N_x and N_xN_5 (for X = 1-5) have been established. Each formula shows that the energy of any two hole state is given as the sum of three types of energies. The major part of this energy is due to sum of the energy of single hole states, the second major part is the spin orbit interaction energy between two holes, and the third contribution is due to weak electrostatic interaction energy between two holes. To calculate the total energy for a particular state, the energy of single hole states like L_3, N_x and N_5 have been taken from the tables of Bearden and Burr [6] to calculate spin orbit interaction energy, the values of spin orbit constants have been taken from tables of Larkins [7]. To calculate the values of electrostatic energies, the values of all Slater integrals F's and G's have been taken from the tables of Mann [8]. In these tables the values of Slater integrals are given in atomic units, therefore, all these values have been changed to eV by multiplying the values of integrals by 13.6. The energies for all possible J values for each two hole configuration, under study have been calculated.

Using the energies of two hole states, the energies of various transitions of arrays L_3N_x - N_xN_5 have been calculated. Only those transitions have been considered which are allowed according to the dipole selection rules applicable for transitions in two electron system.

The averages of five arrays L_3N_x - N_xN_5 have been calculated. Energies of all the five arrays L_3N_x - N_xN_5 (x = 1 to 5) are mutually very much close with one another and hence simple averages of transitions have been calculated. Their overall average has then been calculated and is taken as the theoretical energy of the $L\beta_2^0$ satellite.

Shirley [9] in 1973 had suggested that when two electrons are removed from the atom, the orbits of the atom relax and adjust to the new potential field and this relation modifies the energy of the two hole states. This modification in the energy is named as 'adiabatic relaxation energy' (ARE). The adiabatic relaxation energy (ARE) for a transition can be written as k(initial) - k(final) = Δk and is calculated as the difference between the theoretically calculated transition energies in intermediate coupling using HFS formulae with the experimentally measured energies of satellites. The difference for each transition and satellite energy is found to be nearly constant and an average of the difference is used as the correction term Δk. Using this method Shahlot and Soni have assigned the transition scheme L_3N_x - N_xN_5 (x=1-5) to the satellite $L\beta_2^0$. The theoretical values of the energy of the satellite $L\beta_2$ as reported by Shahlot and Soni are given in Table 1 for elements with Z= 72, 74, 78, 82, 88. The values given in Table 1 for other elements are the values estimated by us by using linear interpolation method. It is seen from this table that the agreement between the theoretical and experimental values is not good in some of the elements.

3. Results and discussions

The results are given in table 1. It can be seen from the table that our calculations show better agreement with the experimental values than that obtained from the values of Shahlot and Soni. However, in some cases, our calculations also do not yield good results. The reason for this may be that the procedure used for grouping of the transition schemes, as devised by Shahlot and Soni, may not be the correct procedure and that we have also used the same transitions for our calculations. The best alternative is to obtain theoretical satellite spectra. But this is not possible at the moment, because relative intensities of many transitions are not available in literature.

References

[1] Cauchois Y and Senemaud C 1978 *Wavelengths of X-ray emission lines and absorption edges* (Pergamon press, Oxford).
[2] Shahlot S 2005 *Ph.D. thesis*, J.N.V. University, Jodhpur
[3] Parante F, Chen M H and Craseman B 1981 *At. Data. Nucl. Data Tables* **26** 383
[4] Huang K N 1976 *Phys. Rev. A* **14** 1311
[5] Mohr P J 1974 *Ann. Phys.* (N.Y.) **88** 52 ; 1975 *Phys. Rev. Lett.* **34** 1050
[6] Bearden J A and Burr A F 1965 *Atomic energy levels* (U.S. Atomic Energy Commission, Oak Ridge Tennessee)
[7] Larkins F P 1977 *At. Data Nucl. Data Tables* **20** 311
[8] Mann J B 1967 *Atomic structure calculations I* (Los Almos Scientific Lab, Report no. LA, 3690)
[9] Shirley D A 1973 *Phy. Rev. A* **7** 1520

IOP Conf. Series: Journal of Physics: Conf. Series **836** (2017) 012052 doi:10.1088/1742-6596/836/1/012052

Thermal transport properties of bulk and monolayer MoS_2: an ab-initio approach

Amreen Bano[a], Preeti Khare and N K Gaur

Department of Physics, Barkatullah University, Bhopal-462026, M.P. INDIA

E-mail: [a)]banoamreen.7@gmail.com

Abstract. The transport properties of semiconductors are key to the performance of many solid-state devices (transistors, data storage, thermoelectric cooling and power generation devices, etc). In recent years simulation tools based on first-principles calculations have been greatly improved, being able to obtain the fundamental ground-state properties of materials accurately. The quasi harmonic thermal properties of bulk and monolayer of MoS_2 has been computed with *ab initio* periodic simulations based of density functional theory (DFT). The temperature dependence of bulk modulus, specific heat, thermal expansion and gruneisen parameter have been calculated in our work within the temperature range of 0K to 900K with projected augmented wave (PAW) method using generalized gradient approximation (GGA). Our results show that the optimized lattice parameters are in good agreement with the earlier reported works and also for thermoelastic parameter, i.e. isothermal bulk modulus (B) at 0K indicates that monolayer MoS_2 (48.5 GPa)is more compressible than the bulk structure (159.23 GPa). The thermal expansion of monolayer structure is slightly less than the bulk. Similarly, other parameters like heat capacity and gruneisen parameter shows different nature which is due to the confinement of 3 dimensional structure to 2 dimension (2D) for improving its transport characteristics.

1. Introduction

Molybdenum disulfide MoS_2 is a 'layered' transition metal dichalcogenide (TMD) in its bulk state with an indirect band gap has attracted many of the researchers for its various electronic and catalytic applications [1-5]. The crystal structure of MoS_2 is similar to graphene, having Mo and S atoms stacked together forming a sandwich where S-Mo-S sandwiches are bonded with weak Van-der-Waals forces. The 2D materials possess unique properties. Hexagonal boron nitride (h-BN), graphene, MoS_2, $MoSe_2$, WSe_2 , WS_2 etc TMDs offered promising candidature for novel nanoelectronics, large band gap insulators, semiconductors and semimetals [6-10]. In particular, 2D semiconductors show high bendability, high on-off current ratio that offers low power consumption and flexible electronics for future generations [11-15]. The confinement of bulk MoS_2 from ploy layer to monolayer has shown drastic improvements in its transport mechanism as its monolayer gives a direct semiconducting band gap which makes it an efficient candidate for thermoelectric, optical, and many other application areas. However, a great amount of work has been carried out in the research filed for its electronic, optical and vibrational properties but very less is available regarding its thermodynamic properties under the temperature change. In the present work, we have explored the temperature dependent thermodynamic nature of MoS_2 in bulk and monolayer both. In the following sections, the results obtained in our study for isothermal bulk modulus (B), specific heat (C_p), thermal expansion coefficient (α) and gruneisen parameter with temperature variation has been discussed.

IOP Conf. Series: Journal of Physics: Conf. Series **836** (2017) 012052 doi:10.1088/1742-6596/836/1/012052

2. Computational details

In the present work, all the calculations are performed using PWSCF code [15] based on DFT. Using the plane-wave pseudopotential method and employing the generalized gradient approximation (GGA) [16]. The all electron projector augmented wave (PAW) method [17]was employed with a plane-wave cutoff energy of 55 Rydberg. The k-point samplings with 11x11x7 for bulk and 9x9x1 for monolayer structures has been used after convergence test in the Brillouin zone using the Monkhorst-Pack scheme [18].The total energy convergence tests showed that convergence to within 1 meV/atom was achieved with the above calculation parameters. To simulate the monolayer of a vacuum of 20Å is included along Z direction to avoid any interactions between the periodic images. In order to investigate the thermodynamic properties, we have used quasi harmonic approximation (QHA) as implemented in Phonopy code [19], which includes the phononic effects. This model is very quick and easy, and it has been successfully applied to predict the thermodynamic properties of some materials [20, 21].

The isothermal bulk modulus (B), specific heat (C_p), thermal expansion coefficient (α) and gruneisen parameter (γ) in QHA are defined by the following relations [22, 23]

$$B(V) = V\left[\frac{d^2E(V)}{dV^2}\right] \quad (1)$$

$$\alpha_V = \frac{1}{V}\left[\frac{dV}{dT}\right] \quad (2)$$

$$C_p - C_v = \alpha_v^2(T)B_0VT \quad (3)$$

$$\gamma = \frac{\alpha B_T V}{C_v} \quad (4)$$

where, E is the total energy per unit cell, V is the unit cell volume, $C_v$is the specific heat at constant volume, and T denotes the temperature.

3. Results and discussions

The crystal structure shown in figure 1, of MoS_2 is hexagonal (honeycomb). The space group of bulk MoS_2 is P3m1 (point group D_{6h}) and that is of the single layer is P6m2 (point group D_{3h}). To obtain the ground state structure of both (bulk and monolayer MoS_2), we have fully relaxed the lattice constants and atomic coordinates by using a DFT code Quantum Espresso [16], and obtained a=3.17 Å and c=11.79 Å for bulk and a= 3.15 Å for monolayer structure, which is in good agreement with the available literature [24, 25]. Phonopy [19] based on QHA has been used to calculate the thermodynamic quantities. The variation of isothermal bulk modulus (B) with temperature is shown is figure 2(a). It can be seen that with upto 300K, B is nearly unchanged for both the structures, but on further increasing the temperature, for bulk structure, B start to decrease and in monolayer structure this decreasing trend is very nominal. This is due to the change in the equilibrium volume of the compound. The different variation trends in bulk and monolayer reveals the fact that the rigidity in bulk structure is along the z-direction, where atoms are interacting under Van-der-Waals interactions [26]. This provides hardness to the bulk and high bendability to the monolayer structure of MoS_2. One of the most important thermodynamic properties is specific heat of the material, as it provides information about many applications and vibrational properties. The response of $C_p$with temperature can be seen in figure 2(b). Upto 400K, $C_p$increases with increasing temperature and it is noteworthy, for T>400K, it is tending to a converging value which is called as Dulong-Petit limit. For low temperatures i.e. T<100K, it follows the T^3 law. In the QHA model, the Gruneisen parameter γ describes the anharmonic effects of the crystal lattice thermal vibration [24, 25]. It is observed from figure 2(c) for both structures, the effect of temperature over γ is not much significant. At T> 150K, γ reaches to a constant value in both structures, i.e. 1.25 for bulk and 1.19 for monolayer. The temperature dependence of thermal expansionα, is presented in figure 2(d). It is observed that it increases rapidly with T^3 upto 400K. And after that, it gently increases at further high temperatures. The values obtained of each thermodynamic quantity at 300K, is listed in table 1.

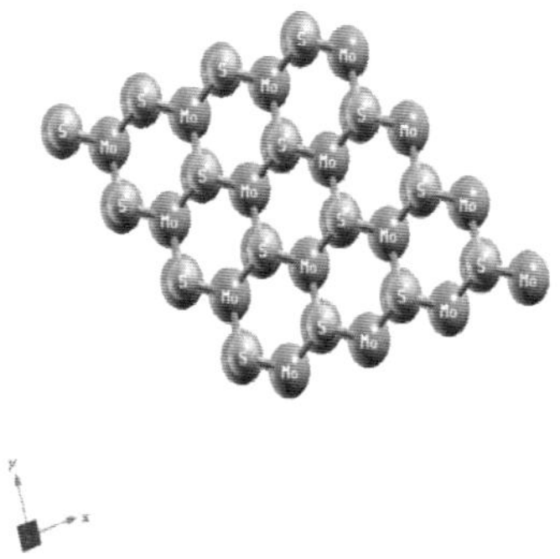

Figure 1. Error! No sequence specified. crystal structure of MoS_2 (top view)

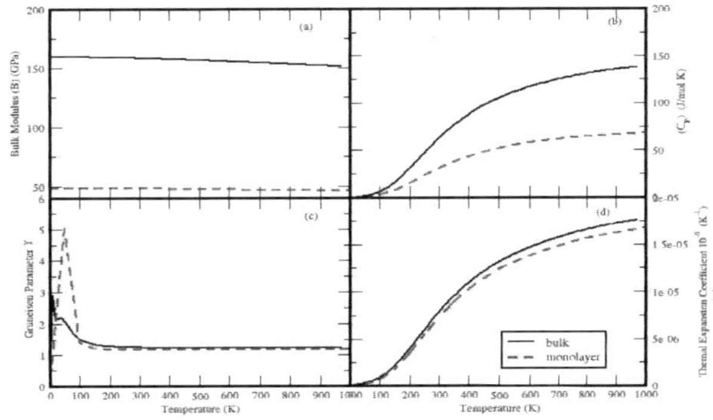

Figure 2. Thermodynamic response of bulk and monolayer of MoS_2 in the temperature range 0K<T<900K

Table 1. Thermodynamic quantities of bulk and monolayer MoS_2 at 300 K

Quantities	Bulk	Monolayer
B (GPa)	158.3	47.9
C_p(J/mol K)	62.23	30.76
γ	1.25	1.17
$\alpha(10^{-6}\ K^{-1})$	7.8	7.2

Acknowledgment
The authors highly acknowledge University Grants Commission (UGC) Delhi, for providing financial assistance in our work by Maulana Azad National Fellowship (MANF).

References

[1] Byskov L S, Norskov J K, Clausen B S and Topsoe H 2000 *J. Catal.* **187** 109
[2] Raybaud P, Hafner J, Kresse G, Kasztelan S and Toulhoat H 2000 *J. Catal.* **189** 129
[3] Galea N M, Kadantsev E S and Ziegler T 2009 *J. Phys. Chem. C* **113** 193
[4] Vojvodic A, Hinnemann B and Norskov J K 2009 *Phys. Rev. B* **80** 125416
[5] Li T and Galli G 2007 *J. Phys. Chem. C* **111** 16192
[6] Kim K K, Hsu A, Jia X, Kim S M, Shi Y, Hofmann M, Nezich D, R-Nieva J F, Dresselhaus M S, Palacios T and Kong J 2011 *Nano Lett.* **12** 161
[7] Shi Y, Hamsen C, Jia X, Kim K K, Reina A, Hofman M,Hsu A L, Li K, Juang Z Y, Dresselhaus M S, Li L J and Kong J 2010 *Nano Lett.***10** 4134.
[8] Chen C H, Wu C L, Pu J, Chiu M H, Kumar P, Takenobu T T and Li L J 2014 *2D Mater.* **1** 034001
[9] Liu K K, Zhang W, Lee Y H, Lin Y C, Chang M T, Su C Y, Chang C S, Shi Y, Zhang H, Lai C S and Li L J 2012 *Nano Lett.* **12** 1538
[10] Li X, Cai W, An J, Kim S,Nah J, Yang D, Piner R, Velamakanni A, Jung I, Tutuc E, Banerjee S K, Colombo L and Ruoff R S 2009 *Science* **324** 1312
[11] Radisavljevic B, Radenovic A, Brivio J, Giacometti V and Kis A 2011 *Nat. Nanotech.* **6** 147

[12] Fang H, Chuang S, Chang T C, Takei K, Takahashi T T and Javey A 2012 *Nano Lett.* **12** 3788
[13] Huang J K, Pu J, C-H.Hsu, Chiu M H, Juang Z Y, Chang Y H, Iwasa Y, Takenobu T and Li L J 2013 *ACS Nano* **8** 923
[14] Pu J, Yomida Y, Liu K K, Iwasa Y and Takenobu T 2012 *Nano Lett.* **12** 4013
[15] Baroni S *et al.* see ***http://www.pwscf.org/*** for more information about the quantum-simulation software
[16] Perdew P, Burke K and Ernzerhof M 1996 *Phys. Rev. Lett.* 77 3865
[17] Kresse G and Joubert D 1999 *Phys. Rev. B* **59** 1758
[18] Monkhorst H J and Pack J D 1976 *Phys. Rev. B* **13** 5188
[19] Togo A, Oba F and Tanaka I 2008 *Phys. Rev. B* **78** 134106
[20] Blanco M A, Francisco E, and Lua-na V 2004 *Comput. Phys. Commun.* **158** 57
[21] Zhang X, Chen L, Ma M, Zhu Y, Zhang S and Liu R 2011 *J. Appl. Phys.* **109** 113523
[22] Zhang X , Qin J , Ning J , Sun X , Li X, Ma M and Liu R 2013 *J. Appl. Phys.* **114** 183517
[23] Zhang X, Chen L, Ma M, Zhu Y, Zhang Sand Riping 2011 *J. Appl. Phys.* **109** 113523
[24] Grimme S, Antony J, Ehrlich S and Krieg H 2010 *J. Chem. Phys.* **132** 154104
[25] Kadantsev E S and Hawrylak P 2012 *Solid State Comm.* **152** 909
[26] Luo F, Cai L C, Chen X R, Jing F Q and Alfe D 2012 *J. Appl. Phys.* **111** 053503

IOP Conf. Series: Journal of Physics: Conf. Series **836** (2017) 012053 doi:10.1088/1742-6596/836/1/012053

Transition probabilities for explaining the origin of L-emission satellite

Renuka Kendurkar[a] and B D Shrivastava

School of Studies in Physics, Vikram University, Ujjain, M.P.- 456010, India

Email: [a)] renukakendurkar@gmail.com

Abstract. The X-ray satellites are caused by transitions in atoms which are doubly ionized or multiply ionized. To explain the origin of X-ray satellites, the relative intensities of the various possible transitions between the doubly ionized initial and final states of an atom, have to be calculated. The intensity can be calculated from the probability of creation of the double vacancy states. These probabilities are due to Coster-Kronig and Auger transitions and can be calculated using the tables of McGuire. These tables are based on non-relativistic calculations. The latter calculations of transition probabilities are of Chen *et al.*, who have done relativistic calculations in elements with $18 \leq Z \leq 96$. The aim of the present work is to study the variation with atomic number (Z) of the transition probabilities for the different Coster-Kronig transitions as given in McGuire's tables and Chen et al's tables and to discuss the difference between the two calculations. The present study shows that for calculation of the intensity of a satellite, the tables of Chen *et al.* should be preferred over the tables of McGuire while investigating the origin of L-emission satellites.

1. Introduction

When X-rays are generated by electron impact on a target in an X-ray tube, the radiation consists of both the continuous X-rays and characteristic X-rays. The characteristic X-ray spectrum can be classified into two broad group, namely, diagram lines and non-diagram lines or satellite lines. The diagram lines are those which can be represented by transitions in the normal energy level diagram of a singly ionized atom. The X-ray satellites are caused by transitions in atoms which are doubly ionized or multiply ionized. Though intensive work on the theories of their origin has been published, the theories regarding their origin do not appear to have as yet reached a final decisive stage. Hence, we have been interested in finding possible explanations of the origin of the satellites in the region of the L-emission spectra, especially because, several new satellites have been reported experimentally from our X-ray laboratory.

Any transition taking place between the energy levels in a multiply ionized atom will be associated with an X-ray emission line of slightly different energy than the corresponding diagram line arising as a result of transition in a singly ionized atom. After the creation of a single-hole state, it is the probability of a particular subsequent process that leads to the formation of initial two-hole state, required for the emission of a satellite. The single-hole state can get converted into the two-hole states by two processes:

(i) Auger transition: The process of radiationless reorganization of an atom ionized in an inner shell is known as the Auger effect. After an Auger transition an atom may be left doubly ionized in inner shells.

(ii) Shake-off process: A vacancy in the outer shell of the atom may be created by shake-off process when an electron from the outer shell of the atom may escape out simultaneously to the formation of an inner shell vacancy. Photo ejection of an inner shell electron in an atom results in a sudden change of atomic potential. This sudden change can excite other electron to an unoccupied bound state (electron shake-up) or eject into the continuum (electron shake-off).

2. Calculation of transition energies and transition probabilities in multiply ionized atoms

To investigate the origin of the X-ray satellite lines, the requirements are: (i) identification of the double vacancy initial and final states, (ii) the energy difference between these states and (iii) the relative intensities of the transitions between these states. To calculate the energy of a satellite theoretically, the difference in energies of the double vacancy initial and final states has to be calculated. This simple explanation becomes complicated when one has to take into account the RS or LS, jj or intermediate coupling between the two states which leads to the multiplicity of levels. Since, we have been interested in assigning transitions in doubly ionized atom which give rise to the various L-emission satellites; this can be done by calculating the theoretical X-ray satellite spectrum. To obtain such a spectrum, we require energies of different transitions which are allowed between the initial and final double vacancy states and relative probabilities of such transitions. Ab initio calculation of energies and intensities of various double hole state transitions, considering multiplicity of energy levels, is a difficult and laborious task. It is also not necessary because, alternatively, calculations can be done with the help of several tabulations, in which necessary data are available.

For energy calculations, one can use either the tables of Larkins [1] which contain theoretically calculated Auger-electron energies for a wide range of transition processes in all the elements from Z = 1 to 100 or the tables of Parente *et al.* [2] which contain the energies of L-series X-ray satellites that arise from the presence of one 'spectator hole' in the M-shell or N- shell of the emitting atom, for 11 elements, namely $_{65}$Tb, $_{67}$Ho, $_{70}$Yb, $_{74}$W, $_{78}$Pt, $_{80}$Hg, $_{85}$At, $_{88}$Ra, $_{90}$Th, $_{92}$U and $_{95}$Am.

For calculations of intensities of satellites, one can use the tables of McGuire [3] which contain non-relativistic calculations for Coster-Kronig transition probabilities for creation of double vacancy states or the tables of Chen *et al.* [4] which contain relativistic calculations of the rates of all energetically possible radiationless transitions with intensities above 10^{-5} milli atomic units with initial K- and $L_{1,2,3}$ - shell vacancies in elements with $18 \le Z \le 96$.

To calculate the probability of creation of the initial double vacancy state by the process of Coster-Kronig transition and/or the process of shake-off, the cross section for creation of single vacancy state has to be first calculated. This can be done by using the formulae given by Moores *et al.*[5]. This cross section is then multiplied by the transition rates for the Coster-Kronig transition. After multiplication, one gets the Coster-Kronig transition probability for the creation from the single vacancy state to the double vacancy state. The singly ionized state may get converted to doubly ionized state also by an associated shake-off of an electron. Hence, to calculate the probability of creation of the initial double vacancy state by the shake-off process, the cross section for creation of single vacancy state is multiplied by the shake-off probability. After multiplication, one gets the shake-off probability for the creation from the single vacancy state to the double vacancy state. Finally, the total probability of creation of doubly ionized initial state can be determined by adding both the cross sections calculated above, i.e., by Coster-Kronig transition and by shake-off process. From these probabilities of creation of doubly ionized initial states, the probability of transition between doubly ionized initial and final states can be calculated by assuming the cross section for a set of double vacancy levels to be the total probability of all the transitions from this set. This probability can be distributed statistically among all the allowed transitions from this set of levels for relative probabilities of the transitions, taking into consideration all the multiplets of supermultiplets from various $^{(2S+1)}$(L) levels of the set.

In the present paper, we are concerned with the calculations of the intensities of L-emission

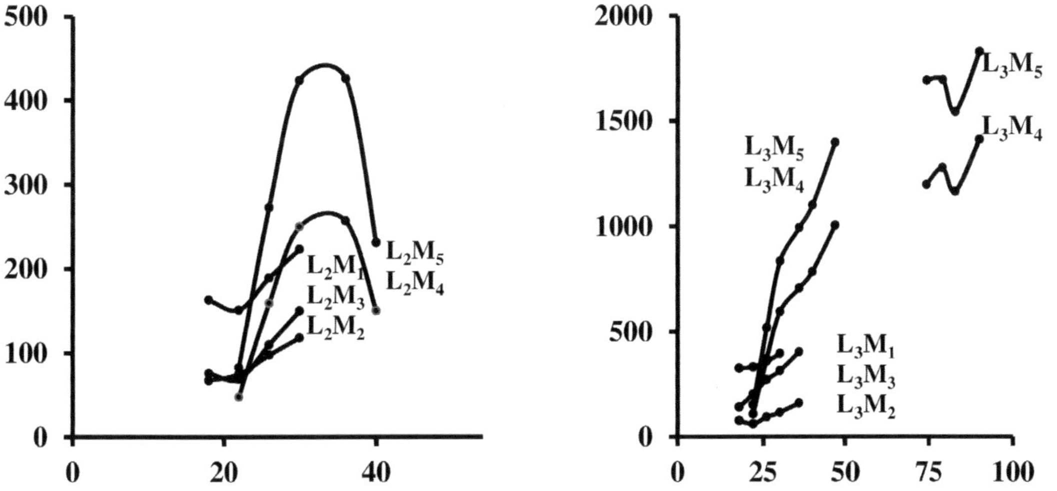

Figure 1. Variation of transition probability of elements (along Y-axis) with increasing atomic number (Z) (along X-axis) as given in the tables of McGuire, for transition arrays L_2M_x (for X= 1-5) (left figure) and L_3M_x (for X= 1-5) (right figure).

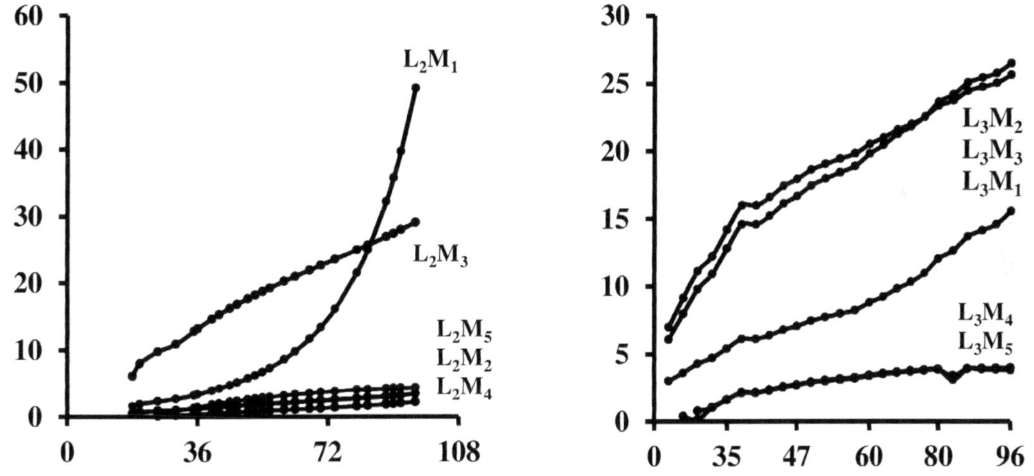

Figure 2. Variation of transition probability of elements (along Y-axis) with increasing atomic number (Z) (along X-axis) as given in the tables of Chen *et al.*, for transition arrays L_2M_x (for X= 1-5) (left figure) and L_3M_x (for X= 1-5) (right figure).

satellites. For such calculations, one can use the tables of McGuire or the tables of Chen *et al.* Both of these tables have been used by several authors for explaining the origin of the satellites. Our aim is to compare these tables and to know which table is better to be used for assigning transitions to the satellites. Hence, the details of the tables are first described here in brief.

3. Coster-Kronig transition probabilities using McGuire's and Chen's tables

McGuire has calculated the probability of creation of double vacancy states due to Auger and Coster-Kronig transitions in j-j coupling scheme and has tabulated the values in the form of tables. The calculations have been done using Harman-Skillman one-electron eigenvalues, radial matrix elements and oscillator strengths. They have tabulated the transition probabilities only for 14 elements, namely, $_{18}$Ar, $_{22}$Ti, $_{26}$Fe, $_{30}$Zn, $_{35}$Br, $_{40}$Zr, $_{47}$Ag, $_{54}$Xe, $_{60}$Nd, $_{67}$Ho, $_{74}$W, $_{79}$Au, $_{83}$Bi and $_{90}$Th.

The latter calculations of Coster-Kronig transition probabilities are of Chen *et al.*, who have done relativistic calculations of the rates of all energetically possible radiationless transitions with intensities above 10^{-5} milli atomic units with initial K- and $L_{1,2,3}$ - shell vacancies in 25 elements, namely, $_{18}$Ar, $_{20}$Ca, $_{25}$Mn, $_{30}$Zn, $_{35}$Br, $_{36}$Kr, $_{40}$Zr, $_{42}$Mo, $_{45}$Rh, $_{47}$Ag, $_{50}$Sn, $_{52}$Te, $_{54}$Xe, $_{56}$Ba, $_{60}$Nd, $_{63}$Eu ,$_{67}$Ho, $_{70}$Yb, $_{74}$W, $_{80}$Hg, $_{83}$Bi, $_{88}$Ra, $_{90}$Th, $_{92}$U and $_{96}$Cm. The Auger transition probabilities have been calculated from perturbation theory, in *j-j* coupling, assuming frozen orbitals. The wave functions were generated according to the Dirac-Hartree-Slater approach for configurations that contain one initial inner-shell vacancy. The continuum wave functions were obtained by solving the Dirac-Slater equations with the same atomic potential as that used for the initial state. With this treatment, the orthogonality of the wave functions is assured, and the approximation is good for all but the lightest elements. A general relativistic Auger program was used to compute the radiationless transition probabilities.

4. Results and discussions

The difference between McGuire's tables and Chen et al.'s tables is that Chen et al.'s tables are based on relativistic calculations and have been published much after the publication of McGuire's tables, which are based on non-relativistic calculations. Also, for using McGuire's tables, one has to use the probabilities of the creation of single hole states in the L_1 or L_2 or L_3 subshell, while for using Chen et al's tables, one has to use the probabilities of the creation of single hole states in the K- shell.

The main difficulty experienced in the use of both the tables is that both of them give the values of transition probabilities for selected elements. For explaining the origin of satellite in any element, the transition probabilities are required for that element, which may not be available in any of the tables. The only alternative is to estimate the values of that element using linear interpolation method. In the present work, we have explored this possibility by plotting the tabulated values of transition probabilities versus the atomic number Z. The obtained plots of atomic number Z versus the transition probabilities for the different Coster-Kronig transitions as given in McGuire's tables and Chen *et al.*'s tables are shown in Figs. 1 and 2 respectively. These figures clearly demonstrate the difference between the two calculations. If one uses the values given in the McGuire's table for linear interpolation, then one may not get proper values of transition probabilities for intermediate elements, since the curves do not have regular variations. On the other hand, the plots of values of Chen *et al.*'s tables show regular variation with Z. Hence, if they are used for linear interpolation, one is likely to get the correct values of transition probabilities for intermediate elements. From the present study, it is clear that, since Chen *et al.*'s tables are based on relative calculations and since the method of linear interpolation can be used for estimating the values of transition probabilities for those elements, for which values are not given in the tables, these tables should be used for calculations of intensities of the satellites and thereby for explaining the origin of the L-emission satellites.

References

[1] Larkins F P 1977 *At. Data Nucl. Data Tables* **20** 311
[2] Parante F, Chen M H and Craseman B, 1981 *At. Data. Nucl. Data Tables* **26** 383
[3] McGuire E J 1971 *L-Shell Auger, Coster-Kronig and radiative matrix elements, and Auger and Coster-Kronig transition rates in j-j coupling* (Report No. Sc-RR-710075) (New Mexico: Sandia Laboratories)
[4] Chen M H, Crasmann B and Mark H 1979 *At. Data. Nucl. Data Tables* **24** 13
[5] Moores D L, Golden L B and Sampson D H 1980 *J. Phys. B* **13** 385

IOP Conf. Series: Journal of Physics: Conf. Series **836** (2017) 012054 doi:10.1088/1742-6596/836/1/012054

Studies on ternary zinc blend based semiconducting quantum dots for hybrid solar cell applications

Harit Kumar Sharma[1,a], S K Pandey[2] and S L Agrawal[1]

[1]SSI Laboratory Department of Physics Awadhesh Pratap Singh University, Rewa (M.P.) 486003, India
[2]Govt. Engineering College, Rewa (M.P.) 486002, India

Email: [a)]haritsharma2@gmail.com

Abstract: A systematic study is presented in the present work on the fabrication of hybrid solar cell using doped zinc blend based semiconductor quantum dots (QDs) prepared by a simple wet chemical precipitation method. Formation of quantum dot has been ascertained by X-ray Diffraction (XRD), UV-visible Spectroscopy and Impedance Spectroscopy measurements. XRD studies established the zinc-blende phase in quantum dot regime. UV-visible studies reveal decrement in optical bandgap of the QDs with co-doping of Cd ion in ZnS lattice. Mott-Schottky analysis revealed n- type conductivity with increasein band bending from binary to ternary configuration respectively. Bulk heterojunction hybrid solar cells fabricated in 1cm X 1cm size in conjunction with p type Polypyrroleunder AM 1.5 illumination resulted in highest conversion efficiency of 1.6 %.

1. Introduction

One of the major concerns for solar cell technology is to find novel materials that have appropriate properties for solar energy conversion. In this perspective synthesis and characterization of semiconductor nanoparticles have attracted much interest in recent times due to their novel size dependent properties as a consequence of the large number of surface atoms and the three dimensional confinement of electrons [1-2]. Among the different semiconductor materials, binary metal chalcogenides of group II-VI have been extensively investigated in recent times [2]. In the case of ZnSchalcogenides, the band gap varies in the range of 3.7 to 5.5 eV depending upon the particle size and finds application in variety of opto-electronic applications [2-4]. It has been well recognized in recent years that optical absorption and photoluminescence properties of semiconductor nanoparticles are strongly modified due to size quantization [3-5]. Alternatively, ternary semiconducting nanoparticles formed by two binary semiconductors can also be used to tailor these optical properties. Such semiconducting particles are expected to improve the performance of thin film solar cells and photo electrochemical energy conversion cells. In recent report energy band gap has been found to vary from 5.5 to 2.5 eV at room temperature in $Zn_{1-x}Cd_xS$ solid solution semiconductors [5] which favours their candidature for solar cell applications. Bearing these aspects in mind, an attempt has been made in the present work to dope ZnS with Cd^{2+} ions leading to formation of ternary system $Zn_{1-x}Cd_xS$ and characterize them for hybrid solar cell application.

2. Experimental details

Zinc blend based semiconductor quantum dots have been synthesized at room temperature using co-precipitation method. The chemical reaction involved in the synthesis of ZnS, $Zn_{0.75}Cd_{0.25}S$ and $Zn_{0.5}Cd_{0.5}S$ nanoparticles has been described elsewhere [5]. The resulting precipitates were filtered several times with methanol to remove impurities and dried at room temperature. X-ray studies were conducted on Shimadzu make X-ray diffractometer (model 6100) in 2θ range 20° to 70°C having Cukα radiation wavelength 1.5406Å. Band gap of all the samples were estimated from UV-Vis data obtained with the help of Systronics make UV-visible spectrophotometer (model 2203) in 200nm-1100nm wavelength range. Mott-schottky analysis was carried out with the help of LCR meter (Hioki Japan, model 3520). With these nanopowders and as synthesized p type polypyrrole a bulkheterojunction solar cell was fabricated employing doctor blade technique. Electrical characterization was performed using a computer-controlled Keithley 2400 Source Meter in the dark and under illumination to assess the performance of solar cell. The photovoltaic devices were illuminated using a metal-halide lamp calibrated to an intensity of 1 kW/m^2 (in accordance with standard testing conditions AM 1.5G). All measurements were performed at a temperature of 295 K (room temperature) and the temperature of the device was kept constant.

3. Results and discussions

3.1 X-ray diffraction studies

Figure1 depicts the XRD pattern of wet-chemical derived $Zn_{1-x}Cd_xS$ nanoparticles for different mole percentages of cadmium (x). Polycrystalline nature of $Zn_{1-x}Cd_xS$ QDs is apparent from these X-ray diffraction patterns. Further, these patterns exhibit three prominent peaks around 2θ values 28.54°, 47.62°, and 56.28° corresponding to (111), (220), and (311) planes of the cubic phase of ZnS with the lattice constant a = 5.4Å (JCPDS card no. 05-0566). These results reflect zinc blend structure of as synthesised powders. The gradual shift of diffraction peak positions towards lower diffraction angles with increase of Cd concentration (from 0 to 0.5) indicates that the synthesised crystals are not a simple mixture of pure ZnS and CdS rather a solid solution of $Zn_{1-x}Cd_xS$ [5-6]. The small difference in the bond length of ZnS and CdS and the very close electronegativity value of Cd and Zn allow the formation of such a solid-solution.

Table 1. XRD data of $Zn_{1-x}CdxS$ nanocrystallites

Sample description	Angle (2θ)	d-spacing (Å)	Peak width (2θ in°)	Identified possible h k l	Crystalline size (mm)	Lattice constant (Å)
ZnS	28.54	3.1250	4.39	1 1 1	1.75	5.406
	47.62	1.9080	2.63	2 2 0	2.76	
	56.28	1.6332	2.58	3 1 1	2.71	
$Zn_{0.75}Cd_{0.25}S$	27.57	3.2318	2.97	1 1 1	2.59	5.613
	45.67	1.9849	2.74	2 2 0	2.67	
	53.84	1.7012	2.57	3 1 1	2.75	
$Zn_{0.5}Cd_{0.5}S$	26.50	3.3608	1.36	1 1 1	5.68	5.716
	43.95	2.0582	1.56	2 2 0	4.72	
	52.15	1.7523	1.81	3 1 1	3.94	

The crystallite size corresponding to the observed $Zn_{1-x}CdxS$ peaks have been calculated from the Debye- Scherrer relation [7]. It is apparent from these calculations (table 1) that crystallites size of $Zn_{1-x}CdxS$ Nanocrystallites increases with increasing Cd concentration.

3.2 Optical studies (UV-visible absorption study)

Band gap of synthesized nano-powder is estimated using thewell-known Tauc's relation.

$$(\alpha h\nu) = K(h\nu - E_g)^n (1)$$

where K is the edge width parameter, n is a parameter which depends on the type of transition, i.e. allowed direct, allowed indirect, forbidden direct and forbidden indirect and can have values ½, 2, 3/2, and 3 respectively. The direct band gap value of $Zn_{1-x}Cd_xS$ nanoparticles was evaluated from the plot of a graph between $(\alpha h\nu)^2$ versus hν as shown in figure 2.

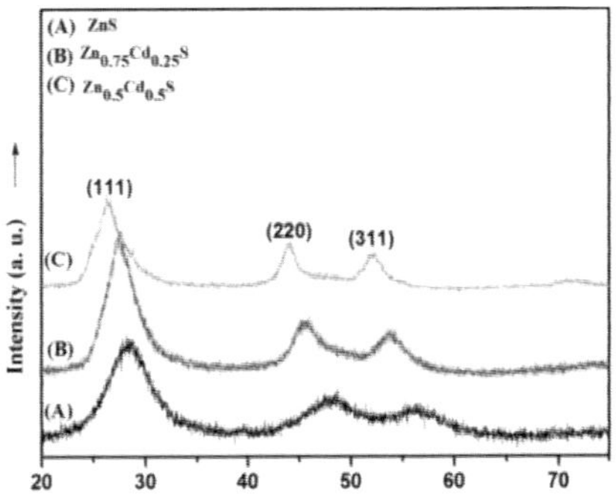

Figure 1. X-ray diffraction pattern of $Zn_{(1-x)}Cd_xS$ ($0 \leq x \leq 0.5$) semiconductor quantum dots

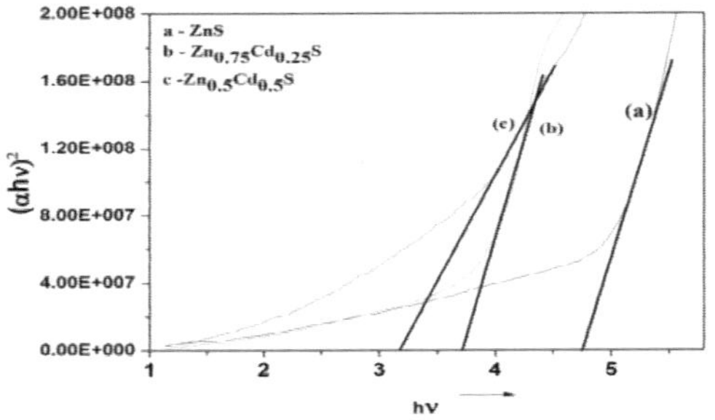

Figure 2. Shows optical band gap graph of $Zn_{(1-x)}Cd_xS$ ($0 \leq x \leq 0.5$) semiconductor

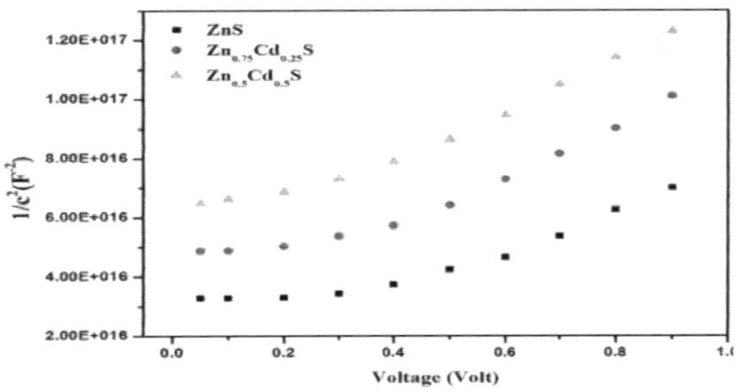

Figure 3. Mott-schottky study of $Zn_{1-x}Cd_xS$: Agsystem samples

The obtained experimental band gap energy 4.7 eV, 3.7 eV and 3.2 eV for ZnS, $Zn_{0.75}Cd_{0.25}S$ and $Zn_{0.5}Cd_{0.5}S$ respectively reveal decrement of optical band gap with increasing Cd^{+2} concentrations in $Zn_{1-x}Cd_xS$ QD's. This red shift of band gap can be attributed to quantum confinement effect which takes place essentially due to growth of grain size and decrease in defect states near the band edges.

3.3 Mott-Schottky study

Mott-schottky plots have been drawn (in dark conditions) to evaluate the semiconductor parameters as shown in figure 3. The value of flatband potential (V_{fb}) was calculated using the relation [8]. the positive slope of the Mott-Schottky plot indicates n-type conductivity and decrement of flat band potential 0.29, 0.21 & 0.15 as for increasing Cd ions in ZnS lattice results increase in doping density of as prepared $Zn_{1-x}CdxS$ samples.

3.4 Solar cell characteristics study

Figure 4 shows the photo response curve of as prepared Polypyrrole/ZnS and Polypyrrole/$Zn_{0.5}Cd_{0.5}S$ solar cells with respect to variation in voltage. The obtained efficiency of ZnS/Ppy structure is 0.01% while for $Zn_{0.5}Cd_{0.5}S$/Ppy structure it is 1.6%.Such poor efficiency for ZnS/Ppy structure can be attributed to the fact that ZnS being wide band gap semiconductor there is improper band alignment between polypyrrole and ZnS which in turn causes inefficient excitation dissociation and thus less efficient device formation. On the other hand, the devices with Polypyrrole/$Zn_{0.5}Cd_{0.5}S$ are efficient.

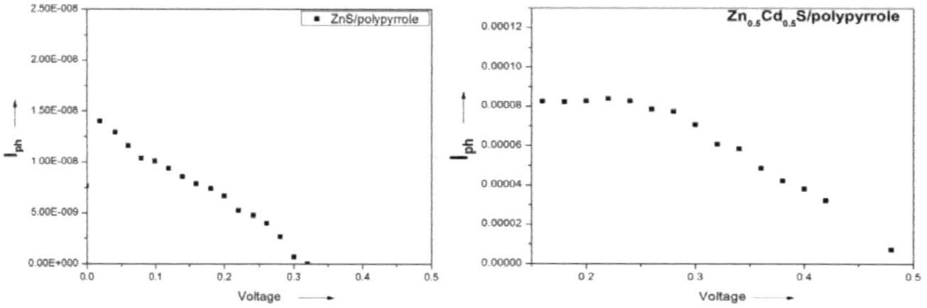

Figure 4. Photocurrent versus voltage graph of ZnS/Ppy and $Zn_{0.5}Cd_{0.5}S$ hybrid heterojunction solar cell

It can be expected that cell efficiency of Polypyrrole/$Zn_{0.5}Cd_{0.5}S$ results from the good charge balance of hole and electron at the interface and forming of a well-ordered morphology. The reason of the dramatic increase in efficiency could be attributed to the possibility that energy transfer rather than charge transfer could occur from the donor (polypyrrole) to acceptor (using ternary and quaternary QD's) in polypyrrole/$Zn_{0.5}Cd_{0.5}S$ resulting in a sufficient generation of free charge carriers.

4. Conclusions

$Zn_{1-x}Cd_xS$ QD's ($0 \leq x \leq 0.5$) have been prepared by simple wet chemical precipitation method without using any capping agent or surfactant. X-ray diffraction study confirmed cubic phase formation and nanometric size of particles around 2to 5nm. Due to the increasing of Cd content, all the XRD peaks shifted towards lower diffraction angles which show alloy formation of nanoparticles of $Zn_{1-x}Cd_xS$ samples. Optical study revealed that the optical band gap of materials decreased with the increase of Cd content which shows quantum confinement effect of QDs.Mott-schottky plot of prepared samples shows n-type of conductivity with decreasing flat band potential for their binary to ternary conversion. Doctor blade technique has been deployed for deposition of active layer onto commercially available SnO_2 coated glass substrate for the preparation of hybrid solar cell structure. The characterization and findings on hybrid solar cell designed around inorganic semiconducting QD's and polymeric semiconductor shows better photoresponse with ternary quantum dots with highest conversion efficiency of 1.6%. Present studies reveal importance of II-VI group of semiconductor in the design and performance of hybrid solar cell structure.

References

[1] Kosyachenko L and Toyama T 2014 *Sol. Energ. Mat. Sol. C.* **120** 512
[2] Dennler D, Scharber M C and Barbec C J 2009 *Adv. Matter* **21** 1323
[3] Sharma H K, Shukla P K and Agrawal S L 2015 *J. Optoelect. Adv. Mater* **17** 608
[4] Wang L, Y Jiany, Wang C, Wang W, Cao B, Niu M and Qian Y 2008 *J. Alloys Compd.* **545** 255
[5] Mohapatra J 2013 *Intern. Nano Lett.* **3** 1
[6] Zu S, Wang Z, Liu B, Fan X and Qian G 2009 *J. Alloys Compd.* **476** 689
[7] Culity B D and Stock S R 2001 *Elements of X-ray diffraction 3rd ed.* (New Jersey: Prentice Hall)
[8] Fabregat-Santiago F, Garcia-Belmonte G, Bogdanoff P and Zabanc A 2003 *J. Electrochem. Soc.* **150** 293